D1083060

A Guidebook for Technology Assessment and Impact Analysis

NORTH HOLLAND SERIES IN
SYSTEM SCIENCE AND ENGINEERING

Andrew P. Sage, *Editor*

1	**Wismer and Chattergy**	Introduction to Nonlinear Optimization: A Problem Solving Approach
2	**Sutherland**	Societal Systems: Methodology, Modeling, and Management
3	**Šiljak**	Large-Scale Dynamic Systems
4	**Porter et al.**	A Guidebook for Technology Assessment and Impact Analysis

A Guidebook for Technology Assessment and Impact Analysis

Series Volume 4

Alan L. Porter
Frederick A. Rossini
Stanley R. Carpenter
Georgia Institute of Technology

and

A. T. Roper
Rose-Hulman Institute of Technology

with

Ronal W. Larson and **Jeffrey S. Tiller**
Solar Energy Research Institute — Georgia Institute of Technology

NORTH HOLLAND
New York • Oxford

Elsevier North Holland, Inc.
52 Vanderbilt Avenue, New York, N.Y. 10017

Distributors outside the United States and Canada:

Thomond Books
(A Division of Elsevier/North-Holland Scientific Publishers, Ltd.)
P.O. Box 85
Limerick, Ireland

Library of Congress Cataloging in Publication Data

Main entry under title:

A Guidebook for technology assessment and impact analysis.
(North Holland series in system science and engineering)

Includes bibliographical references and index.
1. Technology assessment. I. Porter, Alan L.
T174.5.G84 301.24'3 79-12699
ISBN 0-444-00314-2

Desk Editor Michael Gnat
Design Series
Production Manager Joanne Jay
Compositor Lexigraphics Photocomposition, Inc.
Printer Capital City Press

Manufactured in the United States of America

Contents

Preface xiii
Acknowledgments xvii
Glossary xxi

1. Introduction 1

1.1 The Context 2
1.2 TA and EIA: Descriptions and Distinctions 3
1.3 The "What" and "Who" of TA/EIA 4
1.4 Brainstorming 6
1.5 Concluding Observations 7
Exercises 8

2. Technology and Society 10

2.1 Introduction 11
2.2 What is Technology? 12
2.3 Technology in the Social Context 14
2.3.1 Setting the Scene, 14
2.3.2 Technology Causes Social Change, 16

2.3.3 Society and Its Values Cause Technological Change, 19
2.3.4 Technology and Society—All of a Piece, 21
2.4 Concluding Observations 24
Recommended Readings 24
Exercises 25

3. Institutionalization of TA/EIA 26

3.1 Introduction 27
3.2 The National Environmental Policy Act of 1969 27
3.3 Creation of the Office of Technology Assessment 30
3.3.1 Legislative Research Service Study: Technical Information for Congress, 31
3.3.2 National Academy of Engineering Study, 32
3.3.3 National Academy of Sciences Study, 32
3.3.4 Follow-Up and Establishment of the OTA, 33
3.4 Other Instances of TA/EIA in the United States 34
3.4.1 TA/EIA in Other Federal Agencies, 34
3.4.2 TA/EIA at the State Level, 36
3.4.3 TA in the Private Sector, 36
3.5 TA/EIA Outside the United States 37
3.5.1 Sweden and Japan, 38
3.5.2 Common Market Countries, 38
3.5.3 Eastern European Countries, 39
3.5.4 Elsewhere, 40
3.6 Concluding Observations 40
Recommended Readings 41
Exercises 41

4. Basic Features of an Assessment 42

4.1 Introduction 43
4.2 Assessment Objectives 43
4.2.1 The Large Picture, 43
4.2.2 Validity, 45
4.2.3 Utility, 46
4.2.4 Improving Assessment Methodology, 47
4.3 Types of Assessment 47
4.3.1 Distinctions Between TA and Other Studies, 47
4.3.2 Distinctions Between TA and EIA, 50
4.3.3 Three Types of TA/EIA, 51
4.4 Assessment as Program: Possible Variations 52
4.5 Components of an Assessment 54
4.5.1 Problem Definition, 54
4.5.2 Technology Description, 55
4.5.3 Technology Forecast, 55
4.5.4 Social Description, 58
4.5.5 Social Forecast, 58
4.5.6 Impact Identification, 58
4.5.7 Impact Analysis, 59
4.5.8 Impact Evaluation, 60
4.5.9 Policy Analysis, 60
4.5.10 Communication of Results, 60

4.6 Fitting the Components Together 60
4.7 Concluding Observations 61
Recommended Readings 62
Exercises 62

5. Strategies for Particular Assessments: Bounding and Techniques 64

5.1 Introduction 65
5.2 Problem Definition 65
5.3 Bounding an Assessment 66
5.3.1 General Considerations, 66
5.3.2 Areas for Bounding, 67
5.3.3 The Bounding Microassessment, 69
5.3.4 Interpretive Structural Modeling, 71
5.4 Analytic Strategies 73
5.4.1 Content and Process Dimensions of an Assessment, 74
5.4.2 Knowledge in TA/EIA, 77
5.4.3 Particular Methods and Techniques, 81
5.4.4 Guidelines for Selection of Analytical Techniques, 81
5.5 Concluding Observations 87
Recommended Readings 88
Exercises 89
Appendix: Summaries of Selected Assessments 91

6. Technology Description and Forecasting 97

6.1 Introduction 98
6.2 General Considerations 99
6.2.1 Interdependence: Forecasting Technology and Society, 100
6.2.2 Levels of Emergence and Impact of a Technology, 101
6.2.3 Effects of Subject Characteristics, 103
6.3 Technology Description 104
6.3.1 A Checklist, 105
6.3.2 Gathering Descriptive Information, 110
6.4 Technology Forecasting 111
6.4.1 Basic Considerations, 111
6.4.2 Monitoring, 113
6.4.3 Trend Extrapolation, 115
6.4.4 Expert Opinion Methods, 122
6.5 Concluding Observations 130
Recommended Readings 131
Exercises 132

7. Social Description and Forecasting 134

7.1 Introduction 135
7.2 General Considerations 135
7.2.1 Social Context—Critical for TA/EIA, 135
7.2.2 The Technological Delivery System, 136
7.3 State of Society Description 138

7.3.1 Elements, 138
7.3.2 Social Indicators, 140
7.3.3 Values and Value Change, 143

7.4 Forecasting the State of Society 146

7.4.1 General Principles, 146
7.4.2 Scenarios, 150

7.5 Concluding Observations 151

Recommended Readings 153

Exercises 153

8. Impact Identification and Policy Considerations 155

8.1 Introduction 156

8.2 Impact Identification Strategies 157

8.2.1 Some Dimensions, 157
8.2.2 Criteria for Constructing a Strategy, 161

8.3 Impact Identification Techniques 162

8.3.1 Scanning Techniques, 162
8.3.2 Tracing Techniques, 168
8.3.3 Criteria for Selecting Techniques, 172

8.4 Selectivity 173

8.5 Policy Considerations 176

8.6 Concluding Observations 177

Recommended Readings 179

Exercises 179

9. Impact Analysis 181

9.1 Introduction 182

9.2 General Considerations 182

9.2.1 Model Characteristics, 183
9.2.2 Models in Impact Analysis, 185

9.3 Cross-Effect Matrices 187

9.3.1 Overview, 187
9.3.2 Cross-Impact Analysis, 190

9.4 Simulation Models 195

9.4.1 Overview, 195
9.4.2 KSIM: A Simple Cross-Impact Simulation Model, 196
9.4.3 Physical Models, 203
9.4.4 Planning Models, 203
9.4.5 System Dynamics, 204
9.4.6 Stochastic Modeling Approaches, 207

9.5 Sensitivity Analysis 208

9.6 Concluding Observations 209

Recommended Readings 212

Exercises 212

10. Environmental Analysis 215

10.1 Introduction 216

10.2 General Considerations 217

10.2.1 Procedural Issues, 217

10.2.2 Substantive Issues, 218
10.2.3 Range of Methods and Approaches, 219
10.3 Ecological Systems 222
10.3.1 General Considerations, 222
10.3.2 Ecosystem Parameters and Analysis, 223
10.4 Land 230
10.5 Water 232
10.6 Air 238
10.7 Noise 242
10.8 Radiation 245
10.9 Other Issues 247
10.10 Concluding Observations 249
Recommended Readings 250
Exercises 251

11. Economic Impact Analysis 253

11.1 Introduction 254
11.2 Cost–Benefit Analysis 255
11.2.1 The Cost–Benefit Framework, 255
11.2.2 Definition of the Subject of the Analysis, 256
11.2.3 Identification of Costs and Benefits, 257
11.3 Measurement of Costs and Benefits 258
11.3.1 Direct Costs and Benefits, 264
11.3.2 Externalities, 265
11.3.3 Public Goods, 268
11.4 Evaluating Economic Impacts 269
11.4.1 Discounting, 270
11.4.2 Alternative Decision Criteria, 272
11.5 Variations on the Cost–Benefit Framework 278
11.5.1 Risk–Benefit Analysis, 278
11.5.2 Net Energy Analysis, 281
11.6 Analysis of Secondary and Higher-Order Impacts 283
11.6.1 Macroeconomic Concerns, 283
11.6.2 Economic Base Models, 284
11.6.3 Input–Output Models, 289
11.6.4 Regression Analysis, 291
11.7 Concluding Observations 294
Recommended Readings 296
Exercises 296

12. Analysis of Social and Psychological Impacts 300

12.1 Introduction 301
12.2 SIA Strategies 302
12.3 Profiling and Projection 308
12.3.1 Profiling, 308
12.3.2 Data Gathering and Use, 309
12.3.3 Projection, 314
12.4 Identification, Description, and Display of Social Impacts 316

12.4.1 Strategies, 316
12.4.2 Some Specific Social Impacts, 318
12.4.3 Other Considerations, 319
12.5 Impacts on Values 320
12.6 Psychological Impacts 323
12.7 Concluding Observations 325
Recommended Readings 326
Exercises 327

13. Technological, Legal, and Institutional/Political Analyses 329
13.1 Introduction 330
13.2 General Methodological Considerations 331
13.3 Technological Factors 333
13.4 Legal Factors 334
13.4.1 Implications of the Law for Technological Development, 334
13.4.2 Legal Analysis, 337
13.4.3 Legal Impacts, 337
13.5 Institutional /Political Factors 339
13.5.1 General Considerations, 339
13.5.2 Institutional Analysis, 343
13.5.3 Institutional /Political Impacts, 344
13.6 Concluding Observations 344
Recommended Readings 345
Exercises 349

14. Impact Evaluation 350
14.1 Introduction 351
14.2 The Role of Values in Impact Evaluation 352
14.2.1 Value Issues Arising from the Societal Context for TA /EIA, 352
14.2.2 Value Issues Internal to TA /EIA, 354
14.3 Impact Evaluation Strategy 355
14.3.1 Who Is to Be Involved?, 356
14.3.2 What Criteria Are to Be Used? The Concept of Utility and Alternatives to It, 357
14.3.3 What Is to Be Evaluated?, 359
14.4 Evaluation Techniques 360
14.4.1 Dimensionless Scaling, 362
14.4.2 Decision Analysis, 363
14.4.3 Policy Capture, 367
14.5 Concluding Observations 371
Recommended Readings 375
Exercises 375

15. Policy Analysis 378
15.1 Introduction 379
15.2 Policy and Policy Studies 379

15.2.1 A Description of Policy, 379
15.2.2 Policy Studies, 381
15.3 Models of the Policy Process 382
15.4 Policy Analysis in TA /EIA 384
15.4.1 Objectives of Policy Analysis, 384
15.4.2 Policy Analysis Parameters, 385
15.4.3 A Policy Analysis Model for TA /EIA, 388
15.5 Some Lessons from Actual Experience 394
15.6 Concluding Observations 397
Recommended Readings 397
Exercises 398

16. Communication of Results 400

16.1 Introduction 401
16.2 Two Communities 401
16.3 Uses of TA /EIA 403
16.4 The Communication Process 404
16.4.1 Before the Study, 404
16.4.2 During the Study, 406
16.4.3 The Report, 411
16.4.4 After the Study, 412
16.5 Innovative Approaches 416
16.6 Concluding Observations 417
Recommended Readings 418
Exercises 418

17. Project Management 420

17.1 Introduction 421
17.2 Management from the Sponsor's Perspective 421
17.3 Structural Features of the Assessment 423
17.3.1 Boundary Conditions, 424
17.3.2 Project Team Characteristics, 425
17.4 Process Features of the Assessment 427
17.4.1 Project Scheduling, 428
17.4.2 Communication Patterns Among the Project Team, 432
17.4.3 Integration, 435
17.5 Concluding Observations 438
Recommended Readings 439
Exercises 439

18. Evaluation of Technology Assessments and Environmental Impact Statements 442

18.1 Introduction 443
18.2 Criteria and Purposes for the Evaluation of TA /EIAs 444
18.3 Design of Evaluations 446
18.4 Practical Considerations: The Human Element in Evaluation 450

18.5 Current Evaluation Practice 451
18.6 Concluding Observations 452
Recommended Readings 453
Exercises 453

19. Critiques of TA/EIA 455

19.1 Introduction 456
19.2 Substantive Critiques of TA /EIA 456
19.2.1 Attitudes Toward Technology, 457
19.2.2 Relation of TA /EIA to the Political Process, 459
19.3 Methodological Critiques of TA /EIA 460
19.4 Critiques of TA /EIA Institutions 464
19.5 Concluding Observations 467
Recommended Readings 468
Exercises 469

20. Future Prospects 471

20.1 Three Scenarios 472
20.2 General Observations 474
Exercises 475

References 476

Index 505

Preface

Technology Assessment (TA)? Environmental Impact Analysis (EIA)?

What are they?

Why do them?

How are they done?

Who does them?

What are their faults?

This guidebook is our response to the above questions. It is particularly intended for use by practitioners of technology assessment and/or environmental impact analysis (TA/EIA) and for college students studying these topics. It is suitable for graduate or undergraduate students interested in the design and conduct of impact assessments. A quantitative background is not required.

We aim to acquaint the reader with a growing number of systematic tools to anticipate the indirect, and sometimes unwanted, effects of a new or modified technology or technology-based project. The practice of this "look before you leap" analysis has developed rapidly since the U.S. Government adopted the National Environmental Policy Act of 1969 requiring environmental impact statements. TA/EIA is essentially a decade old as we write this book. Much has been written on what it should be, but relatively little has been written on what has been learned from practice to date. We attempt to present a balance between conceptual and practical facets of TA/EIA, leaning a bit toward the latter.

The idea of combining TA and EIA, as implied by our use of the term TA/EIA, emerged during the preparation of the book drafts. Until now, these activities have been largely separate practices with distinct sets of professionals writing about and performing them. As we attempted to determine the boundaries between them, we saw great benefit in considering these activities together. We want to emphasize strongly that EIA is not restricted to consideration of effects on the natural environment; it is mandated to consider the "human environment." This means that the scope of coverage of an EIA can be fully as broad as a TA. (Hence, Chapter 10 on environmental analysis is not equivalent to EIA; it is only one part of EIA.) Establishment of a greater dialog between the practitioners of TA and EIA is one goal of this book.

A word about the order of presentation of material that we have followed: Based on teaching experience with the guidebook drafts, we believe that the best strategy is to get quickly into the "how" of doing impact assessment. To this end the first three chapters describe what TA and EIA are, and place them in present societal context. Chapters 4 and 5 set forth the general strategy and essential steps in doing TA/EIA. Chapters 6–16 then provide the meat to put on the skeleton of the steps—the methods by which one can produce a TA/EIA.

In Chapters 17 and 18, we discuss another group of "how to" issues involving managing and evaluating assessments. These place the analytic activities in the context of practical project considerations. The final two chapters first raise the broad range of critiques of TA and EIA, and offer our suggestions for the future of impact assessment.

Individual chapters can be used out of sequence. We have attempted to make material easily accessible via the index and considerable cross referencing in the text. In addition, the main analytical techniques are summarized and indexed in Table 5.3 (p. 82). TA/EIA is not a linear process that step by step takes one to a quality assessment; rather it is a circular process demanding continual iteration of analytical portions based on new insights gained by other analyses. We try to suggest important feedback points within a study. We also caution the reader that this (and any other) book has a seductive linearity (Chapter 1 leads to Chapter 2, to Chapter 3, etc.)—one must override this with alert common sense to develop a sound TA/EIA.

We have employed a number of editorial tools to make the book more readable and useful. Each chapter begins with an overview to indicate what the reader can anticipate. Each closes with concluding observations to highlight the "take-home" lessons. *Cameos* are presented throughout as "asides" and examples of the topics under discussion; these can be identified by a surrounding border. Also, the chapters contain exercises designed to complement the text material and recommended readings to supplement it. A Glossary of Abbreviations and an Index may be of use. Some references are flagged to indicate the type of material they represent.

We invite reader comment on the content, style, recommendations, and exercises.

Acknowledgments

Ron Larson was deeply involved in the conceptualization and initial structuring of this effort. After moving to the Solar Energy Research Institute in 1977, he continued to critique the manuscript.

Jeff Tiller was the primary author of the critical chapter on economic impact analysis (Chapter 11).

We thank Georgia Tech and Rose-Hulman Institute for their aid. Georgia Tech's School of Industrial and Systems Engineering and the Department of Social Sciences compatibly scheduled the technology assessment course. We deeply thank Vicki DeLoach, Gwen Nance, and Anita Bryant, who beautifully prepared draft upon draft from some thoroughly mutilated materials (an unintended consequence of multiple authorship). Shelley Easton and Debbie Turner of Rose-Hulman deserve similar thanks. Any author knows that without the patient aid of a truly competent secretary, naught else is possible.

We especially thank the students who took our jointly listed course through Industrial and Systems Engineering, Philosophy and History of Science, or Electrical Engineering—Winter and Summer 1977, and

Spring 1978 at Georgia Tech. They read and criticized the successive drafts and performed the exercises. Our current readers, too, owe them thanks for trying to get four academics to organize materials, cull jargon, and produce comprehensible text and exercises.

A special note of thanks is due three colleagues at Rose-Hulman. Thad Smith both uncovered problems and suggested their resolution. James Eifert and Thomas Mason added their support and insight.

In addition to subjecting our students and colleagues to the drafts of this book, we have imposed on our other peers as well. We deeply appreciate the time and effort they took to provide us with valuable comments. (Could the reader see the drafts, the enormity of the help would be obvious!) We first thank three colleagues who went over the entire manuscript, sharing their insights as they went:

Kan Chen (University of Michigan)

Chris Hill (Washington University and Office of Technology Assessment)

Andrew Sage (University of Virginia, and our Series Editor)

We also sincerely thank the following professionals who reviewed one or more chapters in depth:

Sherry Arnstein (National Center for Health Services Research)

Mark Berg (University of Michigan)

Margaret Boeckmann (George Washington University)

Barry Bozeman (Syracuse University)

Cliff Bragdon (Georgia Tech)

James Bright (Industrial Management Center)

Stanley Changnon (Illinois State Water Survey)

Joseph Coates (Office of Technology Assessment)

Lawrence deBivort (IR&T)

Jerry Delli Priscolli (Corps of Engineers—Institute for Water Resources)

Ed Dickson (SRI International)

Willis Harman (Stanford University)

Martin Jones (Impact Assessment Institute)

Don Kash (University of Oklahoma)

Dennis Little (Congressional Research Service)

Theodore Lowi (Cornell University)

Donald Michael (University of Michigan)

Ian Mitroff (University of Pittsburgh)

Srikanth Rao (Pennsylvania State University)

Peter Sassone (Georgia Tech)

John Stover (The Futures Group)

Fred Tarpley (National Science Foundation)

Sarah Taylor (Corps of Engineers—Baltimore District)

Albert Teich (George Washington University)

Evan Vlachos (Colorado State University and National Science Foundation)

Carol Weiss (Columbia University)

Jack White (Environmental Protection Agency)

Gene Willeke (Miami University, Ohio)

C. P. Wolf (CUNY Graduate Center)

We also received helpful guidance from the staff of the Atlanta Regional Environmental Protection Agency and Department of Interior. Obviously, we alone are responsible for final interpretations and any errors.

We would also like to thank Elsevier North Holland and our patient editors, Ken Bowman and Michael Gnat.

Above all, we owe our appreciation to our wives (Claudia, Tina, Dorothy, and Karen) and children for their long-suffering patience.

Glossary of Abbreviations

CEQ	The President's Council on Environmental Quality (they review EISs)
EIA	environmental impact analysis (or, equivalently, environmental impact assessment)
EIS	environmental impact statement
EPA	U.S. Environmental Protection Agency
EPISTLE	Economic, psychological, institutional/political, social, technological, legal, and environmental impacts (all forms of societal impacts)
NEPA	National Environmental Policy Act of 1969 (required preparation of EISs)
NSF	National Science Foundation
OTA	Office of Technology Assessment, U.S. Congress
SIA	social impact assessment
TA	technology assessment
TA/EIA	generic term for impact assessment
TA/EIS	technology assessment/environmental impact statement
TDS	technological delivery system
TIE	The Institute of Ecology

A Guidebook for Technology Assessment and Impact Analysis

1

Introduction

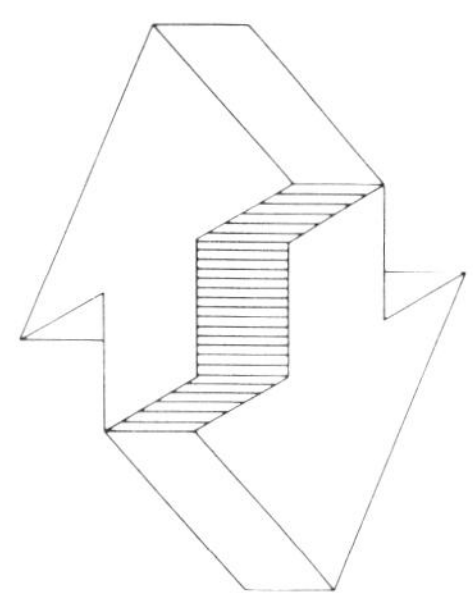

This chapter introduces the concepts of technology assessment (TA) and environmental impact analysis (EIA). It first relates these to the societal context that spawned them. Then TA and EIA are described and compared with each other and with other sorts of studies. Both differences and similarities between TA and EIA are noted. A brief description is presented of what is involved in TA/EIA, who does it, who uses it, and for what purposes it might be used. Finally, the technique of brainstorming is explained.

1.1 THE CONTEXT

Humans address the problem of survival in two basic ways. The first is to modify the effects of a hostile environment through the construction and use of artifacts such as shelters, clothing, and dams. At the same time, ways are found to force that environment to yield the materials necessary for human survival. This approach is technological because it makes use of tools, artifacts, and all the components of a material technology to alter the world in ways beneficial to mankind.

The second approach to survival is, by contrast, behavioral. Proponents of this approach assert that the human mind, within limits, is capable of redefining needs and of adopting lifestyles in which the dependence on material technology is minimized. The first strategy embraces the material world and tries to alter it. The second employs the considerable spiritual and psychological capacities of our species to adapt to the environmental realities of this world. Although no society has ever adopted one approach to the exclusion of the other, a variety of mixtures of these approaches have been attempted.

For a variety of complex reasons, Western civilization has elected the technological approach as the premier survival strategy. The results have been spectacular, especially during the past two centuries. This period has witnessed an industrial revolution powered in large part by newly developed sources of nonrenewable energy. New "social technologies," such as banking systems, industrial arrangements, and management information systems, have complemented the physical advances. Within this framework, items for material comfort have emerged in a steady flow. Living shelters have become artificial environments nearly independent of their surroundings. Additionally, human alteration of the biosphere—the life-sustaining mantle of the earth—has begun to rival in scale and power the forces of nature itself.

Why have we been so successful? Hans Jonas, a leading philosopher of technology, suggests that it is due in part to the analytic character of the scientific method that can reduce the complexities of nature to a few fundamental principles (e.g., Newton's laws and the laws of thermodynamics). Science provides a way to selectively focus our efforts; technology then produces, tests, and diffuses innovations, often based on scientific principles, to solve specific problems.

The pressures of economic competition reinforce the reductionist nature of the analytic approach to lend a discrete, and sometimes piecemeal, character to our technological solutions. Thus, whereas the

Western brand of survival strategy has been particularly adept at reducing practical problems to manageable size, it has done so at the expense of the "big picture." Although the analytic method abstracts and simplifies, the whole of nature remains complex, interconnected, and, as we are beginning to perceive, somewhat fragile. A kind of "tyranny of small decisions" has too often determined the drift of technology, and too few provisions have been made for evaluating the sum total of such decisions. Technology assessment and environmental impact analysis can be seen as attempts to modify this situation.

1.2 TA AND EIA: DESCRIPTIONS AND DISTINCTIONS

Joseph F. Coates (1976b: 372), a guiding force in the TA movement, offers the following definition of TA as

> a class of policy studies which systematically examine the effects on society that may occur when a technology is introduced, extended or modified. It emphasizes those consequences that are unintended, indirect, or delayed.

EIA owes its emergence as a major activity to the U.S. National Environmental Policy Act [NEPA (1969)]. Section 102C of that act requires the preparation of an environmental impact statement (EIS):

> for legislation and other major Federal actions significantly affecting the quality of the human environment, a detailed statement by the responsible official on—
>
> (i) The environmental impact of the proposed action, ...

In response to this mandate, and under the general guidelines of the Council on Environmental Quality, the federal government has institutionalized EIA. From that base, EIA has been adopted by other countries and many states. It has led to the advance in sophistication of the analysis of impacts on the natural environment, such as changes in water quality and ecosystem stability. EIA also treats human environments in terms of economic costs and benefits, as well as social and aesthetic considerations.

TA/EIAs range in scope from broad *problem-oriented* studies, which may examine a major societal problem (e.g., energy shortage), through *technology-oriented* assessments intended to anticipate the effects of a particular technology (e.g., videophone), to the most specific form, the *project* assessment, which focuses on a single application of technology (e.g., a power-plant siting).

TA and EIA have differences but also share similarities. Although various distinctions between TA and EIA are discussed throughout this book, it is appropriate to emphasize two at this point—geographical extent and emphasis on policy. TA usually deals with a technology that could be geographically situated almost anywhere, whereas EIA is more likely to consider a site-specific technology (say, a particular dam or highway segment). In terms of policy, TA is likely to explore a wide range of possible subsidies, incentives, regulations, and so forth. EIA, on the other hand, is more often focused on the impacts of "go/no go" decision alternatives relative to a particular project. Another distinction between the two assessment types is that EIA is procedurally structured (by NEPA requirements for federal EIAs), whereas TA is not. Chapter 4 explores the distinctions between TA and EIA and their implications for assessment practice.

The differences just noted are not always clear-cut. Program EIAs address issues of national extent and, often, heavy policy concern (e.g., the plutonium recycle EIA described in the Chapter 5 Appendix). The scope of a given assessment can vary widely depending on such factors as the funding level and time frame of interest. In general, however, both TA and EIA are broader than activities such as market research, program planning, and cost-effectiveness studies—all of which display some similarities to TA/EIA. Both TA and EIA focus on technological issues. They involve more than technological forecasting, however, because of the systematic consideration of such matters as environmental and social impacts. They also share a futuristic orientation, consistent with the notion of planning. But because this orientation is technology specific, they are not so broad as some futuristic studies that attempt to produce comprehensive scenarios of the future.

This book attempts to establish a combined perspective on "TA/EIA." In so doing, we hope to consolidate the distinctive knowledge gained in each of these areas. For instance, the broad TA perspective and the effective EIA analytical work complement each other. Concurrently, we shall contrast TA and EIA where appropriate, with a particular interest in real operational differences. For instance, social impact analysis proceeds quite differently in a local project assessment than in one of national scope.

1.3 THE "WHAT" AND "WHO" OF TA/EIA

What are the elements that comprise a thorough TA/EIA? This guidebook attempts to provide a workable answer to that question and to indicate how these elements might be executed and assembled to form a final product (see Chapter 4). A TA/EIA should begin with a careful definition of the problem. It must accurately describe the state of the art of the relevant technology and forecast the changes that the technology is likely to undergo. It should also describe the relevant features of the society, its values, and ways of managing the technology, and should anticipate how these factors might limit and shape the context in which the technology must function in the future. It must identify areas of the social and physical environments that are likely to be impacted by the technology, and analyze the character and extent of significant impacts. Policy options open to decision makers for the modification of expected impacts must also be identified. Finally, since no TA/EIA can be effective unless its results are communicated to the user, it should be presented in a clear, logical, and effective manner.

Who performs TA/EIA studies? Typically, a team of technically trained individuals. One might expect to find environmental specialists, engineers, economists, systems analysts, social scientists, and physical scientists on an assessment team. Lawyers, humanists, and representatives of interest groups may also participate. Interdisciplinary research is a difficult task (see Chapter 17), but since TA/EIA is intended to provide a holistic picture of technological change, extensive interaction among a range of professionals is necessary. Assessors may typically work for an agency responsible for the assessment or for a contract research firm or university hired to work on the study.

Who uses the results of TA/EIA? A U.S. Senate study (Senate Rules and Administration Committee 1972) listed among those groups with an understanding and inclination to use TA: (1) most federal executive departments, (2) state legislatures, (3) universities, and (4) many "high-technology" industries throughout the developed countries. Not surprisingly, user interests vary considerably (see Chapter 16). EIA is, of course, required for major U.S. federal projects and also by many states and other countries. Other agencies and interested publics use the EIA process as a vehicle to understand and comment on proposed developments. Most importantly, the agency preparing the EIS is the primary user of the assessment, ideally incorporating insights gained into its planning process.

Finally, what are the uses of a TA/EIA? These depend on the particular needs of the users. Naturally, a proprietary study by a firm would not be handled in the same way as a public study. Arnstein and Christakis (1975: 16) suggest a range of possible outcomes of a well-executed assessment:

1. Modify the project.
2. Specify a program of environmental or social monitoring.

3. Stimulate research and development, particularly to deal with adverse effects of the technology.
4. Stimulate research to specify or define risks.
5. Develop latent benefits.
6. Identify regulatory and legal changes to promote or control the technology.
7. Define institutional arrangements appropriate to the technology.
8. Define intervention experiments to reduce negative or enhance positive consequences.
9. Delay the project or technology until some of the preceding steps are completed.
10. Stop the technology.
11. Provide a reliable base of information to parties at interest.

In sum, this new form of policy analysis has gained wide visibility through federal government assessment sponsorship and EIS requirements. In concept, the notion is as sensible as the adage "look before you leap." In practice, TA/EIA is an uncertain activity demanding broad-ranging, yet careful, analyses, but promising both immediate and long-range benefits in return.

Socrates claimed that the unexamined life is not worth living. TA/EIAs are positive attempts to examine the context of human life in a technology dominated world and to determine future options that are, in the best sense, desirable.

1.4 BRAINSTORMING

Before delving into the context of TA/EIA (Chapters 2 and 3) and "how to do TA/EIA" (beginning with Chapter 4), it is useful to get a feel for what a TA/EIA is. Toward that end we now introduce one of the techniques potentially useful in TA/EIA, brainstorming. This technique can then be used in Exercise 1-1 or 1-2 to explore the content of a sample assessment topic.

Brainstorming is an idea-generation technique (Osborn 1957). In essence, it is nothing more than an effort to stimulate creative thinking on a topic by explicitly removing censorship of ideas. The "cameo" entitled "brainstorming rules" presents the ground rules.

Brainstorming can be conducted either in groups or by individuals who then pool their ideas. The literature indicates no clear relationship between group size or group interaction pattern and the number and richness of ideas generated (Lewis et al. 1975).

By nature, TA and EIA find use for brainstorming in a variety of ways, such as identification of alternative technologies, and identifica-

BRAINSTORMING RULES

1. Criticism of any sort is barred, both of one's own ideas and those of others. Avoid "killer phrases" such as "that's ridiculous" and "it won't sell."
2. Quantity of ideas is a primary objective.
3. Unusual, remote, or wild ideas are sought—freewheeling is welcome.
4. Combinations, modifications, and improvements on ideas are encouraged. It is therefore important that every idea offered is kept visible to the participants (e.g., on a blackboard) to generate additional ideas.

tion of ways in which effects of a development may interact with each other to produce what are called "higher-order" effects.

1.5 CONCLUDING OBSERVATIONS

This chapter has presented a brief rationale for the development of TA and EIA in our technological society. It has gone on to note several attributes of TA/EIA:

They are both systematic, prospective studies of the range of potential effects of proposed technological developments.

They can differ on several dimensions, particularly in: (1) geographical localization, where TA is typically location independent and EIA is location specific, (2) the range of allowable policy considerations, where TA is typically broad and EIA narrow, and (3) their procedural ground rules, where EIA is fixed by law and TA is not.

They consist of identifiable study components (see Chapter 4).

They are performed by teams of professionals with varied backgrounds, working sometimes for governmental agencies, private firms, or universities.

They can be used by policy makers and interested parties in a number of different ways.

Finally, the technique of brainstorming is described to enable the reader to get a taste of TA/EIA in the following exercises.

EXERCISES

1-1. (Project-suitable) This might be called a "brainstorm TA."[1] One person (e.g., class instructor) should research a particular technological issue beforehand and come prepared to lead a group brainstorm on a specified topic.[2] The topic should be carefully circumscribed so that it can be satisfactorily explored in the time available (about 2 hours is recommended). The following steps are suggested:

1. Leader presents the issue [e.g., the adoption of electronic point-of-sale monetary transfers on all retail transactions—see Arthur D. Little, Inc. (1975).]

2. Leader briefly describes the basic technology involved, its current state of development, and essential social context (e.g., electronic fund transfer technology is available; various institutional factors are important, such as banking regulations).

3. Group brainstorms the issue, the technology, and the context loosely. Each participant then suggests a factor or so worthy of consideration in formulating an assessment in this area. The leader takes these ideas into account to specify a tightly bounded issue for further group assessment (e.g., electronic point-of-sale fund transfers for the state of Iowa over the next decade).

4. Group brainstorms potential *impacts* of the development in question (e.g., electronic fund transfers could eliminate informal "float" credit and raise security dilemmas).

5. Group brainstorms possible *policy options* to mitigate negative impacts and accentuate positive ones (e.g., Iowa imposes security requirements on electronic fund operations).

6. Discussion of major points of emphasis for a TA on the issue addressed and likely uses of such a study.

1-2. (Project-suitable) Apply logic similar to Exercise 1-1, but to an EIA. Modify the sequence to conform to EIA practice (e.g., see Chapter 10):

[1]This exercise is modeled on the successful practice of Joseph Coates of leading "mini-TAs" that last several hours on a given topic.

[2]Topical ideas may be generated through scanning the references at the end of the book for actual TAs. These are excellent sources, in turn, for issue identification and technical background.

1. Description of the action and the setting (note land use plans, institutional authorities, etc.).
2. Alternative actions.
3. Potential impacts of the proposed action and of the alternatives.
4. Irreversible commitments and countervailing policy considerations that offset the adverse environmental effects identified.

2

Technology and Society

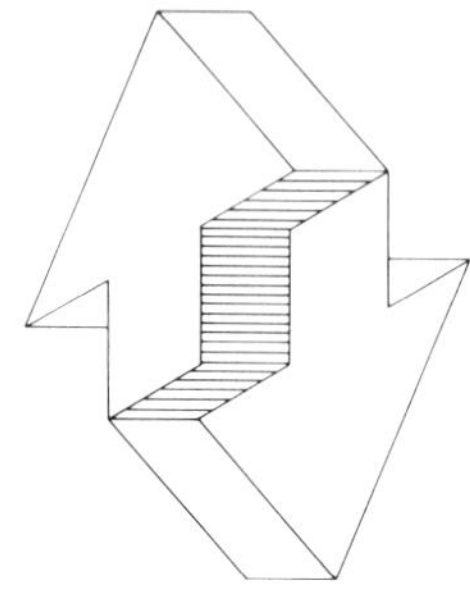

This chapter deals with the relationship between technology, taken broadly, and society. It is intended to provide a background against which the development and performance of TA/EIA can be considered in the remainder of this text. The chapter emphasizes the causal relationships existing between technology and society and the role of values in modern technology and its assessment. The work of a number of thinkers is summarized so that the reader may gain insights from a diversity of views on these questions. One's perspective on these larger issues may significantly influence how one perceives a particular technological development under assessment.

2.1 INTRODUCTION

Technology is a pervasive factor in modern society. It is often taken for granted as an unobtrusive, but necessary, backdrop to life, like the air we breathe. It is generally conceded that ours is a technological society, with the technological component and its ties to human society intensifying with the passage of time.

However, the technological society is not without its problems. In many cases technological developments have been seen as causing social problems. Lawless (1977) lists 45 cases of "social shock" caused by technological developments that went wrong (Table 2.1).

The notion of assessing technology has originated from the convergence of two observations: (1) that technology is a crucial force in modern society and (2) that technological development can go awry.

The assessor's conception of the character of technology and its rela-

TABLE 2.1. Technological Case Studies[a]

1. Human artificial insemination	25. The rise and fall of DDT
2. Oral contraceptive safety hearings	26. Asbestos health threat
3. Southern corn-leaf blight	27. Taconite pollution of Lake Superior
4. The great cranberry scare	28. Foaming detergents
5. The diethylstilbestrol ban	29. Enzyme detergents
6. The cyclamate affair	30. Nitrilotriacetic acid (NTA) in detergents
7. Monosodium glutamate (MSG) and the Chinese Restaurant Syndrome	31. Saltville—an ecological bankruptcy?
8. Botulism and "bon vivant"	32. Truman reservoir controversy
9. The fish-protein-concentrate issue	33. Plutonium plant safety at Rocky Flats
10. The flouridation controversy	34. Storage of radioactive wastes in Kansas
11. Salk polio-vaccine-hazard episode	35. Amchitka underground nuclear test
12. The thalidomide tragedy	36. The Dugway Sheep Kill incident
13. Hexa-, hexa-, hexachlorophene	37. The nerve-gas-disposal controversy
14. Krebiozen—cancer cure?	38. Project West Ford (orbital belt of copper needles)
15. Dimethylsulfoxide (DMSO)—suppressed wonder drug?	39. Project Sanguine (giant underground transmitter)
16. X-ray shoe-fitting machine	40. Project Able (space orbital mirror)
17. Abuse of medical and dental X-rays	41. The chemical mace
18. X-radiation from color TV	42. The bronze horse (debate over antiquity of a work of art)
19. Introduction of the lampreys	43. Synthetic turf and football injuries
20. The Donora air-pollution episode	44. Disqualification of Dancer's Image
21. The Torrey Canyon disaster	45. The AD-X2 battery-additive debate
22. The Santa Barbara oil leak	
23. Mercury discharges by industry	
24. The mercury-in-tuna scare	

[a] From Lawless (1977). Copyright © 1977 by Rutgers, The State University of New Jersey. Reprinted by permission of Rutgers University Press. Another 56 cases are also briefly discussed in the same work.

tionship to society underlies the performance of any TA/EIA. It is, therefore, valuable to explicitly examine these matters before getting into assessment specifics.

This chapter considers broad issues in the relationship between technology and society that affect how TA/EIA is conceptualized and performed. It begins with an attempt to define the complex notion of technology and proceeds to discuss various views of the relationships between technology and society. In this treatment we focus on the mutual influences between technology and society and the consequences of these views for TA/EIA. The subject is a rich and complex one with important contributions from incredibly diverse perspectives. This chapter can merely touch on the most significant concerns that underlie TA/EIA.

2.2 WHAT IS TECHNOLOGY?

There is ambiguity in the use of the word technology. This ambiguity is compounded by historical changes in the extent and character of activities considered to be technological. Therefore, before offering an explicit definition, we shall explore the evolution of modern technology.

Before tracing the development of technology, it is necessary to draw a critical distinction. In one usage "technology" refers to technological ways of doing things. Thus "Western technology" refers to the totality of technological activities used by the Western nations. But "technology" also possesses a less encompassing sense, as in "solar-energy technology" or "pencil-manufacturing technology." The term "technology" in "technology assessment" refers to the second, more restricted usage. That is, it pertains to specific instances of technological development. However, in this chapter we consider technology in its first (broader) meaning to set the stage for the assessment of specific technologies.

Until fairly recently, technology could generally be differentiated from other forms of human activity by its direct involvement with the material world. Persons involved in the practice of technology shaped and altered their physical environment with artifacts fashioned to meet their needs. Other modes of human activity, such as politics, were distinguishable from technology. Furthermore, technological activities did not have high status in society. From classical times a disparagement of the practical and glorification of the contemplative relegated even the most skilled artisan to an inferior position.

By the end of the 19th century, however, important trends were in motion that would change the basic character of technology. In particular, the physical sciences gradually began to influence technological practice. The laboratory began to rival the shop or farm as a source of new inventions and techniques. The firm grounding of technology in scientific theory in fields such as chemistry and electricity provided the technologist a focus vastly more powerful than trial and error approaches. Coincidently, the scientist came increasingly to depend on the technologist for the design and construction of experimental apparatus necessary to verify scientific theories. Thus science and technology became partners and both prospered as a result.

As the 20th century progressed, new institutions were organized to exploit the merger of scientific theory and technological know-how. Large laboratories, often supported by the government, employed large numbers of scientists and technologists. These new "big science" organizations represented an increase in the commitment of national resources to scientific inquiry. In view of this commitment, they were generally directed toward the solution of specific classes of problems, often military, deemed to be in the national interest. Such mission-oriented organizations represented a major institutionalization of science and technology as a formal tool for achieving governmental policy and corporate development. They helped lower the boundaries separating government, industry, and the university.

Whereas the latter part of the 19th century marked the beginning of technology grounded in physicochemical theory, the post-World War II era witnessed the birth of what Bunge (1966: 331) has termed "operational technologies." At the heart of these new technological developments were advances in applied mathematics that facilitated operations research, information theory, and systems engineering. The mushrooming power of the electronic computer greatly aided these developments. These techniques sparked what may be considered a second industrial revolution in which information replaced energy as the key resource.

With the advent of the information-processing revolution, the line dividing technology from other forms of human activity has become much less clear. Queuing theory, for example, can be applied as easily to people passing through toll gates as to machine parts moving through inspection points. Systems models facilitate placing a man on the moon as well as suggesting optimal pricing arrangements between oil-producing and -consuming nations. Virtually every area of activity in modern society has been affected by the information-processing revolution. Communications, business inventory control, banking, military defense strategy, medicine, and government record keeping, for example, have all been directly shaped and altered.

Any adequate definition must accurately reflect the modern,

broadened spectrum of technological activity, not just the making of artifacts. Gendron's definition (1977: 23) adequately delineates the current meaning of technology:

> A technology is any systematized practical knowledge, based on experimentation and/or scientific theory, which is embodied in productive skills, organization, or machinery.

Although this definition includes a broad spectrum of activities, it excludes techniques based on faith or mysticism, productively useless skills, and knowledge that, although possessing an abstract potential for practical utility, has not been utilized for production, organization, or machinery.

2.3 TECHNOLOGY IN THE SOCIAL CONTEXT

2.3.1 Setting the Scene

Although nearly everyone possesses a mental model of how society is structured and functions, few of us in our everyday lives have reflected on the subject to any great extent. The objective of this section is to highlight fundamental assumptions about the relationship between technology and society. For example:

What is the proper role of technology in society?

Does technology cause social change?

Does society cause technological development?

Is technology a neutral instrument in societal process, or are some of its forms of organization inconsistent with the changing goals, values, and aspirations of society?

Is a maximization of the values intrinsic to modern technology consistent with the broader ideals of a free society?

These are questions of great scope to which a bewildering variety of answers has been offered. In the following paragraphs some of these are surveyed to sharpen perceptions of the modern technological society. The discussion focuses on the question of the causal influences between technology and society. Two other important recurrent issues cutting across these concerns are treated. The first is the question of whether *technology is in itself neutral* with respect to human values. The second issue is whether a *value-free analysis of technology* (in either the broad or the narrow sense of "technology") is possible. These two issues are not independent, and the second is an ongoing concern of this book.

The issue of the neutrality of technology revolves around the claim that technology should be viewed as morally neutral in and of itself. Proponents of the neutral view assert that technology is an enterprise with potential for ill as well as good. A hammer, for example, can be used to build a cathedral or to crack a human skull. Therefore, it is not technology that is to blame if things go wrong. Rather the responsibility lies with the human actors—the designers, implementors, users, and regulators. The neutrality argument effectively shifts attention from the technology to the intentions of those who apply it. If the assertion of technological neutrality is accepted by the assessor, there are obvious implications for the formulation of policy to deal with new or expanded applications of technology.

The second issue concerns the possibility of an assessment free from value intrusions. The notion of value-free assessment springs from a transference of the scientific approach that claims to yield objective information free of personal wishes, feelings, or prejudicial judgments. Patterning their methods after those of the physical sciences, social scientists have generally adopted the scientific approach. Political scientists, the argument goes, can look at voting trends or the performance of government organizations as objectively as a physicist describing particle traces in a cloud chamber. Sociologists can document the values of others without the fear that personal norms will jeopardize their objectivity. Economists, especially, rely on the scientific method and the analytic and quantitative character of economic theory to enable them to do their work in a "value-free" manner. The extension of the value-free concept to the analysis of highly technological issues holds that the assessor can remain sufficiently aloof from partisan positions to generate a range of objective and balanced technological options.

In the following sections we present three views of the relationship between technology and society. The first emphasizes the causal influence of technology on society and its values, with the effect of society on technology treated as a less important "feedback loop," as diagrammed in Figure 2.1a. The second view reverses the primary influence pattern; society and its values primarily determine technological development (Figure 2.1b). The third view emphasizes the mutual causal relationships between technology and social forces. Whether societal or technological influence is dominant varies with time and context. Figure 2.1c illustrates this view. The issues of the neutrality of technology and value-free assessment are significantly affected by the view of the society–technology relationship held by the assessors, and they are discussed as these views are presented.

In actuality, the following three sections juxtapose a variety of

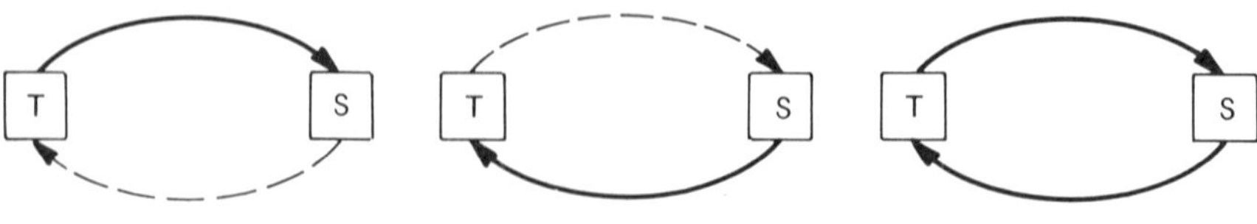

(a) Technology causes social change

(b) Society and its values drive technological change

(c) Technology and society are related by mutual causality

FIGURE 2.1. Causal relationships between technology and society.

deeply thoughtful viewpoints. The categorization of these into three perspectives is a simplification. Furthermore, any partition between technology and society should not be taken as literally true in our "technological society."

2.3.2 Technology Causes Social Change

The first of the three views has been advanced by diverse thinkers. A frequently cited statement of this position is that of Emmanuel Mesthene, who expressed the causal linkage in four steps (1968: 135–143):

1. Technological advance creates a new opportunity to achieve some desired goal;
2. this requires (except in trivial cases) alterations in social organization if advantage is to be taken of this new opportunity,
3. which means that the functions of existing social structures will be altered,
4. with the result that other goals that were served by the older structures are now inadequately achieved.

Mesthene sees society as basically reacting to technology, rather than leading it. New technology alters the range of available choice. It provides new ways of doing old things, as well as totally new options. Yet at the same time new technologies have a way of removing old options. Modern plumbing, for example, displaced the way of life that surrounded the village pump. Mesthene, however, stops short of the suggestion that particular technologies control lifestyle and seriously restrict human freedom. He calls his point of view *soft determinism*, and supports the view of historian of technology Lynn White, Jr. (1966: 28), who observed "a new device merely opens a door; it does not compel one to enter." Mesthene, however, admits that political pressures and blind faith in progress can sometimes make the lure of new technology virtually irresistible.

A central feature of Mesthene's picture of technological change is optimism. This optimism is not shared by the French philosopher–theologian Jacques Ellul. According to Ellul (1964), in the current situation technology is completely out of human control. It has a life of its own. It is the self-propelled driving force that determines the structure of society and can be brought under control only with incredible difficulty in improbable circumstances.

Perhaps an analogy can help to suggest the force of Ellul's position that seems, on its face, so implausible. If the reader has had occasion to confront some agency of the government, been shuttled from department to department, and faced the shrugs of a string of bureaucrats, he or she may well conclude that no one is really in charge. The bureaucracy has an inertia, a resistance to change, that seems to defy human purpose. The organization, with its self-protective mechanism, appears beyond human control. That, Ellul pessimistically concludes, describes modern technological society.

An interesting extension of Ellul's position has recently been provided by Langdon Winner (1977). Two concepts, *technological imperative* and *reverse adaptation*, are central to his viewpoint. The technological imperative holds that the kinds of decisions required by the technological order, rather than consciously selected ends directed toward human purposes, dominate in modern society. Thus specific technological objectives (e.g., more oil, better roads, and higher buildings) become the dominant policy concerns in place of more basic human needs. Such redefinition of human demands in terms of sophisticated and intertwined technologies causes mankind, ostensibly the master, to acquire a pathological dependence on technology, supposedly the servant.

In explaining "reverse adaptation," Winner wrote (1977: 227) that "technical systems become severed from the ends originally set for them and in effect reprogram themselves and their environments to suit the special conditions of their own operation." Thus, for example, tomatoes are bred for picking and shipping rather than for eating. Reverse adaptation arises in a perfectly logical manner when viewed within a technological frame of reference. A goal such as communication or transportation is pursued by specific means. These means require their own means, such as government subsidy, right-of-way legislation, foreign policy, and regulation. As a result, decisions are made to optimize the means without regard to the original purpose they were intended to serve. Thus reverse adaptation suggests that means may not only specify ends, but in some cases become ends themselves. A growing economy, for example, may become the goal to which human labor and lifestyles must adapt.

Another famous observer of human society saw technology as a dominant, though not independent, social force, but in a somewhat different light. Karl Marx (1964) argued that all social and economic organization rested on three factors: (1) the *means* of production—raw materials. land, and energy, (2) the *forces* of production—factories, machinery, technology, industrial knowledge, and the skills of those who labor, and (3) the *social relations* of production—ownership patterns, divisions of labor, and sources of political and economic power. Marx viewed the political state as the instrument through which the most powerful elements in society, those who own the means of production (industrial technology), maintain their privileged status and exert control over a vast underclass. Under feudalism, ownership of land created a class of lords and a subordinate class of serfs. Similarly in the industrial system, where capital rather than land creates political power, owners of capital control the state and define the rights and obligations of the underclass of workers.

Marx held that the forces of production tended to change faster than social relationships, which are by nature conservative of the status quo. Therefore, an instability would be created within the social order that would eventually lead to its breakdown. In the main, he saw the destabilizing factor to be technology. The emergence of industrial technology led to the concentration of workers and their subjection to the "discipline of the machine." As these tendencies intensified, workers would become progressively more alienated from their work. This alienation, exacerbated by the increasing substitution of machinery for human labor, would create a militant class of workers that would eventually rebel against capitalism. The result would be a new social order in which the means of production were owned communally and where industrial production would be based on need rather than the promise of economic gain.

David Noble's recent book, *America by Design* (1977), is an extension of Marx's critique of capitalism. Marx was primarily concerned with the forces that made the industrial revolution possible: (1) destruction of mercantilist restraint of trade, (2) emergence of a bourgeoisie who owned industrial capital, (3) creation of landless proletariat, (4) separation of the workers from their tools, and (5) the creation of the factory system. Noble argues that these developments have been superseded by a new stage of the industrial revolution wherein worker skills and technical knowledge have been expropriated exclusively for the service of corporate capitalism. Noble (1977: xxiv) traces the

> capitalist wedding of science to the useful arts along three intersecting paths—the rise of science-based industry, the development of technical

> education, and the emergence of the professional engineers—showing how each of these nineteenth-century developments reflected and contributed to the social process of technology as corporate social production.

Control of each of these developments has been secured through: (1) consolidation, patent monopoly, and merger (especially in the electrical and chemical industries) resulting in a relatively few extremely powerful corporations; (2) licensing and definition of appropriate technical education; and (3) professional engineering gradually gaining a monopoly over the practice of scientific technology.

In its various versions the view in which the causal influence of technology dominates is extremely influential. In effect, it may be interpreted as saying that technology is a necessary and, in some cases, sufficient condition for social development. (Some of the theorists perceive this for the better; others, for the worse.)

2.3.3 Society and Its Values Cause Technological Change

What of the role that society and its values play in encouraging certain technologies while discouraging others? Willis Harman examines this question in his book, *An Incomplete Guide to the Future* (1976). He concludes that values that have served society reasonably well in the past now seem to be causing problems. In fact, the entire complex of accepted beliefs, values, and implementation strategies—what he calls the *industrial state paradigm*—may be becoming disfunctional. At the very least, technology seems to be producing dilemmas (Harman 1976: 9).

> The basic system goals that have dominated the industrial era and that have been approached through a set of fundamental subgoals have resulted in processes and states which end up counteracting human ends. The result is a massive and growing challenge to the legitimacy of the basic goals and institutions of the present industrial system.

Harman specifies four particular dilemmas. First is a *growth dilemma*. Growth in the creation of goods and services seems essential to the economic health of the social system. However, society seems to be reaching physical limits in many areas, including energy, nonrenewable resources, and human population size. How can growth be maintained in light of such limits? Some claim this dilemma can be resolved through a new kind of growth that is coordinated and planned. But this leads to the second dilemma, that of *control*. If control or planning is required, it appears to violate Western society's long-standing commitment to free enterprise. Classical economic theory holds that the free market, for example, cannot regulate and self-correct if government planning interferes. Yet how can growth be coordinated and planned without such interference?

The third dilemma Harman calls the *distribution dilemma*. A healthy, growing economy generates benefits that trickle down to even the poorest members of society. Goods are thus distributed to a wider group than those responsible for their production. Yet as this distribution spreads to the poorest members of the world community, it would seem that the earth's resources are insufficient to sustain industrialized society's levels of material consumption throughout the world.

Finally, Harman suggests a *work-role* dilemma. The Puritan ethic of Western society generates strong feelings that human worth is closely tied to having a job. Yet modern industrial practices, including the preference for capital-intensive over labor-intensive alternatives, seem a cruel hoax, especially for the least skilled members of the work force.

Harman sees technology as shaped by dominant societal values. Effective technical change will thus require value shifts at all levels of society.

Major reorientations of societal values can occur rather rapidly during times of crisis. Normally, however, such alterations take place gradually over extended periods of time. Elgin and Mitchell note one such incremental change when they claim that the most rapidly growing section of today's consumer market is "people who don't want to buy much" (1977: 4). There are estimated to be from four to five million young, active, educated men and women within the United States who have adopted this lifestyle, which Elgin and Mitchell characterize as *voluntary simplicity*. They identify another eight to ten million as partial adherents to this lifestyle. At the heart of the concept of voluntary simplicity are five values: (1) material simplicity, (2) human-scale living and working environments, (3) self-determination rather than dependence on large complex public or private institutions, (4) ecological awareness, and (5) personal growth in psychological or spiritual directions.

Voluntary simplicity could be expected to have a number of effects on current technological practices. First, more concern could be anticipated for socially appropriate technology. Technologies may also be more closely scrutinized. Further, acceptable processes and products, in addition to the requirements of efficiency and productivity, may have to be demonstrably safe, nonlittering, energy conserving, nonpolluting, and made of renewable or recyclable raw materials. Thus Elgin and Mitchell have painted a picture of potentially dramatic technological changes flowing from shifts in societal values. They cite poll data to buttress this conclusion (Table 2.2).

TABLE 2.2. Shifts in Consumer Values[a]

92 % of Americans are willing to eliminate annual model changes in automobiles.
91 % are willing to have one meatless day per week.
90 % are willing to do away with annual fashion changes in clothing.
82 % are willing to reduce the amount of advertising urging people to buy more products.
73 % are willing to wear old clothes (even if they shine) until they wear out.
73 % are willing to prohibit the building of large houses with extra rooms that are seldom used.
57 % are willing to see a national policy that would make it much cheaper to live in multiple-unit apartments than in single-family homes.

[a] From Harris (1975); cited by Elgin and Mitchell (1977:11).

From a different perspective, the historian Lynn White, Jr. (1967) makes an analogous point regarding the effect of Judeo-Christian values on the development of technology. These beliefs make a sharp distinction between man and nature. They also refute the beliefs that nonhuman entities possess spirits. This reduces nature to a mere resource of man and removes important psychic barriers preventing its careless exploitation.

> The spirits in natural objects, which formerly had protected nature from man, evaporated. Man's effective monopoly on spirit in this world was confirmed, and the old inhibitions to the exploitation of nature crumbled (White, Jr. 1967: 1205).

Just as the views of Harman, Elgin and Mitchell, and White reverse the causal emphasis from the first perspective; so, in general, they inject societal values more significantly into technology and its analysis. Since technology is a direct function of social values, it is not neutral with respect to these values—it is their product. According to this view, the value-free character of assessment is less certain because of the dominant causal role values play in technological development. Indeed, assessment may take on a normative or planning character—by establishing a technological path from the present to a state in which goals relating to the social values are achieved.

2.3.4 Technology and Society—All of a Piece

The final view of the relationship between technology and society suggests a more interdependent connection than either of the previous two positions. All human experiences, basic perceptions, definitions of the good life, expectations for the future, and technological developments are shaped by the social, political, and technological milieu in which they exist. Thus, as shown in Figure 2.1c, the reciprocal causal links between technology and society are of equal importance. All arguments of this form reject as naive the suggestion that technology is merely a neutral instrument. Whether technology is seen as the most basic factor in defining and limiting the forms of social life or is tied to something even more basic (e.g., ownership patterns), it is never neutral, never simply an amoral instrument. Nor can the assessment of its consequences be free from the values of the social milieu.

Max Weber, a major figure in the establishment of sociology, emphasized the influence exerted mutually by social and technological factors. He supplemented Marx's causal argument by arguing not only that values result from the forces of production, but also that religious (Calvinist) values facilitated the rise of modern capitalism and its teachings.

Schumacher (1973) also stressed the importance of both linkages between society and technology. He argued that materialistic values lead to the large-scale production of goods. This development wastes material and human resources, which deteriorates the quality of society. Value shifts away from a materialistic ethic can induce a more benign technology. His "Buddhist economics" (1973: 54) conveys this contrast:

> While the materialist is mainly interested in goods, the Buddhist is mainly interested in liberation . . . . The modern economist is used to measuring the "standard of living" by the amount of annual consumption, assuming all the time that a man who consumes more is "better off" than a man who consumes less. A Buddhist economist would consider this approach excessively irrational: since consumption is merely a means to human well-being, the aim should be to obtain the maximum of well-being with the minimum of consumption.

Schumacher faults modern industrial practice for its tendency to place profit ahead of the right of everyone to have a proper job—one that yields sufficient livelihood, provides the setting for human cooperation, and offers the chance for the development of one's skills. He also criticizes the tendencies of industrial technology toward massive factories that require raw materials and finished products to be shipped over long distances and encourage centralized management bureaucracies and careless waste-disposal practices. He sees size itself as a problem. An emerging group in our society who call themselves variously "careful," "soft," "frugal," or "appropriate" technologists share his view. They emphasize small-scale, localized production, minimal dependence on daily motor travel, regional self-sufficiency, and low-impact approaches to growing food, home building, and community development. Such individuals regard the modification of technological practices and the creation of an alternative society as two sides of the same coin.

The mutual causality involved in this third view is quite complex. Technology, with its attendant goals of efficiency and rationality, sets a milieu in which these qualities permeate and influence many human activities, such as household routines and scheduling childrens' activities. Witness, for example, value conflicts surrounding genetic engineering developments. On the one hand, there is a scientific/technological pressure toward enhanced capability to reduce birth defects, and so on. On the other are grave concerns about preserving the sanctity of human life.

Higher-order cause–effect relationships generate complex social and technological changes. For instance, the automobile has contributed to air pollution. This has led to pollution-control regulations and development of pollution-control technologies. At the same time, Western society's eagerness for the automobile has led to increased fuel consumption. This has prompted energy-conservation measures, some of which have directly conflicted with the air-pollution-reduction measures (both social and technological).

In summary, a relationship of mutual causality exists between technology and society. This complex relationship leads to the creation of background conditions, immediate or first-order effects, and higher-order effects. Social values held by differing groups vary widely and are often conflicting. Likewise, the technologies available at a given time may not fit together consistently. Assessment involves isolating a part of this complexity for a closer look. But the very act of assessment is part of the sociotechnical environment and thus is potentially a link in the causal chain. Assessments may be performed from a variety of value-laden perspectives. Each assessment represents a point of view that is somewhat unique. Thus, in this view, technology is not free of social values, nor is its assessment ever value-free. What can be hoped for is to make explicit, as far as possible, the nature of this involvement. Because of its similarity to the problem of the priority of the chicken and the egg, the third view is not very comforting. It displays the full range of the potential complexity of our technological society. We believe, however, that it serves as the safest point from which to view the technology–society interaction. The first two views can be seen as limiting cases where one or the other type of causal influence is dominant.

2.4 CONCLUDING OBSERVATIONS

This chapter considers broad questions relating to the interaction between technology and society. In the process it considers "technology" to refer to a global phenomenon rather than to a single technology, as it does in the term "technology assessment."

We adopt Gendron's definition (1977: 23):

> A technology is any systematized practical knowledge, based on experimentation and/or scientific theory, which is embodied in productive skills, organization, or machinery.

We consider two important questions as background to the discussion of technology and society:

Is technology morally neutral in and of itself?

Can an assessment of technology be performed which is free of social values?

The viewpoints of a number of authors on the relationship between technology and society are contrasted. These are categorized according to three general perspectives:

Technology is a prime determinant of social change. Hence technology can be assessed relatively objectively, free of social values.

Society and its values steer technological development. This implies that social values enter into the assessment of technology (e.g., in the sense of planning for the future).

Technology and society are tightly intertwined by complex, mutual causal relationships. Since the components cannot be easily separated, the notion of value-free assessment of technology is not tenable.

All three perspectives lend useful insights; the last is most general.

RECOMMENDED READINGS

Gendron, B. (1977) *Technology and the Human Condition.* New York: St. Martin's Press.

Offers three views of technology in human society—utopian, dystopian, and socialist.

Harman, W. (1976) *An Incomplete Guide to the Future.* San Francisco: San Francisco Book Company.

Shows values underlying current technology and outlines value shifts the future may require.

Noble, D. (1977) *America By Design*. New York: Knopf.
A neo-marxist analysis of capitalist society's stimulation and control of technology.

Winner, L. (1977) *Autonomous Technology*. Cambridge, MA: MIT Press.
An answer to the question "What does it mean to say technology is out of control?"

EXERCISES

2-1. Select one or more of the following statements for discussion. Organize a debate pro or con, or write an essay taking a pro or con position. Be sure to cite specific examples to support your position.

1. Technology is never neutral. It defines new freedoms for us, yet it also cuts us off from other options we might wish to pursue.
2. Technology creates political power. It makes society less democratic.
3. Technology strengthens the control of an elite minority within the society.
4. The values embraced by our technological society may turn out to be counterproductive to the goal of human survival.
5. Technologies that served us well in the past may become disfunctional.
6. No values, not even efficiency or productivity, should be free from reflective assessment.
7. Technological change is not always progress.

2-2. Technology is one important way to solve human problems. What are some other ways? How do such solutions differ from technological solutions?

3

Institutionalization of TA/EIA

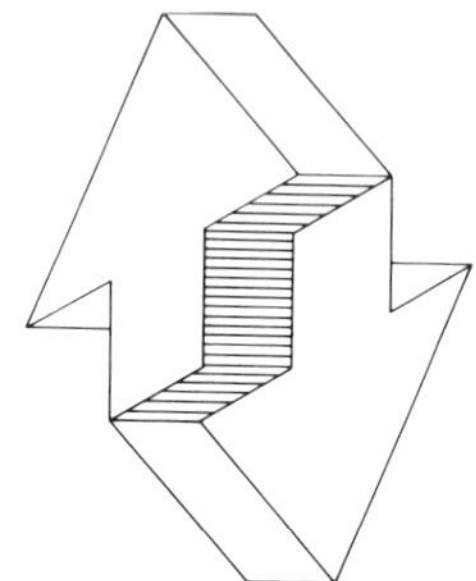

This chapter discusses the steps that have been taken to translate the concerns about technology, discussed in the previous chapter, into institutional form. In the United States, institutions have been created at the federal and state levels, and a few steps in this direction have occurred in the private sector. Primary emphasis is given to the passage of the National Environmental Policy Act of 1969 (NEPA) and the creation of the Office of Technology Assessment (OTA) in 1972. International developments in TA/EIA are also briefly surveyed, as is progress in state government and the private sector.

3.1 INTRODUCTION

The late 1960s were a traumatic period in U.S. history. During that period, long accepted beliefs about the best way to supply goods and services were reexamined. Conventional wisdom concerning economic growth seemed at times in conflict with impulses toward conservation and calls for a cleaner environment. A minority viewpoint that suggested the end of the era of plentiful raw materials and cheap energy gained a wider hearing. Perhaps, it was argued, the economics of the marketplace had been too frequently geared to short-range considerations. A longer-range approach was needed, requiring an increased understanding of technological change. Such an understanding would entail a broad base of environmental and social data in addition to the more customary economic information.

As a result of the concerns of the previous decade, the 1970s witnessed the institutionalization of new methods, organizations, and laws intended to extend the horizon of technological planning. The present chapter outlines the emergence of the TA/EIA movement, focusing specifically on the passage and implementation of the National Environment Policy Act of 1969 (NEPA) and the establishment of the U.S. Office of Technology Assessment (OTA). It also describes attempts to introduce TA into private industry, cites its growing acceptance by state and local governments, and discusses international developments in TA/EIA.

3.2 THE NATIONAL ENVIRONMENTAL POLICY ACT OF 1969

Lynton Caldwell (1970: 43) speculated that "Had there not been an Apollo Program there might not have been a National Environmental Policy Act in 1969." He suggested that the space experience may have convinced us as a nation that the earth is indeed a spaceship on which we travel. A growing consensus that the environment must be protected may reflect a sense of stewardship combined with more prosaic factors. As Americans acquired increasing mobility and leisure time, remaining areas of natural beauty seemed increasingly precious and desirable.

However complex the root causes of changing attitudes toward the environment, the 1960s saw a gradual shift to a more holistic concern for environmental policies. Although conservation had emerged as a national movement in the 19th century, major disagreements over goals prevented establishment of a single, comprehensive environmental policy. Some conservationists argued for more rational utilization of natural resources; others emphasized the aesthetic values of landscape, waters, and wildlife. Still others saw in the flora and fauna a great laboratory for science and fought its elimination under the rubric of progress.

The legislative history leading to NEPA demonstrates the advance of these concerns toward effective policy. The first attempt at federal legislation to provide comprehensive environmental policy was introduced by Senator Murray of Montana in 1959 (Senate Bill No. 2549). It sought "to declare a national policy on conservation, development, and utilization of natural resources . . . to meet human, economic, and national defense requirements, including recreational, wildlife, scenic and scientific values, and the enhancement of the national heritage for future generations." Hearings were held and the bill was endorsed by the Democratic platform in 1960, but it was later abandoned by President Kennedy. President Johnson's preoccupation with the Vietnam War delayed environmental policy formulation. As the decade neared its end, Congress took the lead in responding to public pressure. Over 50 bills relating to environmental policy were introduced in the Ninetieth and Ninety-First Congresses. In 1968 Senator Jackson proposed S. 1075, the forebear of NEPA. It was passed July 10, 1969. An amended version passed the House of Representatives on September 23. Following agreement reached in conference in October and passage by both houses, President Nixon signed the bill into law on January 1, 1970.

NEPA (1969):

1. declares that the policy of the Federal Government is to use all practicable means to create and maintain conditions under which man and nature can exist in productive harmony and fulfill the social, economic, and other requirements of present and future generations of Americans;
2. establishes within the Executive Office of the President a full-time, three-man Council on Environmental Quality which is appointed by the President and subject to Senate Confirmation;
3. requires the President to submit to Congress environmental quality reports annually. These reports are to set forth the status of the Nation's various environmental programs and review their impact on the environment, and on the conservation, development, and use of our national resources;
4. defines the duties of the Council on Environmental Quality (CEQ) to (a) assist and advise the President in the preparation of the annual report; (b) develop and recommend to the President na-

tional policies which promote environmental quality; and (c) accumulate necessary data for a continuing analysis of changes or trends in the national environment.

One of the most significant operational components of NEPA requires the preparation of EISs. (Interestingly enough, this requirement was a last-minute provision.) Specific wording of this requirement is as follows [NEPA (1969), title I, Section 102(c)]:

> All agencies of the Federal Government shall include in every recommendation or report on proposals for legislation and other major Federal actions significantly affecting the quality of the human environment, a detailed statement by the responsible official on
>
> (i) the environmental impact of the proposed action,
> (ii) any adverse environmental effects which cannot be avoided should the proposal be implemented,
> (iii) alternatives to the proposed action,
> (iv) the relationship between local short-term uses of man's environment and the maintenance and enhancement of long-term productivity, and
> (v) any irreversible and irretrievable commitments of resources which would be involved in the proposed action should it be implemented.
>
> Prior to making any detailed statement, the responsible Federal official shall consult with and obtain the comments of any Federal agency which has jurisdiction by law or special expertise with respect to any environmental impact involved. Copies of such statement and the comments and views of the appropriate Federal, State, and local agencies, which are authorized to develop and enforce environmental standards, shall be made available to the President, the Council on Environmental Quality and to the public as provided by Section 552 of Title 5, United States Code, and shall accompany the proposal through the existing agency review process.[1]

Since its passage, NEPA has occasioned deep changes in general awareness and agency practices concerning the environmental impli-

[1]The EIS preparation process itself has been a significant institution. In general, for the Environmental Protection Agency (EPA), that process has included (1) an applicant's environmental assessment, (2) a regional planning review (A-95), (3) a state agency environmental review, (4) an EPA Regional Office review to determine whether an EIS is required (if it is not, an environmental appraisal and negative declaration must be prepared and made available for comment and the decision must then be reevaluated), (5) a preliminary draft EIS for in-house EPA review, (6) a draft EIS for external review, (7) an evaluation of review comments, after a minimum of 45 days, prior to preparation of (8) the final EIS, (9) which is distributed to the Council on Environmental Quality, and commenting agencies, and (10) administrative action on the applicant's proposal after a minimum of 30 days for comment on the final EIS.

cations of technological developments. Despite much initial uncertainty and criticism from various quarters, the provision for environmental impact assessment (EIA) has taken hold effectively [U.S. Council on Environmental Quality (CEQ), 1976c, 1977]. In Chapter 19 we review the performance of NEPA, particularly its requirement for EIS preparation. Now, we turn to the second landmark in the institutionalization of impact assessment in the United States—the establishment of the Congressional Office of Technology Assessment (OTA).

3.3 CREATION OF THE OFFICE OF TECHNOLOGY ASSESSMENT

In the year 1830, prompted by a series of boiler explosions aboard steamboats, the federal government set about to acquire information on boiler-construction practices. The intent was both to spot flaws in existing design practices and to develop a more adequate theory of steam boilers. Congressman Emilio Q. Daddario (1967) cited this as an early federal attempt to evaluate a particular technological enterprise. Since that time, he continued, the government has performed many similar investigations, however sporadically. A more systematic approach was needed. Daddario called for a new form of policy research that he termed *technology assessment* (TA). TA was intended to provide policy makers with a novel and powerful tool for coping with the dynamic and pervasive impacts of technology on the fabric of society. It would systematize the identification of positive payoffs of technological innovations and foster their transfer into practice. At the same time it would isolate potential negative consequences, thus serving as an early warning system. Daddario (1967: 9) noted that "[A]gain and again, this country has moved to assess technology after some major crisis or catastrophy." He chose as examples the sinking of the U.S.S. *Thresher*, the Apollo spacecraft fire, the grounding of the tanker Torrey Canyon, and the four-state electric power blackout.

Additionally, Daddario continued (1967: 9):

> [T]echnical information needed by policymakers is frequently not available, or not in the right form. A policymaker cannot judge the merits or consequences of a technological program within a strictly technical context. He has to consider social, economic, and legal implications of any course of action.

TA was to remedy this situation.

Two factors lent urgency to the development of TA according to Daddario (1967: 3). First, the scale and variety of man's alterations of the biosphere has reached a level that rivals the forces of nature. Furthermore, biological, chemical, radiation, and energy effects resist local confinement. These problems are exacerbated by a second factor, an increased worldwide population that overtaxes the resilience of natural systems.

To expedite the development of a TA capability, H.R. 6698 was presented before the Congress on March 7, 1967. Congressman Daddario made it clear that he introduced this bill primarily as a stimulant to discussion. The Subcommittee on Science, Research, and Development of the House Committee on Science and Astronautics, which Congressman Daddario chaired, subsequently commissioned independent studies by the National Academies of Science and Engineering and the Legislative Reference Service of the Library of Congress. Because each of these studies significantly influenced the conceptualization of TA, brief highlights are provided in the following sections.

3.3.1 Legislative Reference Service Study: Technical Information for Congress

This study (U.S. Congress, Legislative Reference Service 1969) examined 14 cases with substantial technical aspects that required governmental action (e.g., the AD-X2 battery-additive case, the Mohole undersea drilling project, the test-ban treaty, the Camelot episode in which the military attempted to influence the behavior of South Americans, and the Salk polio vaccine development). In each case the mismatch between scientific and political behavior and the decision making process was in evidence (1969: 507–509):

> Among the obstacles to congressional treatment of technical issues were: political acceptance of invalid hypotheses, sensationalism, the improper use of outstanding personalities, and the raising of questions that cannot be answered scientifically . . . there is a mistaken tendency for politicians to expect scientific findings to be absolute and "beyond the peradventure of a doubt."

To aid in better communication between technical experts and politicians, the report recommended more cross-disciplinary exchange among the technical experts, integration of findings by scientific generalists, and greater input from social scientists informed on the political factors of technological change. The idea of an adversary process among technical experts was mentioned several times. [This concept remains controversial in the technical community; witness the recent debate over a "science court" (Kantrowitz 1975).] Finally, the report recognized the inherent tension between the aims of sound technical analysis and effective political action. For example, thorough analysis requires time, yet social costs may rise catastrophically if decisions are too long delayed. Although there is no adequate resolution of this problem, TA as an "early warning" procedure could serve to pinpoint potential problems before they become irreversible.

3.3.2 National Academy of Engineering Study

This study by the Committee on Public Engineering Policy (COPEP) of the U.S. National Academy of Engineering (1969) involved the actual conduct of three small-scale TAs on (1) the technology of teaching aids, (2) subsonic aircraft noise; and (3) multiphasic health screening. It was concluded that TA could perform a useful social function for both the Congress and the public in four ways. First, it could clarify the nature of technology-driven social problems and propose corrective legislation. Second, TA could help establish a list of long-term priorities for public policy and offer guidance in resource allocation. Third, it could recommend incentives to both public and private sectors that would encourage and assist development of socially desirable technologies. Fourth, TA could serve as an educational tool for both the public and the government decision makers.

COPEP was first to suggest that TA addresses two different forms of problems: those dealing with problem-oriented cases in which the issue could be focused and those dealing with an expanding set of variables stemming from a growing use of technology in the future. The former is closely allied to the analytic methods underlying engineering endeavors. In the latter, technology-initiated situations, a complex pattern of consequences evolve that can be foreseen with diminishing certainty as the time horizon increases. The committee observed that "because of the uncertainty and potentially broad scope of the impacts of new technologies, their assessment is probably of most concern to Congress. The availability of appropriate methodology to perform these is more limited."

COPEP deserves special mention for providing the first codification of a TA methodology in seven steps, of which the 10 step process used in this guidebook is a descendant.

3.3.3 National Academy of Sciences Study

The most general and reflective among the TA studies was produced by the Committee on Science and Public Policy (COSPUP) of the National Academy of Sciences (U.S. National Academy of Sciences 1969). This study began by cataloging prevailing attitudes toward technology, ranging from the claim that it is an unmitigated boon to mankind to the position that it is destructive of both the biosphere and personal freedom. It then argued that TA should strive to provide policy analysis leading to more systematic and effective maximization of gains and minimization of losses.

The uncertainty that has persisted concerning citizen participation in TA/EIA was first raised by COSPUP. The panel acknowledged that the essential points of technological debates should be open to the public, but it also pictured public reaction as often irrational, emotional, and fraught with bickering.

Finally, COSPUP recommended that a government assessment agency be established that could, without regulating technology, catalyze the cooperation of professional and corporate communities, inform legislative and executive decision makers, sponsor basic research, and offer policy advice. The report suggested formation of either a new joint committee or a separate OTA.

3.3.4 Follow-Up and Establishment of the OTA

The three reports just described provided the basis for a conference attended by government, academic, and technical personnel at Andover, New Hampshire, in 1969. A significant polarization of viewpoints regarding the desirability of interference, governmental or otherwise, in the free processes of technological innovation occurred at the Andover meeting. Phrases such as "technology harrassment" or "technology arrestment" were used by those fearing a straitjacket effect of regulations on innovation. This group felt that the process of technological innovation, if left relatively free of regulative interference, could and would correct most mistakes or oversights. Others at the conference argued for a larger role for government in technological-innovation processes. This position reflected the concerns of Daddario and others who gave the initial impetus to a TA movement.

Congressional hearings were held in 1969 and 1970 culminating in a bill to create an OTA. This bill was reported out of the House Committee on Science and Astronautics in September 1970. An amended version passed the full house in February 1972.

One amendment to the bill was quite consequential. It replaced the proposed 11-member Technology Assessment Board (TAB), of which the President would appoint six, with a 12-member board consisting of an equal number of Senators and Representatives. This change signaled a clear intention to create a source of technical information for the Congress, independent of the Executive Branch. A modified bill passed the Senate in September 1972; on October 13, 1972 the President signed the Technology Assessment Act of 1972.

Public Law No. 92-484 established the OTA, which was to be governed by the TAB, consisting of three majority and three minority members from each house. The Director of OTA served as a nonvoting member of the TAB. A Technology Assessment Advisory Council

(TAAC) was created, consisting of 10 private citizens (appointed by the TAB) plus the Director of the Congressional Research Service and the Comptroller General ex officio [for detail on the institutional structuring, see U.S. Congress, Congressional Research Service (1973)].

The procedure to be followed by OTA was envisioned as follows (U.S. Congress, House 1972: 4,5):

1. Requests for assessments are submitted to OTA by Congressional committee chairmen.
2. OTA assigns priorities among the requests and formulates the assessment task.
3. A contractor is selected by OTA, or more typically these days, OTA staff are assigned to direct the assessment.
4. The TA is performed, and then OTA prepares a summary report.
5. OTA transmits a summary report, with or without recommendations, to the original requesting committee.

The first TAB was appointed by April 1973, and initial funding for the office was made available in November of the same year. Former Congressman Daddario was selected as the first director. The TAAC met for the first time in January 1974, and OTA produced its first TA in July of that year.

During its brief existence, OTA has generated a number of effective studies. However, its development has involved controversy over Senatorial influence and effective management. Further discussion of OTA's performance appears in Chapter 19.

3.4 OTHER INSTANCES OF TA/EIA IN THE UNITED STATES

The most visible instances of the institutionalization of TA/EIA activities are represented by OTA and the federal EIA effort, but other examples should also be mentioned. A number of agencies of the Executive Branch have been involved in TA. The National Science Foundation (NSF) has played a major role, both in terms of funding specific studies and coordinating efforts of other agencies of the government. At the state level, broad-gauge TAs are still rare, although EIAs are ongoing in many states. Within the business community, there have been a relatively small number of TAs performed by an even smaller number of firms. Other nations have undertaken TA/EIA activities within the traditional planning function.

3.4.1 TA/EIA in Other Federal Agencies

EIA has become a pervasive element of planning by federal agencies. EIS procedural requirements have been formulated to meet agency conditions with specific guidelines and manuals. Most importantly, consideration of environmental consequences appears to be increasing and becoming an integral part of the planning of technological developments (U.S. CEQ 1976a, 1977).

H. Guyford Stever, then director of NSF, noted that a number of federal agencies were carrying on environmental impact studies, national assessments, future studies, planning studies, social impact analyses, and so forth, although not consistently using the term "technology assessment" to describe their efforts (U.S. Congress, OTA 1976b: 23). As early as 1972 Vary Coates identified 86 offices within federal agencies concerned with technological programs and projects, with 11 claiming TA to be the major function of their office. However, as she notes in a later statement (V. Coates 1976: 5):

> In many agencies, concerned with a technological development, economic analysis is done in one part of the agency, environmental impact analysis in another part, social impact analysis in a third part, and policy formulation somewhere else entirely. Until these activities are integrated to inform the policy-maker, there is no Technology Assessment no matter how much that term may be used or insisted upon.

Notable efforts to institutionalize TA are underway in several federal agencies. For instance, the Department of Energy and the Health Systems Research Center (in the Department of Health, Education, and Welfare) have begun TA programs. EPA has supported a series of "integrated technology assessments," very large studies on such topics as regional energy development in the West and in Appalachia.[2]

The NSF has played a leadership role among the Executive agencies. As Stever described it, the NSF approach is three-pronged. First, the NSF has supported at least 30 comprehensive TAs intended to provide substantive policy information. These studies have been undertaken to identify new technological possibilities and to examine technologies that cross-cut or fall outside the purview of existing agency missions. Second, the NSF has instituted several studies of TA methodology itself to assess the present state of the art, pinpoint methodological weaknesses, and suggest ways in which TAs may be more effectively accomplished. Third, the organization has sponsored programs to stimulate and enhance the capability to conduct and utilize TAs by means of workshops, public meetings, seminars, and cooperative ventures in TA.

A notable development in federal technology policy was the passage

[2]Vary Coates was completing an update and expansion of her previous survey of federal TA activities as this book went to press.

of the National Science and Technology Policy, Organization and Priorities Act of 1976 (P. L. 94-1046). This law established the Office of Science and Technology Policy within the Executive Office of the President. (The former Office of Science and Technology had been disbanded several years before.) Technology assessment is explicitly noted as one function of the office.

3.4.2 TA/EIA at the State Level

There is very limited and relatively recent experience with TA in state government in the United States (Kelly 1976: 4). Four states—California, Georgia, Hawaii, and New York—have undertaken assessments of one form or another. Both Georgia and Hawaii have established centers for carrying out TAs, although their present status is unclear.

Unlike TAs performed in business, which are nearly always technology driven, state-sponsored TAs tend to focus on issues or problems. The short time frame reflected in the concerns of legislators has presented difficulty for state TA efforts, no less than for OTA. Since TA is meant to inform a policy process, the deadline imposed by next month's vote or a proposed bill is critical. Such constraints are inimical to the idea of long-range assessments.

EIA has received substantial attention at the state level. As of May 1977, a large number of states had acted legislatively or administratively to establish requirements. Fifteen had adopted NEPA-like guidelines (so-called "little NEPAs" or "SEPAs"), four had comprehensive executive or administrative orders, and seven had special or limited EIS requirements (U.S. CEQ 1977).

3.4.3 TA in the Private Sector

Although TA originated in the public sector, at the 1976 hearings before OTA's TAB, 18 private companies were identified as conducting TA in one form or another (U.S. Congress, OTA 1976b: 98). What are some of the factors that influence private industry in the decision to invest funds in a TA study?

In these OTA hearings, Congressman Brown of California observed that the investment of industry's resources to implement strategies compatible with the public welfare might reduce the need for government intervention. To this, Dr. Dale Compton, Vice President for Research of the Ford Motor Company, responded, "If one could be assured one's competitors would be doing the same, and, if not, that there would be not net disadvantage to you, then of course . . . the free market will operate properly" (U.S. Congress, OTA 1976b: 127).

Day (U.S. Congress, OTA 1976c: 21) cited certain pressures that prompt introduction of TA activities in the private sector. It may be used as a defense mechanism by firms in response to the growing number of legal suits related to product safety. Here TA is viewed as one way of anticipating such problems before too much has been spent in developing and introducing a new product. A second pressure is that for increasingly sophisticated long-range planning. In these activities, TA may be viewed as a new and useful tool. Those businesses in which government regulation is an already accepted fact of life appear most disposed toward TA. "Regulated industries," Day observed, "have experience in dealing with governments on a routine basis and are not quite as caught up in the standard government vs business rhetoric that occurs elsewhere" (U.S. Congress, OTA 1976b: 336). Finally, Day noted the pressure for corporate social responsibility. A firm engaged in TA can provide tangible evidence of good faith in this regard.

Whereas business response has been generally slow, one organization, Monsanto Chemical Company, is a notable exception. Over the past 6 years this corporation has carried out TAs concerning 35 large and small business projects. Kenneth Craver and Monte C. Throdahl emphasize their commercial development group's early awareness of societal interest in environmental protection, consumerism, and energy conservation (U.S. Congress, OTA 1976b: 33):

> In terms of providing products to meet market requirements, these three forces have all pushed in the same direction. At times some of these forces seem to be in opposition, but these differences can be resolved. We have found that TA is a technique both for resolving the differences and for measuring progress.

Monsanto has ventured into the area of public participation in TA, holding symposia in conjunction with the assessment of a new beverage container, and has fostered the development of cross-impact analysis (see Chapter 9).

3.5 TA/EIA OUTSIDE THE UNITED STATES

Concerns about undesirable change produced by technology, including damage to the biosphere, have been voiced by a number of the industrialized nations. International emphasis seems clearly placed on environmental impact studies. In Europe particularly, international cooperative efforts are strong. The U.S. political system differs from the more common parliamentary forms of government, and there has been an historical bias in the United States against long-range planning. Thus what has emerged as TA in this country is often seen abroad as nothing radically new. At the present time only the United States has created a separate TA office outside existing agencies, although France has considered the possibility. Hence one might argue that TA has retained a distinctly American flavor, especially in its institutionalization (Vlachos 1976a).

The Organization for Economic Co-operation and Development (OECD), encompassing some 24 of the major developed, noncommunist nations, has had a long-standing interest in TA. It has coordinated joint national studies on various technological developments and has contributed significantly to TA methodological development and practical application concerns (Hetman 1973: OECD 1975).

3.5.1 Sweden and Japan

According to Hahn (1977), Sweden and Japan currently have the widest range of TA/EIA activities outside the United States. In Sweden TA is conducted by the national government as part of the Secretariat for Future Studies, Office of the Prime Minister. A subsidiary of the Swedish Academy of Engineering Sciences, the Swedish Association of Future Studies, has conducted a range of assessments beginning in 1971. A major Association effort has been a joint industry and government future study titled "Sweden 2000." A third locus of TA/EIA activity is the Swedish National Board for Technical Development.

Japan recognized the potential of TA as early as 1969. Initial studies were conducted by a Subcommittee of Technology Advancement of the Economic Council. By 1971 the Science and Technology Agency began studies of computer-aided instruction, pesticides, and tall buildings. Subsequent TA/EIA studies have been actively supported by the Ministry of Trade and Industry (MITI) and the Japan Techno-Economics Society. Public and private research institutes such as Nomura, Mitsubishi, and the Institute for Future Technology of Nippon Telephone and Telegraph Public Corporation have also engaged in TA activities. The Japanese view themselves as in a third state of industrialization (Hahn 1977: 37). First came industrial revolution, then technological innovation, followed by technological reexamination—TA. A fourth stage—the age of civil participation—has not yet been entered.

3.5.2 Common Market Countries

A brief survey of Common Market countries yields the following TA/EIA activities (Hahn 1977: 38–44). Although Great Britain has no equivalent of the U.S. OTA, technological impact analyses have been performed by the Programmes Analysis Unit of the Parliamentary Select Committee on Science and Technology of the House of Commons. Ongoing studies of environmental impacts of nuclear energy are being conducted by the Royal Commission on Environmental Pollution. Related university programs exist at Sussex and Manchester.

In France, activities have focused on environmental impact studies, with most analyses sponsored by the Bureau du Plan. To date France has shown the most interest in establishing a national office of assessment and has studied the organization of the U.S. OTA as a possible model. A bill to establish such an agency was introduced in 1976.

In the Federal Republic of Germany, a variety of studies have been conducted dealing with impacts of nuclear power, water supply, and pollution. Nonprofit groups in Karlsruhe, Heidelberg, and Frankfurt (Battelle) have performed TAs, and many large industries have executed systems analyses that resemble TAs in many respects. For political reasons, the party in power has resisted the establishment of a national office of assessment. It is argued that such an office would shift too much power to the opposition in a parliamentary system (Paschen 1977: 14). This same pattern of opposition has apparently developed in the Netherlands as well (Hahn 1977: 43).

In Finland, the Science Policy Council has commissioned a TA feasibility study (CITRON) and in Italy, also without a formal TA organization, several agencies are looking at second order impacts of various regional development proposals.

3.5.3 Eastern European Countries

Eastern European countries, the Council for Mutual Economic Assistance nations, have tended to view TA as effective management of science and technology. Technological change is interpreted against a backdrop of historical determinism. Hence TA becomes not the postulation of alternative future scenarios (as a Western-based TA might attempt), but rather the fine tuning of the ongoing historical forces (Chen and Zacher 1977). In the Soviet Union TA/EIA is the concern of the State Committee for Science and Technology of the Council of Ministers. The most important centers conducting this research are located at Moscow, Leningrad, Kiev, and Novosibirsk.

The Polish National Academy of Science is the leading TA center in that country. There are indications that the academy is anxious to increase dialog between TA practitioners in OECD countries and its own members.

In both Czechoslovakia, where it is called "selective goal analysis," and in the German Democratic Republic, TA-like functions are performed as an ongoing planned activity.

3.5.4 Elsewhere

Granting that much more assessment of technology is occurring than is indicated by explicit mentions of TA, there are TA/EIA activities ongoing in Israel, Indonesia, Jamaica, Ghana, Egypt, Mexico, the Philippines, and Pakistan (Hahn 1977: 46). Canada has a small Technology Assessment Division in the Ministry of State for Science and Technology, although the emphasis appears to be shifting increasingly to broad futures research.

Contemporary dialog about technology among Third World countries is likely to center on topics of appropriate or intermediate technology. These approaches call into question many of the underlying technological assumptions of the industrialized nations—such as those concerning the appropriateness of Western models of modernization and development. The adaptation of TA to such settings has been a topic of discussion at the East–West Center in Honolulu (Koppel 1977).

3.6 CONCLUDING OBSERVATIONS

Popular sentiment to protect the environment influenced the U.S. Congress in the late 1960s, leading to NEPA and the virtually concurrent creation of OTA. NEPA, with its Section 102(C) requirement for the preparation of EISs, is the dominant bastion of TA/EIA and has led to increased state attention to the environment.

TA was influenced by three studies conducted for the House Committee on Science and Astronautics by the Legislative Reference Service, the National Academy of Engineering, and the National Academy of Sciences. Subsequent legislation created the OTA as a staff arm of the U.S. Congress.

Federal agency efforts in EIA have been widespread and influential on agency planning processes. Federal TA efforts have been spearheaded by the NSF, with momentum now building in the mission agencies, such as EPA.

There is a largely untapped potential for TA in the private sector as an element in advanced corporate planning.

Developments in nations other than the United States are of a different genre, which is attributable to differences in government structure and historic acceptance of the validity of planning. Various institutionalizations of TA/EIA exist, carrying out impact assessment-like activities under various labels. The emergence of TA and EIA reflects growing appreciation of the long-term and worldwide scale of the effects of technology on the environment.

RECOMMENDED READINGS

Caldwell, L. C. (1970) *Environment: A Challenge to Modern Society*. Garden City, NJ: Anchor Book, Doubleday and Company.

An insightful view of NEPA development.

Casper, B.M. (1978) "The Rhetoric and Reality of Congressional Technology Assessment," *Bulletin of the Atomic Scientists* 34 (February), 20–31.

A lively comparison of the legislative intent and the institutional practice of OTA.

Hahn, Walter (1977) "Technology Assessment: Some Alternative Perceptions and Its Implications Outside the United States." Testimony before the Subcommittee on Science, Research, and Technology of the U.S. House of Representatives Committee on Science and Technology (August 3) (mimeo., 56 pp.).

A fine review of current TA activities outside the United States.

U.S. Council on Environmental Quality (March 1976) "Environmental Impact Statements: An Analysis of Six Years Experience by Seventy Federal Agencies." Washington, DC: U.S. Government Printing Office.

A review of the institutionalization of EIA in the federal government.

EXERCISES

3-1. Which came first, TA or EIA? Consider "first" in this question to refer to a number of significant dimensions such as time, legislation, and institutionalization.

3-2. Compare implications for assessment practice of the manner in which TA and EIA have been institutionalized in the federal government.

3-3. Contrast TA/EIA institutionalization in the federal government, OECD nations, and in the private sector. Consider the differing *objectives* in terms of key assessment questions, time horizons, and users of the information produced. Identify serious *constraints* on how impact assessment can be structured, such as resource limitations, information access, policy maker contact, and so on.

3-4. As consultant to a corporate vice-president, suggest three alternative institutional structures to provide TA for the firm. Recommend one of the three and defend this recommendation.

4

Basic Features of an Assessment

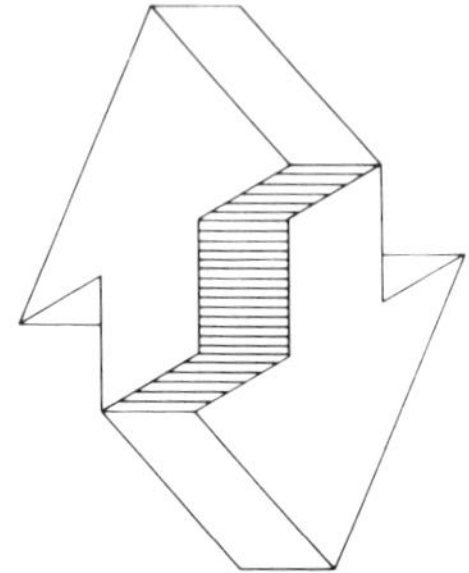

This chapter presents an overall orientation to the TA/EIA process. It describes what TA/EIA is intended to accomplish, for whom, and how it differs from other studies. Two basic objectives are set forth for assessments: (1) to provide valid information as to the likely consequences of certain courses of action and (2) to have this information prove useful to policy making processes. Implications of these objectives and of different user needs are explored. The chapter then differentiates various types of TA/EIAs from each other and interprets the significance of these differences for assessment practice. The latter part of the chapter introduces the 10 essential components of a TA/EIA. These components provide the structure for this book's approach to impact assessment, to be elaborated in the next dozen chapters.

4.1 INTRODUCTION

Previous chapters have introduced TA/EIA, explored the general issues concerning the interaction between society and technology, and described the institutionalization of TA and EIA. We now turn to the assessment process itself, setting forth the basic considerations for a TA/EIA and the components that go together to produce one. This chapter thus lays the foundation for performing TA/EIA.

We claim that TA and EIA are variations of the same activity. This chapter spells out the basic structural features of this common activity. We note differences as well as similarities in explaining the objectives, uses, and components of impact assessments. Differences arise from such factors as the subject, context, and form of a particular study. A balanced perspective for TA/EIA requires consideration of the underlying common aspects of the assessment process along with sensitivity to the idiosyncrasies of any particular study situation.

4.2 ASSESSMENT OBJECTIVES

4.2.1 The Large Picture

The TA/EIA process contributes to society's effort to direct technological development in a beneficial manner. Figure 4.1 presents the notion of the technological delivery system (TDS), which will also appear later in the book as well, indicating a role for TA/EIA in anticipating the consequences of technological development. As implied in Figure 4.1, assessments (1) provide pertinent information to policy makers, (2) alert concerned people who, in turn, influence the policy making process, and (3) may even contribute to serious thought about societal values. Any or all of these three roles may be important in a given TA/EIA. For instance, the TA of life-extending technologies (The Futures Group 1976) provides a categorization of major policy alternatives, directs the attention of potentially concerned parties to the attendant issues, and raises profound concerns about the value of long life.

Each assessment has its own context, sponsor, particular interests, and unique problem definition. However, there are general objectives that apply across all assessments (Rossini et al. 1976). The two most important are validity and utility; a third is improving assessment methodology.

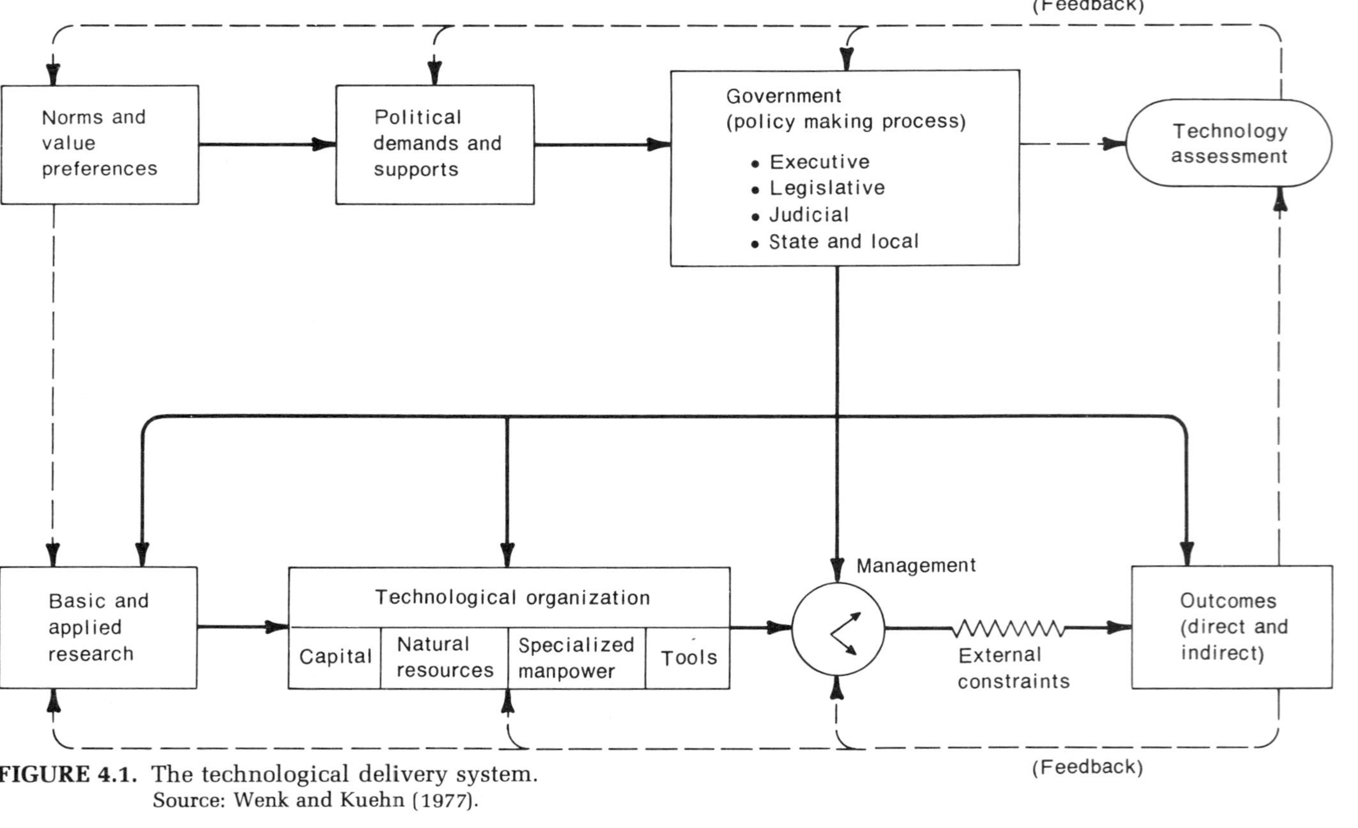

FIGURE 4.1. The technological delivery system.
Source: Wenk and Kuehn (1977).

4.2.2 Validity

"Validity" carries a host of precise connotations. Generally, it refers to being well-grounded in fact and verifiable. In the case of research dealing with the future, we take validity to refer to the congruence between predicted and actual results. Since the results of an assessment apply to the future, and since their correspondence with actual happenings may vary as a function of time, the validity of an assessment will remain uncertain. However, three specific characteristics that can be analyzed in the present pertain to the predictive validity of an assessment.

Cause–Effect Understanding. The underlying assumption in TA/EIA is that a technological intervention as a cause produces impacts as its effects. These impacts, in turn acting as causes, may produce other impacts as (higher-order) effects (see Section 4.5.6). The central task of an assessment is to understand the full set of interactions involved in the coupling between technology and society, including the "feedbacks" involved. This task is not made easier by the fact that existing theories of social change are not adequate bases for treating the complexities of the technology–society interaction (Forester and Rosenthal 1972).

Balance. According to Emilio Daddario (1968), an assessment should provide a balanced appraisal to the policy maker. "Balance" refers to an even-handed treatment of the major assessment issues, both in terms of coverage of all important aspects and in reflection of salient points of view.

There are several approaches to deal with different points of view (i.e., different values). We recommend a *value-explicit* approach in which the assessors attempt to lay out as clearly as possible the divergent value perspectives involved in the issue. Most importantly, the assessors should try to spell out their assumptions and make clear their personal allegiances, so that the users of their study can judge the positions taken. Note that this is *not* a *value-free* assessment, for there is no precisely value-neutral position from which any assessment can be conducted. Any issue worthy of assessment will generate varied perceptions as people relate it to their own interests. Assessors' perspectives will necessarily color the assessment itself.

Alternative approaches to dealing with balance entail deliberate acceptance of different perspectives. For instance, an industrial TA could reflect a company's own interests. Alternatively, adversarial assessments of a given subject can be prepared by different parties at interest, assuming that understanding will emerge through the interaction of opposing positions (Green 1970).

Methodological Soundness. The current state of TA/EIA development does not provide general criteria for judging assessment methods used in a study. Nevertheless, TAs and EISs should use the available, relevant data and follow established scientific principles and procedures (e.g., meteorological principles with respect to air-pollution projections, proper survey techniques). Where reasonable, the reproducibility of results should be ascertained. Review by knowledgeable peers is a highly desirable way to assure methodological soundness.

4.2.3 Utility

The utility of a TA/EIA is determined by how much useful information the study provides to its sponsor, to parties impacted by its subject, and to makers of decisions involving the assessment subject. Utility depends on such factors as the time a study becomes available relative to when a decision must be made (decisions are usually incremental and partial), the study content, and the presentation (see Chapter 16). Utility can be gauged as the difference, in terms of information gained, the study makes to its potential users. Four characteristics that contribute to utility deserve attention.

Relevance. An assessment is irrelevant and hence useless if it does not ask the questions that the sponsors and parties at interest need answered (Mitroff and Turoff 1973). The problem of producing a relevant assessment is illustrated by a TA on the automobile (Grad et al. 1975). Focusing on air-pollution-control issues, the assessors did not emphasize the fuel costs involved in pollution control. The occurrence of an energy crisis, triggered by the 1973 Arab oil embargo, late in the course of the study was not reflected in the assessment. Thus the study's usefulness was reduced as policy makers became concerned about automotive fuel efficiency as well as pollution.

Timeliness. In a policy context, it is imperative that an assessment be available when the time is ripe. Many TAs and EISs were needed before they were begun. Time pressure can also be a major constraint on the depth of analysis possible in a given TA/EIA.

Credibility. Whatever the validity of a report, it is valueless unless the audience believes the report to be worthy of consideration. A fine example of a credible report is the Jamaica Bay/Kennedy Airport Study (Jamaica Bay Environmental Study Group 1971), which concluded that Kennedy Airport should not be extended into Jamaica Bay. This conclusion was implemented the day after public release.

Communicability. Unless the findings are presented in a usable format, the study may receive scant attention. This threat is particularly acute in the case of more quantitative approaches that use elaborate techniques. Those that combine many impacts under a weighted utility function or those that present a multitude of impacts without a viable means to aggregate toward conclusions are particularly troublesome. Clarity of presentation, good writing, appropriate length, and a minimum of jargon are obvious assets that have often been lacking in EISs and TAs.

4.2.4 Improving Assessment Methodology

Advancing the state of the art of assessment methodology, although a subsidiary objective, is also important. It is generally conceded that the products of assessments do not adequately meet the social need for knowledge about the implications of future developments. Given the social importance of TA/EIA, continued methodological advance through development and refinement of study strategies and techniques that lead to valid and useful results should be carefully considered in every significant study. We recommend documentation of the procedures followed in all TA/EIAs and deliberate design of comparative assessments using different methods of analysis (see Chapter 18 and Rossini et al. 1976).

4.3 TYPES OF ASSESSMENT

4.3.1 Distinctions Between TA and Other Studies

TA, more than EIA, is often confused with other studies. Sherry Arnstein (1977) carefully differentiates TA in terms of its (1) comprehensive view of complex issues, (2) requirement of many disciplines, working in an interdisciplinary mode, and (3) component tasks (beginning with a need to structure the problem and continuing through analysis of policy options). Table 4.1 (from Arnstein 1977: 576) summarizes distinctions among various types of study in terms of the questions addressed and the analytical parameters involved. Note that comprehensive TA stands apart for its breadth, interest in higher order effects, and concern for all parties at interest.

TABLE 4.1. Taxonomy of technical studies [a]

TYPE OF STUDY	CENTRAL QUESTIONS TO BE ANSWERED	ANALYTICAL PARAMETERS											
		Technical efficacy	Economic feasibility	Safety risks	Public policy options	All relevent impact domains (e.g., economics, environment, social, psychological, political)	Only selected impact domains (e.g., economic)	Second- and higher-order effects (i.e., unintended, unanticipated, synergistic, cumulative)	Only first-order effects (i.e., intended impacts)	All key parties at interest (e.g., bankers, doctors, consumers, industry, youth, minorities)	Only particular parties at interest (e.g., doctors)	Effects on all relevant systems, (e.g., health, transportation, communications, education, legal)	Only effects on particular system (e.g., health)
FEASIBILITY	Is it a workable and safe design? Can it be made with available resources and techniques? What will it cost to make?	1	1	1	4		X		X		X		X
MARKET RESEARCH	Who will use it? Who will pay for it? How much will they be willing to pay? How can demand be generated?	4	1	2	4		X		X		X		X
CLINICAL TRIAL	How well did it do what it was designed to do? Did it produce clinical side effects? Was the effect beneficial? What regimen is called for?	1	4	1	4		X	X			X		X
COST–BENEFIT [b]	What are the monetary costs? What are the monetary benefits? What is the net difference between costs and benefits? What are the rates of return on investment? What is the benefit: cost ratio? Which alternative will maximize the net value of benefits?	4	1	4	2		X		X		X		X

COST EFFECTIVENESS	What are the costs of alternative ways for obtaining a particular set of outcomes (generally nonmonetary, e.g., reduction in birth rate or improved water quality)? Which alternatives will maximize the desired outcomes for any particular level of resources?	3	2	3	1		X		X		X		X
ECONOMIC IMPACT	How will it affect micro- and/or macroeconomic factors? How will micro- and/or macroeconomic factors affect it?	4	1	4	—		X		X		X		X
ENVIRONMENTAL IMPACT [c]	What are the beneficial and detrimental impacts on the physical environment? Which adverse effects cannot be avoided and why? What alternatives exist? What irreversible and irretrievable resources are committed?	4	3	3	1		X	X			X		X
MACRO-TA (COMPREHENSIVE TECHNOLOGY ASSESSMENT)	What is the state of the art of the technology? Are there better micro and macro alternatives to achieve the objective? What are its potential second and higher order impacts and consequences? How will these impacts and consequences interact with each other? Who are the parties at interest and how will they be affected? Who are the decision makers and what is the potential for public policy to avert or minimize undesirable impacts and/or enhance desirable impacts?	1	1	1	1	X		X		X		X	
MINI-TAS (PARTIAL TA)	(Mini-TAs are about an order of magnitude smaller than macro-TAs and focus on either depth or breadth)												
Selected Impact Scope	Same questions as macro-TA (listed above) but only selected impact and consequence domains are included.	3	3	2	1		X	X		—	—	—	—
Selected Focus	Same questions as macro-TAs, but limited to effects on particular system and/or particular party at interest.	3	3	2	1	X		X			X		X
MICRO TA (HEURISTIC TA)	Micro-TAs are about an order of magnitude smaller than mini-TAs and are generally based on "brainstorming" or nominal group sessions.	5	5	5	5	X		X		X		X	

Note:

Magnitude of analysis:
1. Generally analyzed in depth
2. Generally included but analyzed only moderately
3. Generally included but analyzed superficially
4. Seldom or never included
5. Generally depthless

— Varies widely depending on objectives of the specific study

[a]From Arnstein (1977:576).

Focus of Study:
X Generally primary focus of study.
— Varies widely depending on objectives of the specific study.

Caveats:
[b]Some exceptional cost–benefit studies include nonmonetary values.
[c]Some exceptional environmental impact studies include social effects.

4.3.2 Distinctions Between TA and EIA

Table 4.1 includes major distinctions between TA and EIA. Although TA and EIA have been quite different in practice, we argue that they are quite close in intended purpose and major features. Table 4.2 presents points of similarity and difference. These extend the preliminary description of Section 1.2, which views TA and EIA as generally differing in degree of policy emphasis and localization. The close relationship

TABLE 4.2. Comparison Between TA and EIA

Similarities	
Impact assessment	The core effort of both; both address higher-order impacts in principle, but are weak in practice
Future-oriented	Both intended to anticipate consequences of developments
Technological focus	Both address a range of technological issues—mostly physical in nature, but some social technologies also
Comprehensive	Both are systematic and comprehensive in principle; in practice, both must bound and focus the study, and EIAs have been highly focused on the physical environment
Comparative	Both consider pros and cons of alternatives in principle; often not well done in practice
Scale	Both are highly variable, from small-scale assessments of a couple person-months efforts to large-scale, multi-person-year ventures
Differences	
Geographic locus	EIAs tend to address localized projects; TAs tend to more global (e.g., national) considerations (but program EIAs are national in scope)
Policy emphasis	This is more circumscribed in many EIAs (but the EIA process is becoming institutionalized earlier in the planning process, bringing it to bear while more options are open)
Format	EIA procedures are legally mandated; CEQ regulations provide considerable uniformity that is absent in TAs
Time horizon	EIA typically addresses technologies close to the point of implementation; TAs range in focus from immediate to long-range options (but both address long-term impacts, at least in principle)
Data sources	EIAs entail more direct (primary) data gathering, less reliance on literature
Performers	EIAs are more likely to be a direct agency responsibility, performed in-house; TAs, more likely to be contracted out to research firms or universities (but EIAs can involve outside-agency work and TAs are also done internally)

between TA and EIA suggests a great potential value in cross-fertilization of techniques, information bases, and experiences. On the other hand, there are real differences that should not be ignored.

4.3.3 Three Types of TA/EIA

A threefold distinction can be made among forms of assessment: (1) *project* assessments focus on a particular, localized project such as a nuclear power plant or highway, (2) *problem*-oriented assessments focus on solutions of a specific problem, such as an energy shortage, and (3) *technology*-oriented assessments examine a new technology (e.g., ocean thermal electric conversion) and trace its impacts on the society.

Project assessments study a fairly well-constrained topic, both technically and geographically. They usually address several alternatives, which may have far-reaching impacts. For instance, an assessment of alternative highway configurations is likely to face a spectrum of urban-development issues associated with the new or improved roads. Where feasible, project assessments should consider a wide range of alternatives, including the "no-build" option. In the case of a dam, assessors might consider alternative flood-control possibilities (e.g., restrictions on building in the flood plain), alternative ways to meet water needs (e.g., conservation measures), and alternative energy sources (e.g., conventional power plants).

In a sense, project assessments are special instances of problem-oriented assessments. The latter study physical or social technological alternatives as solutions to a perceived problem. The problem determines the focus of the assessment. The scope of problem-driven assessments is broad in that they address various technologies as possible problem solutions. Such assessments usually have a strong policy content because they study problems of current concern.

The third, technology-oriented type of assessment often deals with innovative technologies. This contrasts with project and problem-oriented assessments, which typically concern adaptation of existing technologies. Constraints arising from specific locations are largely eliminated. Important considerations in a technology-oriented assessment are the forms of the technology, the time frame of the innovation process, and the ways in which implementation of the technology are likely to occur. The time frame covered by such an assessment is typically more open than that for the other assessment types. Consequently, the range of possible societal contexts is generally wider. Finally, the policy content of a technology-oriented assessment is generally more diffuse since the options are great and the uncertainties large. These assessments should strive, however, to assure that their findings have policy relevance, whether immediate or delayed.

The three assessment types can be complementary. For instance, a technology-oriented assessment should consider what societal problems (needs) the technology could serve. To understand impacts, it should consider what forms actual implementation would take, and thus address project-localized impacts. Obviously, the problem-oriented assessment must determine the prospects of each technology viewed as an alternative. In a sense, then, it incorporates several technology-oriented assessments. Well-done project assessments should take note of new technological developments. They obviously must relate the particular development to the larger societal problems and trade-offs in assessing its desirability. In sum, a good assessment will blend some aspects of all three types.

4.4 ASSESSMENT AS PROGRAM: POSSIBLE VARIATIONS

To this point we have discussed TA/EIAs as if they were single, one-shot, studies, as almost all are. Since policy making is usually ongoing and incremental rather than one-time and comprehensive, it is arguable that assessments should reflect this. The trade-off between systematic, comprehensive analyses and timely, practical ones has been clearly laid out in the policy analysis (Archibald 1970) and forecasting (Ascher 1978) literatures. The TA/EIA study must balance these two considerations to meet the dual objectives of validity and utility.

All TA/EIA conclusions are highly sensitive to assumptions about the state of society, technological breakthroughs, and outside influences. The occurrence of an oil embargo, the development of solid-state electronics, or the emergence of an environmental movement can drastically alter forecasts in a short time. Hence, recent assessments are likely to be more valid than older ones.

It may be appropriate to use a range of types of impact assessment study. The cameo entitled "A Family of Assessment Studies" [based on Rossini et al. (1976)] presents such a set. We recommend establishing *assessment programs* for subjects of significant social concern. These would consist of appropriate combinations of assessment studies. Multiple assessments could be performed from different points of view and at various points in time (Rossini et al. 1976). The resultant information should prove more valid and useful than a single assessment, however extensive. Except in cases with a clear "go/no go" decision, such as the supersonic transport (SST), an iterative, incremental assessment program seems preferable to the one-shot TA/EIA.

To illustrate the notion of an ongoing assessment program, consider

A FAMILY OF ASSESSMENT STUDIES
[Based on Rossini et al. (1976)]

Macroassessment (comprehensive, full-scale): Full range of implications and policies considered in depth (on the order of magnitude of 5 person-years work for technology-oriented to 10 person-years for problem-oriented assessments).

Miniassessment: Narrow in-depth, or broad but shallow focus (about an order of magnitude smaller than the macroassessment in work effort).

Microassessment: A thought experiment, or brainstorming assessment exercise, to identify the key issues or establish the broad dimensions of a problem (about an order of magnitude smaller than the miniassessment, say, 1 person-month of effort).

Monitoring: Ongoing gathering of selected information on a topic (e.g., radioactive emissions from a nuclear plant, or industrial energy use profiles). May be done formally or informally; as a result of a prior assessment identifying critical uncertainties; and/or as a way to identify critical changes that warrant a new assessment.

Evaluation: Evaluation of ongoing projects and programs can determine whether alterations or new programs are needed. In addition, these can provide feedback as to the validity of previous TA/EIA predictions.

an imaginary agency whose informal monitoring indicates that a TA on a particular technology would be appropriate. The agency convenes an elite panel to perform a quick microassessment to bound and delineate the assessment task. A contract is then let to perform a macroassessment based on the specifications established. Part of the output of this assessment is the design of a small monitoring program that is carred on in-house by the agency. At some point an EIS is then prepared by the agency leading to construction of a pilot plant. Six months after completion of the plant, an evaluation of the plant's operation is conducted. At this time, two miniassessments on a critical aspect of the technology are prepared, one by the agency and one by a concerned environmental group. A popularized pamphlet contrasts the findings of the two miniassessments. The reformulated monitoring program continues.

Although the usual TA/EIA design is one-shot, this discussion illustrates the flexibility and continuity possible through composite assessment programs. Thus the scope of assessment efforts may broaden to deal with most major societal problems as concern for the future increases, and assessment methodology improves.

4.5 COMPONENTS OF AN ASSESSMENT

Having discussed the basic characteristics of TA/EIA, we now turn to the question of how to do one. Considerable guidance is available on what is to be done, usually presented in the form of lists of activities to be included in a TA or EIA. We offer a set of 10 components that should be accomplished in almost any TA/EIA ("Components of a TA/EIA" cameo). Naturally, the form and emphasis will vary according to factors such as the type of study, scale of effort, and major interests.

The term "components" is used, rather than "steps," because these activities do not follow a simple, linear sequence. They are arranged in a logical progression, but, in practice, it is crucial to iterate (repeat) the individual component analyses based on what is learned in doing the other assessment components.

Table 4.3 compares several lists of TA/EIA components—that of this book, three addressed to TA, and three addressed to EIA. The Jones (1971) recommendations derive from an early TA methodology study. Joseph Coates (1976b), a prime mover behind TA, has compiled his observations over his wide range of TA experience. Armstrong and Harman's (1977) conclusions are based on a National Science Foundation (NSF)-sponsored analysis of TA study strategy. The U.S. Council on Environmental Quality (CEQ) guidelines (1974) were general for all federal EIS preparation. The U.S. Department of Housing and Urban Development (HUD) directives (1974) are representative of how federal agencies implement EIA. Zajic and Svreck (1975) present a generalized (not necessarily federal) perspective on EIA. Note the common emphasis on technical context, impacts, and policy analysis. Social forecasting and communication of results, however, appear underemphasized. The former is particularly difficult, whereas the latter may be too often taken for granted. Despite the sequencing differences and occasional gaps, the overall similarity of these study strategies is striking.

We now introduce each of our 10 components. Each is developed in detail in subsequent chapters.

4.5.1 Problem Definition (see Sections 5.2 and 5.3)

The first activities of a TA/EIA are to determine the nature and scope of the study. The focus must be defined and the breadth and depth of

COMPONENTS OF A TA/EIA

1. Problem definition
2. Technology description
3. Technology forecast
4. Social description
5. Social forecast
6. Impact identification
7. Impact analysis
8. Impact evaluation
9. Policy analysis
10. Communication of results

coverage determined based on objectives and available resources. A preliminary microassessment can help focus and bound the study. Although it is desirable that the study be bounded early and effectively, it should be anticipated that findings in the course of the study will alter the study boundaries.

An important component of problem definition is the identification of the parties at interest to the assessment subject and the nature of their interest. "Parties at interest" are persons or groups impacted by the subject of the assessment who may either gain or lose depending on the nature of the impact. This identification will indicate the range of social and political values involved in the assessment and help to define the important impacts and policy sectors to be considered.

4.5.2 Technology Description (see Section 6.3)

Thorough and accurate description of the technology under assessment is crucial for proper focus on impacts and policy responses. Particularly in technology-oriented assessments, this activity can demand a substantial proportion of the study resources.

A description of a technology should include identification of the major technical parameters, alternative ways in which these can be implemented, competing technologies, and definition of the pertinent technological delivery system (TDS) (Figure 4.1).

4.5.3 Technology Forecast (see Section 6.4)

Technology forecasting (TF) attempts to anticipate the character, intensity, and timing of changes in technologies. In TA/EIA, TF is used to chart the future of the technology under assessment. It can identify key uncertainties, potential breakthroughs, upcoming substitutions of one technology for another, likely cost reductions, and new applications. It provides information on changes over time in technical parameters and in the relative diffusion of the technology.

TABLE 4.3. A Comparison of TA/EIA Study Strategies[a]

This Book	Technology assessment study strategies: Jones (1971)	J. Coates (1976b)
1. Problem definition	1. Define assessment task	1. Examine problem statements 7. Identify parties at interest
2. Technology description	2. Describe relevant technologies	2. Specify systems alternatives 8. Identify macro-system alternatives
3. Technology forecast		
4. Social description	3. Develop state-of-society assumptions	9. Identify exogenous variables or events
5. Social forecast		
6. Impact identification	4. Identify impact areas	3. Identify possible impacts
7. Impact analysis	5. Make preliminary impact analysis	4. Evaluate impacts
8. Impact evaluation		
9. Policy analysis	6. Identify possible action options 7. Complete impact analysis	5. Identify decision apparatus 6. Identify action options for decision apparatus
10. Communication of results		10. Conclusions (and possibly recommendations)

[a] Schemes outlining the activities involved in a TA have also been offered by Hahn (1977) and the National Academy of Engineering (1969), among others. Overall federal EIA guidance comes from CEQ (1978). The individual agencies' own guidelines (Housing

Armstrong and Harman (1977)	U.S. CEQ (1974) Guidelines	U.S. Department of Housing and Urban Development (1974) Format	Zajic and Svreck (1975)
	Environmental impact statement guidelines		
Ib. Bounding the assessment domain	1. Project description	1. Describe proposed project	1. Statement of objectives 3. Proposed actions
Ia. Data acquisition	2. Land-use relationships 4. Describe alternatives	3. Discuss impact of environment on project design 5. Discuss internal project environment 6. Discuss alternatives	2. Technical possibilities for achieving objectives
Ic. Technology projection			
IVa. Whole societal futures IVb. Societal values	1. Describe present conditions	2. Describe existing environment	4. Environment prior to proposed action
IIa. Impact criteria selection			
IIb. Predicting and assessing impacts	3. Describe probable impact 5. Describe adverse impact	7. Discuss short- and long-term impacts 9. Describe reaction to program development	5. Impacts including magnitude and importance
IIc. Impact comparisons and presentations	6. Short- vs long-term impacts and their relationships 7. Irretrievable and irreversible impacts	4. Evaluate impact on environment	6. Assessment of impact
III Policy analysis	8. Other considerations offsetting adverse effects	8. Note actions taken to mitigate impacts	
IVf. Validation, and public participation			7. Recommendations

and Urban Development 1974), and individuals' suggestions (Zajic and Svreck 1975; Andrews 1973; Warner and Preston 1974; Waller 1975) elaborate on EIA preparation.

4.5.4 Social Description (see Sections 7.2 and 7.3)

In a TA/EIA, description of the state of society must concentrate on those aspects of society (economic, political, etc.) that interact with the subject of study. The notion of the TDS incorporates the social elements that interact with a technology of interest. This is helpful in identifying institutional involvements, parties at interest, and social values that may affect, or be affected by, a technological development.

Both quantitative and qualitative social descriptors are useful. Social indicators (Bauer 1966) are aggregate data (e.g., birth and divorce rates) that describe properties of some unit of society. Qualitative observations, often obtained through survey, and dealing with such aspects of a society as values and alienation, are more important in social than in technical description.

4.5.5 Social Forecast (see Section 7.4)

Social forecasting seeks to represent the most plausible future configurations of certain dimensions of a society or to project changes in social parameters. These forecasts depend on the technology forecast because changes in technological systems may alter the relationships between the technology and its social context. Social forecasting is one of the least developed components of the assessment process. Provision of a range of qualitative, alternative futures in the form of scenarios (comprehensive portrayals of future situations) is a typical approach to social forecasting, given the present state of the art.

4.5.6 Impact Identification (see Chapter 8)

"Impacts" refer to the products of the interaction between a technology and its social context. Direct impacts are those effects directly attributable to the technology; higher-order impacts are the products of direct effects. A unique feature of TA/EIA is its emphasis on the higher order or indirect effects. In the long run, these may prove the most significant. In a nonrigorous way, the cameo entitled "The Effects of Technology" shows such a chain of impacts.

An important component of a TA/EIA is the identification of both direct and higher-order impacts. This book categorizes impacts with the aid of the acronym EPISTLE (Environmental, Psychological, Institutional/political, Social, Technological, Legal, and Economic impacts). This classification follows disciplinary lines enabling the assessment to draw on specific expertise (e.g., economists to treat economic impacts). Another approach to impact identification is to categorize impacts by the parties affected.

THE EFFECTS OF TECHNOLOGY
[Excerpted from J. Coates (1971: 228–229)]

At times, technologies can have unintended consequences that combine to have a serious impact undreamed of by the creators of the technology. The following table suggests how television may have helped to break down community life.

Consequences of Television

First-order	People have a new source of entertainment and enlightenment in their homes.
Second-order	People stay home more, rather than going out to local clubs and bars where they would meet their fellows.
Third-order	Residents of a community do not meet so often and therefore do not know each other so well. (Also, people become less dependent on other people for entertainment.)
Fourth-order	Strangers to each other, community members find it difficult to unite to deal with common problems. Individuals find themselves increasingly isolated and alienated from their neighbors.
Fifth-order	Isolated from their neighbors, members of a family depend more on each other for satisfaction of most of their psychological needs.
Sixth-order	When spouses are unable to meet heavy psychological demands that each makes on the other, frustration occurs. This may lead to divorce.

4.5.7 Impact Analysis (see Chapters 9–13)

Impact analysis studies the likelihood and magnitude of the impacts identified. Impact identification may well have produced extensive lists of possible impacts. These must be consolidated to indicate the seriousness and likelihood of the impacts. Noting that a power plant produces both energy and some radiation is one thing; determining the amounts of each over time (e.g., percent of area's energy needs) and their implications (e.g., health hazards from the radiation) is another. The actual effects on the various parties at interest should be determined. Here, perhaps more than in any other TA component, disciplinary expertise plays its role in such areas as cost–benefit analysis and environmental modeling. In addition, techniques developed primarily for studies of the future may be helpful.

4.5.8 Impact Evaluation (see Chapter 14)

With the impacts identified and analyzed, it remains to determine their interrelationships and significance relative to the societal goals and objectives, pertaining to the technology. This component of a TA/EIA draws together the results of the impact analysis. It tries to sensibly aggregate "apples and oranges" to be able to compare alternatives and assist in policy analysis. Impact evaluation involves judgments of importance for which the assessment team bears ultimate responsibility. Assumptions and values held by the assessors should be set forth as explicitly as possible. Impact evaluation procedures should be clear to the user of the assessment to allow for acceptance or rejection of conclusions drawn.

4.5.9 Policy Analysis (see Chapter 15)

Policy analysis relates the impact assessment to the concerns of the society. It compares options for implementing technological developments and for dealing with their desirable and undesirable consequences. The policy making sectors that can deal with the options must be identified. The policy options available to these sectors are then laid out and analyzed. Probable consequences are studied and presented. Explicit policy recommendations may or may not be appropriate, depending on the preferences of the study's sponsors, users, and performers.

4.5.10 Communication of Results (see Chapter 16)

Communication of the results is essential. Since a major objective of TA/EIA is to provide useful information, communication requires identification of the potential assessment users and of the sorts of information and forms of presentation they require. Determination of the appropriate study presentation depends on the aims of the study, the sponsor's desires, and the resources available. Most public TA/EIAs should strive to communicate to all parties at interest. Private sector assessments may or may not seek to reach a broad public audience.

4.6 FITTING THE COMPONENTS TOGETHER

The components of a TA/EIA are presented as a sequence for logical and methodological reasons. Problem definition must precede the study, or else the study will have no focus. The technology and its social context need to be studied before the impacts of the technology–society interaction can be understood. Policy analysis depends on knowing the impacts. Finally, results of the study are required before they can be communicated.

However, feedback from phase to phase, resulting in iteration of phases and even the whole study, strengthens study quality. For example, certain dominant social values or political conditions may preclude the implementation of a technology in a particular mode. Recognition of this would lead to a revision of the technology description and forecast. Likewise, policy options may emerge that require determination of their impacts after the initial sequence of analyses is completed. It may also prove convenient to perform more than one task at the same time. Policy sectors, for example, should usually be identified while impact identification and analysis are in progress.

Integration of the component analyses, especially the impact analyses, is a difficult task. These components are likely to involve disciplinary techniques and assumptions that require serious effort to make them compatible with each other. For example, economic and environmental impacts and policies interact with one another. Integration is necessary to assure the validity of component analyses and to generate a readable, consistent study free from disciplinary jargon (see Chapter 17). The Council on Environmental Quality (CEQ) NEPA regulations stress the importance of interdisciplinary preparation of EISs (U.S. CEQ 1978: 55995).

4.7 CONCLUDING OBSERVATIONS

The first part (Sections 4.2–4.5) of this chapter described general attributes of TA/EIA. The second part (Sections 4.6 and 4.7) presented the essential components of a TA/EIA. Key points raised are as follows:

In the context of society's effort to manage technological delivery systems (TDSs), the role of TA/EIA is to inform policy makers, alert parties at interest, and even contribute to the reformulation of social values.

Three basic objectives for TA/EIAs are validity, utility, and (to a lesser degree) improving assessment methodology: (1) validity requires attention to cause and effect, balance (we recommend a value-explicit approach), and sound methods; (2) utility demands that the assessment be relevant, timely, credible, and communicable; and (3) structured efforts to improve TA/EIA methodology are needed.

TA/EIA differs from other studies by a focus on impacts, comprehensiveness, and interdisciplinary nature.

TA and EIA are similar in their commitment to impact assessment, future orientation, technological subject matter, comprehensiveness, comparison of alternatives, and variable scale of effort.

TA and EIA differ in their typical geographic locus, relative policy emphasis, format, typical time perspective, data sources, and typical performers.

Project, problem-oriented, and technology-oriented assessments have different, but complementary, characteristics.

Macro-, mini-, and microassessments, along with monitoring and evaluation, can be combined into ongoing assessment programs with advantages over one-shot assessments.

Ten essential components of TA/EIA are presented and compared with six other sequences of component tasks (Table 4.3). Each of the 10 components is then defined and introduced.

Integration of the component activities is difficult but necessary. Iteration is an important means of achieving this result.

RECOMMENDED READINGS

Armstrong, J.E., and Harman, W. W. (1977) "Strategies for Conducting Technology Assessments." Department of Engineering–Economic Systems, Stanford University, Stanford, CA. Prepared for Division of Exploratory Research and Systems Analysis, NSF.

An incisive look at study strategy for TA based on experiences on NSF-sponsored TAs.

Arnstein, S. R. (1977) "Technology Assessment: Opportunities and Obstacles," *IEEE Transactions on Systems, Man, and Cybernetics* SMC-7 (August), 571–582.

A clear statement of how TA differs from other studies and an overview of its current state of development.

Federal Agency Guidelines.

Contact the federal agency dealing with the technology of interest (e.g., U.S. Army Corps of Engineers or the Environmental Protection Agency) for their most recent guidelines for preparation of EISs [see also U.S. CEQ (1978)].

EXERCISES

4-1. What is the difference between validity and utility?

4-2. How might one design a TA/EIA program to develop assessment methodology? [*Hint:* see Chapter 18.]

4-3. Categorize the five sample TA/EIAs (Chapter 5 Appendix) as project, problem-oriented, or technology-oriented assessments.

4-4. (Project-suitable) Pick a subject for assessment. Identify the technological delivery system for this subject (see Section 7.2.2 and Figures 4.1 and 13.3).

4-5. Compare the TA study strategies with the EIA study strategies in Table 4.3. How do they differ? Why do you believe these differences are present?

4-6. Pick a topic (solar energy for home heating and cooling, an ongoing project, or another topic). Give short paragraph descriptions of each of the 10 types of study listed in Table 4.1 related to this selected topic. (Alternatively, describe only a macro-, mini-, and microassessment on this topic. Discuss the critical differences.)

4-7. (Project-suitable) Pick a topic (as in Exercise 4-6). Give paragraph-length descriptions of the 10 components for a sample TA/EIA on this topic.

4-8. Obtain a TA and an EIA (see the References at the end of this book). Discuss how well the generalizations of Table 4.2 appear to apply. [Alternatively, select one TA and one EIA from the five sample assessments (Appendix to Chapter 5)].

4-9. Obtain a TA and an EIA (see the References at the end of this book). Relate what was done with respect to each of the 10 components. Which do not appear to have been included from the evidence available?

5

Strategies for Particular Assessments:
Bounding and Techniques

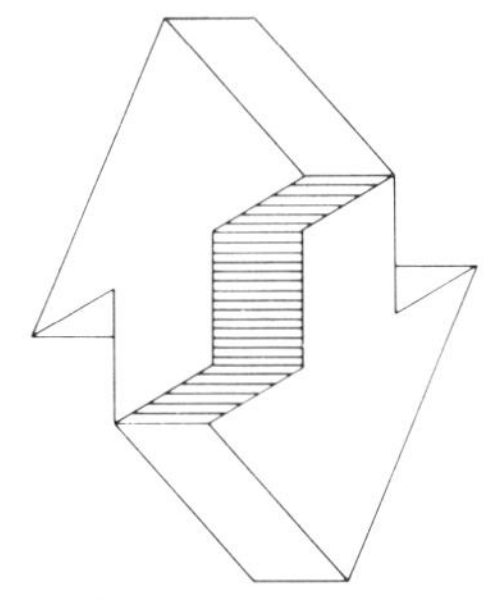

This chapter performs two basic functions relating to the structuring of an assessment. First, it suggests how to define the assessment task. In particular, it explores the issue of bounding the assessment, introducing interpretive structural modeling as a useful technique for that purpose. Second, it addresses the selection of analytical techniques in an assessment. For this purpose important dimensions on which assessments vary are discussed, and characteristics of a variety of techniques are compared. These are combined in the form of guidelines for the selection of analytical techniques in a particular TA/EIA. The Appendix illustrates actual analytical strategies chosen in five TA/EIAs. On completing this chapter, the reader should have insight into the processes of bounding a study and the issues in selecting appropriate analytical procedures to accomplish the assessment.

5.1 INTRODUCTION

Every assessment is unique, and each requires careful attention to the limits of its coverage and the choice of methods to accomplish its objectives. The goals of particular assessments, together with the nature of the subject matter, greatly affect the strategy for performing the work. In Chapter 4 we introduced the generic objectives and types of assessments, together with the components that are common across TA/EIAs. Now by considering content and process dimensions and assessment techniques, we focus on the problem of determining strategies for particular assessments.

We begin with the initial element, problem definition and bounding, to set the context within which a study strategy is developed. As an aid to delimiting a problem area, we introduce the technique of interpretive structural modeling (ISM). Having thus established the outline of an assessment, we discuss important content and process dimensions of TA/EIAs that can differ. As a prelude to knowledge generation and evaluation, there follows a treatment of knowledge characteristics in TA/EIA. Five quite different approaches to knowledge generation are described. Which of these are adopted by an assessment team will have a marked effect on the findings that eventually emerge.

The assessment techniques to be developed in greater detail throughout the book are then introduced. There is a wide range of opinion on the validity and utility of assessment techniques within the TA/EIA community. Considerations, both pro and con, in the use of assessment techniques are discussed and evaluated. A single table presenting the nature, uses, and characteristics of the main assessment techniques is presented to put them in perspective.

With this background, we develop some general guidelines on the relation of techniques to assessment components. Finally, summaries of five diverse TA/EIAs in the Appendix offer concrete illustrations of various assessment strategies.

This discussion of bounding and analytic strategy serves as a jumping-off point for the other components of TA/EIAs that will be treated in subsequent chapters.

5.2 PROBLEM DEFINITION

The first step in contemplating an assessment—whether as sponsor or assessor—is to challenge its existence. Taking as broad and open a perspective as possible, one should pose questions such as:

Why study this technology or project? What can be gained from this assessment?

Is there a core problem reflected in the assessment assignment?

What conditions cause the problem or pose essential opportunities?

What assumptions are being accepted in the TA/EIA formulation? Would reasonable changes in assumptions make a core problem disappear?

Who are the parties at interest to the technology or project? How do their values differ? Are there other social values meriting consideration?

Posing these questions is intended to focus the assessment in its most fruitful direction or to terminate it if it lacks sufficient significance.

Such global questioning can also serve to uncover the value assumptions and world views that affect the study. In so doing, one is opening anything and everything about the TA/EIA to question. This is the ideal starting point for all bounding—a maximum opening of study options before beginning to close them down.

Problem definition in this broad sense is easy to overlook in a TA/EIA. An assessor is usually not asked to raise these questions and, moreover, often faces significant pressures not to do so. A sponsor rarely wants to be told that the study as formulated is inadequate or misdirected. Furthermore, we are all most comfortable with a particular world view and set of approaches to understanding that world. Without a conscious attempt to question and broaden this view, TA/EIA can become a routine task, carried out by technicians representing one narrow viewpoint.

Narrowness is a particular threat to the credibility of EIAs given their focused nature and standardized procedural format. We suggest that a small investment of project resources (even 1 person-day) should be dedicated at the outset to posing the broadest possible questions and perspectives to counter this threat. Options should then be explicitly noted, choices among them made with explanations noted, and underlying assumptions spelled out. Brainstorming (see Chapter 1), followed by critical analysis, is an appropriate procedure for this task.

5.3 BOUNDING AN ASSESSMENT

5.3.1 General Considerations

The previous section suggested questions to pose about the general focus of an assessment. This section carries the issue of bounding to specific domains of concern.

Bounding a TA or EIA, that is, setting its limits, is difficult to accomplish, deeply intertwined with the other assessment tasks, and crucial to the effective conduct and completion of an assessment. The NEPA regulations of the Council on Environmental Quality (CEQ) require such a process, which they refer to as scoping, as soon as possible after the decision to prepare an EIS (U.S. CEQ 1978: 55993). Rossini et al. (1978) found that study depth of analysis and integration were related to the success in bounding. Armstrong and Harman (1977) estimated that delays in settling on study bounds seriously impaired at least half of the 10 TAs they examined. Assessment teams often reported this task surprisingly difficult to achieve, despite extensive effort put into work plans. The main problem arises from the trade-offs involved. Bounding the study early allows the work to proceed in an orderly fashion; keeping the study open allows the assessors to deal with important issues, not foreseen at the beginning of the study. The dangers are, on one hand, a technically competent but irrelevant analysis and, on the other, a rich but unfocused study, not finished on time or within budget.

Bounding an assessment should be an ongoing activity. It depends on constraints set by the study sponsor, and also on characteristics of the development under assessment, the critical impact areas, and the selection of policy options (Figure 5.1). Major bounding decisions should be made relatively early in the assessment. Some of the TA performers Armstrong and Harman (1977) interviewed suggested doing this after about 20% of the study time had elapsed. However, a certain amount of openness allows unanticipated findings to be pursued. For example, energy efficiency emerged as a major automotive consideration after the Grad et al. (1975) TA on the automobile was bounded. One of the advantages of dealing with assessment as a program (Section 4.4) rather than a single study is that bounds may change as external developments unfold.

5.3.2 Areas for Bounding

Berg (1975a) lists six substantive study areas requiring bounding.

Time Horizons. The extent of future projection and the intermediate "viewing times" are central to problem bounding. To deal with currently urgent policy issues, an assessment team may settle on a 10-year projection. If they choose to extend that time horizon to 25 years or more, the technological alternatives and policy orientation of their study would likely be altered significantly.

Spatial Extent. Is the primary emphasis of the study local, regional, state, or national concerns? Spatial bounding is especially pertinent to

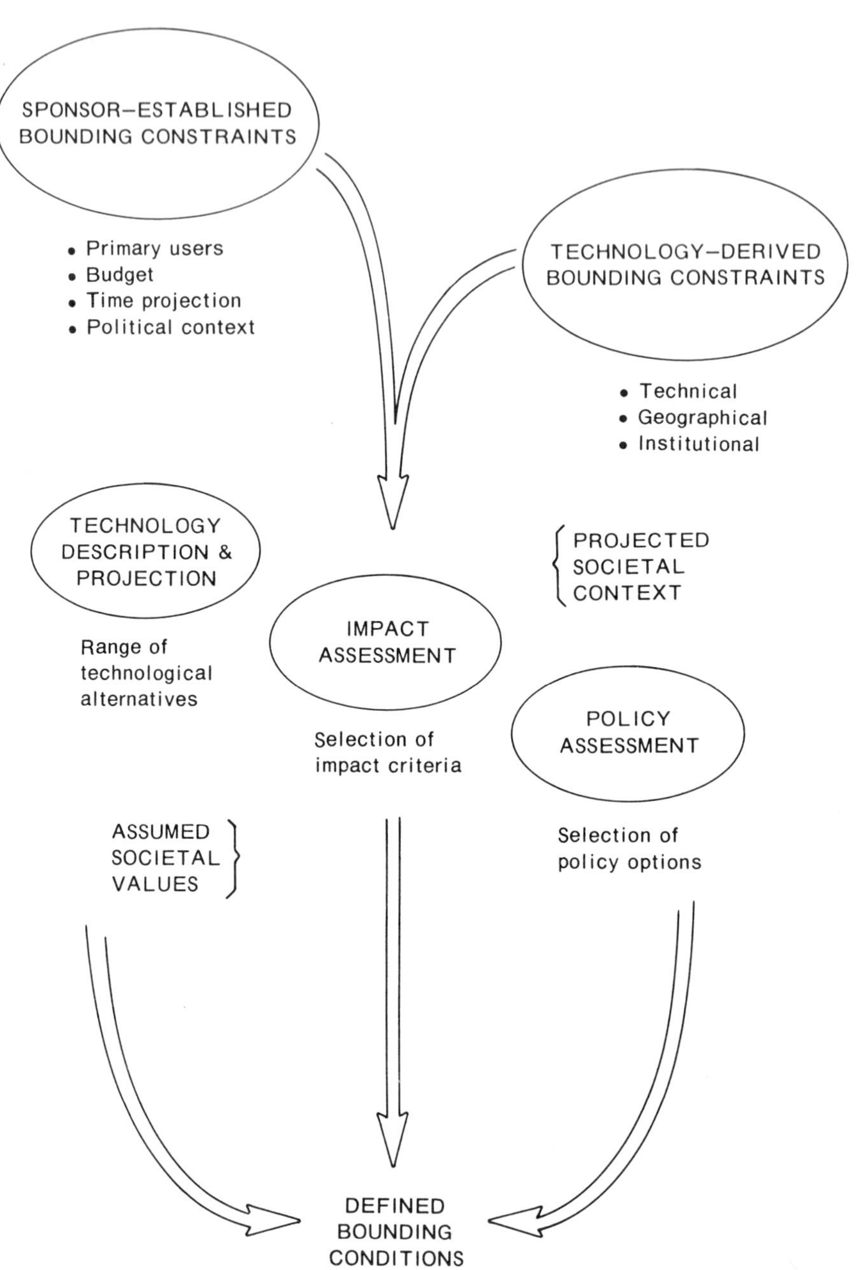

FIGURE 5.1. Critical elements in the process of bounding a technology assessment.
Source: Armstrong and Harman (1977: 30).

what policy makers and policy jurisdictions are to be included. Usually, the more territory considered, the more complex the study.

Institutional Involvements. Institutions considered should include those affecting policy, those likely to use the study, and those impacted by the technology in question.

Technology and Range of Applications. Limiting technological options to a feasible range may be especially important for emerging technologies and for technological solutions to social problems. "Outer bound" alternatives can be set near the feasible technological limits, expecting the comparison to reveal the advantages and disadvantages of various alternatives. Bounding to include only technically conservative and conventional technological options is a frequent source of criticism of TA/EIAs and can constitute the basis of a direct challenge to the objectivity of the study team. Innovative and unconventional alternatives should be included whenever time, funds, and sponsor permit.

Impact Sectors. Criteria for the selection of impact sectors for in-depth treatment should be established to ensure coverage of all areas critical to policy considerations. An initial microassessment (see Section 5.3.3) is the best means of determining which impact areas to assign highest priority.

Policy Options. A wide latitude of policy possibilities exists, especially in emerging and social technologies. Limiting the range must be consistent with the thrust of the assessment. In particular, the sponsor and assessment team must agree upon the range and limits of radical or utopian policy options to be considered.

In addition to these six foci, Berg (1975a) and Armstrong and Harman (1977) note two areas where bounding of a sort is required. The assessors must decide the *sources of input* to the study they wish to accept (e.g., from policy makers or stakeholders). Determination of intended *study users* and their information needs is needed to focus the TA/EIA. The issue of interaction between assessors and such outside parties is treated in Section 16.4.

5.3.3 The Bounding Microassessment

Bounding the study cannot be done without a sound understanding of the technology involved, the most significant impacts, and the main

MICROASSESSMENT

1. A microassessment is any form of study that provides a quick and rough assessment at a level of effort of about 1 person-month or less.
2. It may be accomplished through the efforts of a single person or core team, consulting others as needed; a workshop; a student class project; or some combination of such efforts.
3. This assessment should be relatively comprehensive in its coverage of technological and policy options, possible states of society, and range of impacts. It is not an in-depth analysis of any of these. It assumes that a low effort assessment can identify a sizable proportion, but by no means all, of the impact and policy considerations, and is likely to err in interpretation in a number of respects. A microassessment should, therefore, not be used as a definitive policy tool when other information is available.
4. A microassessment relies upon available literature and opinions to describe the technology. The assessors are likely to speculate about policy options, societal states, and possible impacts—these should at least be tested against the judgment of other individuals.
5. The output can be a *brief* document emphasizing a range of possible developments, impacts, and policy considerations. Available data and literature sources should be noted. Possible courses of action for a more developed assessment may be presented. Note that a microassessment is not an adequate draft EIS.
6. Uses of a microassessment include guidance in bounding a more complete TA/EIA, deciding whether or not a more complete TA/EIA is warranted, or planning an RFP (request for proposals). It may also serve, with appropriate variations in form and emphasis, to elaborate some aspect of a current or previous TA/EIA, to critically review a TA/EIA, or to follow up a completed TA/EIA to determine how recent developments alter previous findings.

areas of policy options (Figure 5.1). To expedite the process, we suggest performance of a "microassessment," that is, a quick attempt at the whole assessment. The results of such a microassessment can be used to set initial boundaries for the TA/EIA that will be revised as the study progresses. The cameo entitled "Microassessment" describes the major features of that process.

5.3.4 Interpretive Structural Modeling

Problem definition and bounding involve the imposition of limits on an initially unstructured milieu. In posing questions about the "core problem" or listing "six areas for bounding," we are trying to derive bases for structuring the assessment task. We noted that a microassessment is one way to get a fuller perspective on the assessment task. In this section we present a relatively simple technique called *interpretive structural modeling* (ISM) that may be useful in the preliminary structuring of the assessment.

ISM is a systematic application of some notions of graph theory to a complex pattern of a contextual relationship between all pairs of the elements in a set. In other words, it helps to identify structure within a system of related elements. It may represent this information either by a digraph (directed graph), which is a set of elements connected by arrows, or by a matrix.

Figure 5.2 shows the logic of constructing an interpretive structural model. One begins with an issue context (e.g., an assessment assignment) that one desires to structure. To do this, one first lists the set of pertinent elements that comprise the issue context. Then one focuses on the key relation of interest among those elements and methodically states whether that relation exists in each of the two directions between every possible pair of elements (see the cameo entitled "An ISM Example."). The relation is expressed simply on a "yes" (1) or "no" (0) basis. Thus ISM is a qualitative, not a quantitative, technique.

As indicated in the cameo, one can manipulate the set of relation-

FIGURE 5.2. The basic conceptual elements of the interpretive structural modeling technique. Arrows denote the presence of activities whereby some elements are examined and elaborated to determine others; the broken arrows denote a feedback, or evaluative, activity.

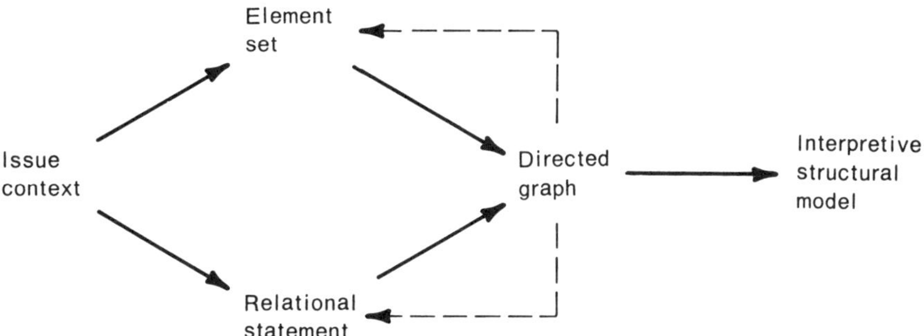

AN ISM EXAMPLE

Suppose we are trying to bound a problem-oriented assessment directed at the possible energy shortage facing the United States in the year 2025. We have limited resources with which to investigate a range of technological alternatives to deal with this problem. Therefore, we should like to decide which ones to include on the basis of how much each is likely to contribute to meeting the energy needs in 2025. Toward this end we have assembled a small group of "experts" and are using ISM to facilitate structuring of the problem.

The issue context (recall Figure 5.2) is as just described. The element set consists of all viable alternative ways to help redress a possible energy shortage. To keep this illustration simple, let us (incorrectly) assume there are only four elements: (1) fossil fuels, (2) solar energy, (3) fusion power, and (4) conservation measures.

Our relational statement is "Is element ——— likely to contribute more to resolving the energy problem in 2025 than element ———?" The group is asked to reach a consensus answer to this question for each pair of elements. The group discussion involved is likely to be very informative—indeed, that may well be the greatest benefit of ISM. Then a 1 (yes) or a 0 (no) is entered in the appropriate cell of the matrix. [The Battelle computer program can help expedite the process by assuming transitive relationships. That is, if (1) contributes more than (2), and (2) more than (3), it will assume (1) contributes more than (3).] Let us assume that the resultant matrix is as follows:

"Row Element Contributes More Than Column Element to Resolving the Energy Problem"

	(1)	(2)	(3)	(4)
(1) Fossil fuels		1	1	1
(2) Solar energy	0		1	0
(3) Fusion power	0	0		0
(4) Conservation measures	0	1	1	

We could construct a digraph to show the same relationships:

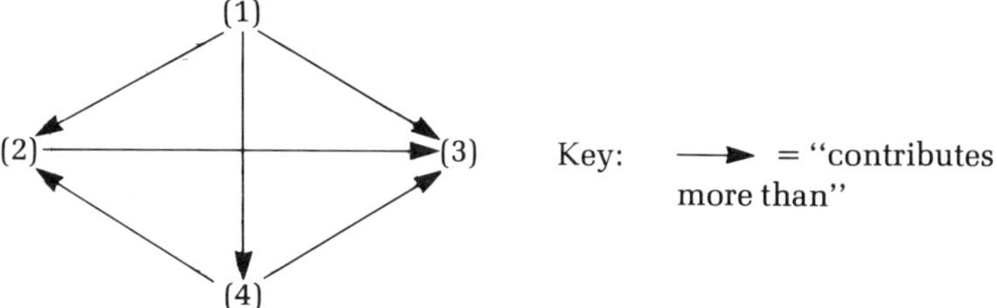

If we are using the computer, it can partition the elements into hierarchical levels and establish a minimal set of linkages to capture the pattern of the relationship. Or, we can do this by hand to yield in this case:

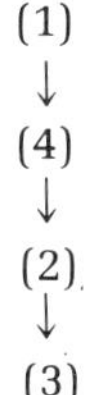

Thus ISM provides what we desired, a ranking to use in limiting the choice of alternatives to be considered. [For further information on ISM, see Malone (1975a, b), Sage (1977), and Watson (1978). Note that the computer program is not necessary to use this simple technique.]

ships to yield a digraph that may be represented hierarchically. A computer program prepared at Battelle Memorial Institute (Columbus, Ohio) facilitates this process. It allows the group or individual to examine the digraph for completeness and allows for possible reinterpretation of particular links. Thus the modelers may revise the set of elements and/or relational statements with ease.

In contrast to models such as KSIM (described in Chapter 9), ISM only describes structural (static) relationships. It is neither dynamic nor quantitative. It can be time consuming if a group ponders a controversial set of elements and relationships. But it is a very flexible means to structure a set of elements (e.g., goals, impacts, or animals) according to any dichotomous relationship among them (e.g., "more important than," "significantly influences," or "eats"). It can be useful in TA/EIA in such tasks as problem definition and bounding, impact evaluation, and identifying relationships among policy sectors.

5.4 ANALYTIC STRATEGIES

With the treatment of study bounding complete, we turn to develop analytic strategies for particular TA/EIAs. This process makes use of the 10 components of an assessment discussed in Chapter 4, a set of content and process dimensions of TA/EIAs, and a classification of assessment techniques in terms of the character of the knowledge they produce. Interrelating these sets of elements leads to some general guidelines for conducting particular assessments.

5.4.1 Content and Process Dimensions of an Assessment

There is a wide variety of emphases in a TA/EIA that affect the choices of specific techniques. Table 5.1 lists selected content and process dimensions of an assessment [Table 15.1 (p. 384) presents the related "policy-analysis parameters"]. The dimensions are dichotomized to allow a more clear-cut discussion of their implications for analytic strategy.

Purpose. A decision document, used to determine the best course of action, will generally require greater emphasis on the identification of key issues, on quantification, and on direct comparison of alternatives. An information document, used to reveal the implications of courses of action, requires a more comprehensive analysis to interpret the significance of a broader spectrum of possible impacts (Warner and Preston 1974: 2).

Focus. In a technology-focused assessment one usually faces considerably greater uncertainty in the technology's development possibilities. Technological description and forecasting should be emphasized. Bounding may also be more difficult. This is likely to be even more the case in a problem-oriented assessment that considers a range of technologies. The alternative problem solutions, as well as the

TABLE 5.1 Content and Process Dimensions of a TA/EIA

Content		
Purpose	1. Decision	2. Information
Focus	1. Project	2. Technology or problem
Spatial inclusion	1. Localized	2. Generalized (regional, national)
Magnitude of intrusion	1. Minor	2. Major
Technological alternatives	1. Restricted	2. Extensive range
Developmental level	1. Existing technology	2. Emerging technology
Time Horizon	1. Short-term	2. Long-term
Process		
Procedural flexibility	1. Guidelines	2. No guidelines
Study resources	1. Low	2. High
Time for study	1. Short	2. Long
Value aspect	1. Value-free (apparently)	2. Value-laden

policies considered, depend on how the problem is viewed and, thus, tend to be less systematic than identification of alternatives in a purely technology-focused assessment (Armstrong and Harman 1977). A project-oriented assessment is more tightly focused with less emphasis on technology forecasting.

Spatial Inclusion. Spatial inclusion is a key distinction between EIAs, as territorially localized, and TAs, as geographically broad. This dimension has methodological implications. For instance, social impact assessment (SIA) of a localized development, such as construction of a highway segment, may rely heavily on personal interviews. In contrast, social impact assessment of national repercussions of a new energy technology may draw primarily upon centralized data accumulations. Familiarity with the technology and the social context is likely to be greater in the localized situation.

Magnitude of Intrusion. In a major intrusion, many factors besides the technological development interact in complex causal patterns to produce direct and indirect effects. Identification and analysis of major impacts is thus both more important and more difficult. Because of the uncertainty involved, efforts to determine the significance of findings, such as sensitivity analyses, are important.

Technological Alternatives. This dimension relates to the first three dimensions (purpose, focus, and spatial inclusion). In the case of a minor, localized project, the study is likely to address several alternatives reflecting closely related technologies or solutions. In contrast, the study of a major technology (or problem) may entail consideration of a wide range of alternatives having repercussions on many secondary technologies. If differences among alternatives are fundamental, such as preventing flood damage by levee construction as opposed to flood plain zoning, impact significance can better be measured against some absolute standard since impacts differ by both kind and size. If differences among alternatives are incremental, direct comparison with the use of quantification is appropriate since they differ in magnitude alone (Warner and Preston 1974: 2).

Developmental Level. A useful distinction can be made between assessments whose subjects are already existing and those where they are emerging. In the former case the form of the technology is relatively fixed and its description may emphasize alternative institutional modes of implementation. In the latter, technology description and forecasting will contain relatively higher levels of uncertainty, and a broader range of alternatives will need to be considered. If an existing technology is controversial, the positions regarding its use may already be polarized, as was the case with No Fault Automobile Insurance (Enzer 1974) when it was assessed. Preexisting analyses may force the assessment to exhibit more precision than if the polarization did not exist, regardless of whether this precision is appropriate to the situation. Impacts of existing technologies tend to be more specific and quantifiable than those of emerging technologies where there is more concern with broad-ranging options.

Time Horizon. A study with a long time perspective, 25 years or more, will face greater uncertainties regarding all major aspects of the assessment—technological and social contexts, impacts, and policy considerations. Forecasting becomes more of a concern and the sensitivity of study conclusions to debatable assumptions becomes critical. Qualitative analysis may be more appropriate than quantitative if the range of considerations is great.

The final four dimensions in Table 5.1 deal with the conduct of the study, rather than the assessment topic.

Procedural Flexibility. This corresponds closely to the distinction between EIAs performed under detailed procedural guidelines and TAs with an open mandate. However, some TAs must prepare an extensive work plan for sponsor approval (e.g., White et al. 1976). In some TAs the sponsor, by specifying in advance the time horizons and range of impacts, may preclude the use of certain methods, thereby incorporating constraints analogous to those in EIAs.

Study Resources. An assessment with limited resources should not simply be a scaled-down version of a major assessment. For instance, small-scale assessments face trade-offs between comprehensiveness and depth. Another serious decision is how quantitative the assessment will be. Quantification, if well done, requires tremendous resources in contrast to a well-thought-out qualitative analysis. For example, Martin Ernst, in guiding the Arthur D. Little, Inc. (1975) study of electronic funds transfer, astutely decided that, given the scope of the topic, its social character, its inherent complexities and uncertainties, and its budget ($220,000), a qualitative assessment was appropriate.

Time for Study. This is a constraint analogous to study resources. A short-time study may not be able to gather primary data, construct sophisticated models, or iterate its findings. Thus such an assessment might choose to use expert panels and qualitative analyses, in contrast to a preferred study strategy for an assessment with equal funding but more time.

Value Aspect. Appropriate analytical means need to be adopted to deal with value-laden issues. These may include a wide range of public inputs to the study, explicit statement of assumptions underlying the assessment, or adversarial TA/EIAs. Certain apparently neutral techniques such as cost–benefit analysis are particularly sensitive to the assessors' values.

5.4.2 Knowledge in TA/EIA

5.4.2.1 Some General Considerations

The content and process dimensions indicate the diversity possible in TA/EIA. Unfortunately, there is no single, focused theory of sociotechnical change to guide TA/EIA. In this situation, assessors have little basis on which to determine the "right" methods to use. It is thus appropriate to step back for a moment to reflect on alternative approaches to knowledge generation. These in turn may offer some guidance in the selection of particular techniques.

We begin by treating general properties that knowledge in a TA/EIA should ideally possess. It should be *systematic*. Like the knowledge in any scientific investigation, it should be coherently interrelated rather than fragmented. It should be generated by the systematic application of sound procedures [as required in EISs; see U.S. CEQ (1978: 55997)]. Reasoning in an assessment should flow without any significant gaps and be as reproducible as possible.

The knowledge should ideally be *comprehensive*, covering all significant impact and policy areas involved in the subject of the TA/EIA. Although all these areas can be covered in depth only in a full-scale study, they can at least be articulated in studies of limited scope.

Knowledge in TA/EIA should extend to *higher-order* effects. "Higher order" refers to effects of effects. An analogy is the collision of balls on a billiard table. The cue ball hitting the first ball is a first-order effect. The first ball hitting a second ball is a second-order effect, and so on. Treating higher-order effects in this manner implies that causality is at work. Thus TA/EIA seeks causal connections between various elements in technological delivery systems.

The TA/EIA process is *future oriented*. It seeks knowledge about a period where no firm data exist. Knowledge of the future in TA/EIA is usually quite uncertain. Very little work has been done to determine the predictive validity of the techniques used in TA/EIA [but see Ascher (1978)].

Finally TA/EIA is *policy-oriented* knowledge. That is, it is knowledge that is going to be used to make real decisions that affect the lives and physical environment of human beings. If it does not provide useful information for decision making, a TA/EIA is worth little.

5.4.2.2 Five Systems for Generating Knowledge

In *The Design of Inquiring Systems*, Churchman (1971) presented a typology consisting of five distinct knowledge-developing, or inquiring, systems that can be used to categorize the wide range of techniques and approaches to knowledge development used in TA/EIA. Mitroff and Turoff (1973) applied this typology to studies of the future. We rely heavily on their treatment.

In presenting the typology one should note that this is not the only possible way to categorize inquiring systems, and that the classification is not exhaustive. None of these five is best in every situation. Each contributes significantly different insights than the others, and the criteria that "guarantee" the knowledge generated differ widely. However, the classification introduces a useful decomposition of knowledge-organizing techniques (Table 5.2).

The first type is the *a priori* inquiring system. This consists of a model of part of the world, typically formal, that does not necessarily depend on any raw data about the external world. Truths about the world are derived analytically from the model. Internal consistency, completeness, comprehensiveness, and the model builder's understanding guarantee the knowledge generated. Examples include operations research, where effort is largely focused on the development and analysis of sophisticated formal models, and the technique of systems dynamics, used in such global modeling efforts as the *Limits to Growth* study (Meadows et al. 1972). A priori inquiry is appropriate in dealing with well-structured problems where an analytic solution is realistic and in which the subject is sufficiently simple and clearly understood for it to be represented by a model.

Empirical inquiring systems treat truth as experiential. A model of a system is developed empirically. Its truth is determined by the ability to reduce complex propositions to simple observations and to verify these observations by the intersubjective agreement of human observers. Empirical systems build up through fact gathering, and results come from induction rather than deduction. Widespread observer

TABLE 5.2. Five Underlying Philosophical Approaches to TA/EIA [a]

Approach	Description	Best suited for problems	Techniques associated with the approach [b]
A priori	Formal models from which one deduces insights about the world, with little need for raw data	Well defined conceptually	Interpretive structural Models Relevance trees Simulation models
Empirical	Beginning with data gathering, one inductively builds empirical models to explain what is happening	Well defined with available data	Monitoring [c] Opinion measurement Probabilistic techniques Policy capture Trend extrapolation
Synthetic	Combines the a priori and the empirical so that theories are based on data, and data gathering is structured by preexisting theory or model	More complex and ill structured	Checklists Cost–benefit analysis Cross-effect matrices Decision analysis Export base models Scenarios Sensitivity analysis
Dialectic	Opposing interpretations of a set of data are confronted in an active debate, seeking a creative synthesis	Ill structured where conflict is present	Adversarial proceedings [c]
Global	A holistic broadening of inquiry by questioning approaches and assumptions	Nonstructured requiring reflective reasoning	Brainstorming

[a] Based on Mitroff and Turoff (1973).
[b] Table 5.3 introduces the techniques.
[c] Not considered in Table 5.3.

agreement on the facts quarantees the knowledge generated. Examples are surveys and simple measurements of phenomena, together with the use of statistical techniques to analyze the data. Empirical inquiring systems are strongest in cases of well-structured problems where there is agreement on the existence and importance of data.

Overemphasis of models with respect to data, or of data with respect to models, can lead to difficulties. The *synthetic* inquiring system takes advantage of the strengths of both models and data. In synthetic inquiry, truth is located both in models (or theory) and in data. Theories may be constructed from data, but at the same time data gathering is structured by some preexisting theory or model. Churchman believes that one of the great strengths of synthetic inquiry is that it offers the possibility of many alternative representations of the problem under consideration. The guarantor of this system is the goodness of fit between the theory or model and the data collected, in terms of that theory. Synthetic inquiry is illustrated by natural science at its best. Theory drives data collection that, in turn, suggests the revision of theory, and then further data collection. Synthetic inquiry can handle more complex and ill-structured problems than either a priori or empirical inquiry.

A fourth inquiring system is the *dialectic*. The dialectic inquiring system depends on the development of two diametrically opposed positions based on a common data set. These are used as vehicles in a debate over the nature of the system being investigated. The object is to develop a position, different from either of the two initial ones, that will reconcile them. In a dialectic inquiring system conflict is the guarantor, since conflict exposes the fundamental assumptions of both positions. The American legal system exemplifies a dialectic inquiring system. Dialectical systems are most appropriate for studying ill-structured problems where plausible arguments can be advanced for diametrically opposing positions. Conflict contributes to "objectivity" by advancing the strongest possible cases for the opposing positions.

The *global* inquiring system is the most complex. The truth of a global inquiring system is measured by its ability to achieve system goals through the creation of alternative approaches and to specify new goals for future inquiry. The system is holistic in that it constantly seeks to expand its area of concern, even to the mode of inquiry. Knowledge so generated is guaranteed only by the success in achieving old goals and identifying new ones. Brainstorming is an example of this approach. The strength of global inquiry lies in its ability to take the broadest possible perspective. Global inquiry seeks the "right" questions to ask. Its weaknesses are complexity and lack of structure.

Table 5.2 lists the five inquiring systems together with some uses and example techniques. Note that techniques associated with TA/EIA distribute among the five approaches. In practice, there is considerable difficulty in distinguishing assessment activities as "purely" associated with any one approach. For instance, one does not expect empirical data gathering without some "model" of what one is seeking. Kan Chen (private communication) has indicated the value of considering the inquiring systems used by members of an assessment team. The presence of individuals operating unawares under different inquiring systems can be a severe barrier to interdisciplinary cooperation.

5.4.3 Particular Methods and Techniques

In this section we gather together the specific methods and techniques that will be systematically discussed in the course of this book. Two techniques, brainstorming and interpretive structural modeling, have already been introduced. Table 5.3 organizes information about these techniques, their characteristics, and some of their potential uses.

Chen and Zissis (1975) point out that the methods used in impact assessments can be dichotomized along two dimensions. First, certain methods require a *group* of participants, whereas others can be performed by an *individual* (or by a group, of course). This may be a factor if resources are limited. Second, they distinguish between those approaches that *scan* for impacts in one fell swoop (e.g., a checklist) versus those that *trace* impacts via some cause and effect sequencing (e.g., relevance tree). The scanning methods are more suitable for a TA/EIA with limited resources, but they, of themselves, are not good for probing impacts in depth and addressing higher-order impacts.

Before delving into the specifics of the various techniques (Table 5.3), it must be noted that the present state of the art is limited. Rossini et al. (1977, 1978) found that the methodological literature that develops techniques usable in TA/EIA has been largely ignored by TA performers. The majority of assessors found the techniques excessively formal and not suited to the work they wanted to do. Another group saw techniques as devices for organizing knowledge without adding any substance. A small minority felt that quantitative techniques increased the quality of the research. Despite all arguments pro and con, it remains that techniques help organize thoughts and data. Used as tools to develop and present analyses more efficiently, we suggest that they deserve serious, but not slavish, consideration.

Systematic validation of particular techniques has seldom been undertaken to a significant extent [see Ascher (1978)], and it is badly needed. If the uncertainty in the knowledge development in TA/EIA is to be decreased, both validation of techniques and continued development of theory are appropriate. In the meantime, Table 5.4. raises 10 points of concern in determining to use particular techniques in a TA/EIA.

5.4.4 Guidelines for Selection of Analytical Techniques

The selection of techniques is context dependent. Our procedure in devising guidelines is to consider the 10 assessment components. We relate them to generic characteristics of techniques and, where appro-

TABLE 5.3 Assessment Techniques

Technique (chapter in which treated)/description	Characteristics	Uses	References
Brainstorming (Chapter 1)/a group or individuals generate ideas with no criticism allowed	Group or individual Scanning Global inquiry Unstructured	Problem definition Generating lists of potential impacts, affected parties, policy sectors, etc. Performing microassessments	Ayres (1969: Chapter 8) Sage (1977: Chapter 5)
Interpretive structural modeling (Chapter 5)/Directed graph representation of a particular relationship among all pairs of elements in a set to aid in structuring a complex issue area	Group or individual Scanning/tracing A priori inquiry	Developing preliminary models of issue areas Impact evaluation	Malone (1975a, 1975b) Sage (1977: Chapter 4) Watson (1978)
Trend extrapolation (Chapter 6)/a family of techniques to project time-series data using specified rules	Individual Tracing Empirical inquiry	Technology forecasting, both parameter changes and rates of substitution Social forecasting	Hencley and Yates (1974) Mitchell et al. (1975) Ayres (1969: Chapter 6) Bright (1978)
Opinion measurement (Chapter 6)/A variety of techniques (including survey, panels, and Delphi) to accumulate inputs from a number of persons, often experts in an area of interest	Group Scanning/tracing Empirical inquiry	Technology forecasting and description Social forecasting and description Impact identification Impact analysis, especially social	Linstone and Turoff (1975)—on Delphi Warwick and Lininger (1975)—on survey
Scenarios (Chapter 7)/composite descriptions of possible future states incorporating a number of characteristics	Individual or group Scanning Synthetic inquiry Largely qualitative	Social forecasting Technology forecasting Impact analysis Policy analysis Communication of results	Hencley and Yates (1974) Bright (1978) Mitchell et al. (1975)

Checklists (Chapter 8) /lists of factors to consider in a particular area of inquiry	Individual Scanning Synthetic inquiry	Impact identification Policy-sector identification	Leopold et al. (1971) Warner and Preston (1974)
Relevance Trees (Chapter 8) /network displays that sequentially identify chains of cause–effect (or other) relationships	Individual Tracing A priori inquiry	Impact identification and analysis	Hencley and Yates (1974) Bright (1978) The Futures Group (1975)—an example
Cross-effect matrices (Chapter 9) /two-dimensional matrix representation to indicate interactions between two sets of elements	Individual Scanning Synthetic inquiry	Impact identification and analysis Analyzing the consequences of policy options	Bright (1978) Kruzic (1974)—on KSIM Linstone and Turoff (1975) Sage (1977: Chapter 5)
Simulation models (Chapter 9) /simplified representation of a real system used to explain dynamic relationships of the system	Individual Tracing A priori inquiry Formal and quantitative	Technology forecasting Impact analysis	J. Coates (1976a) Wakeland (1976)—compares KSIM, QSIM, and Dynamo [a] Sage (1977: Chapter 6)
Sensitivity analysis (Chapter 9) /a general means to ascertain the sensitivity of system (model) parameters by making changes in important variables and observing their effects	Individual Tracing Synthetic inquiry Quantitative	Impact analysis Policy analysis	Leininger et al. (1975)
Probabilistic techniques (Chapter 9) / stochastic properties are emphasized in understanding and predicting system behaviors	Individual Tracing Empirical inquiry Often requires opinions of subjective probabilities	Technology forecasting Impact analysis Impact evaluation	Gordon and Stover (1976) Gohagen (1975)

(continued)

TABLE 5.3 Assessment Techniques *(continued)*

Technique (chapter in which treated)/description	Characteristics	Uses	References
Cost–benefit analysis (Chapter 11)/a set of techniques employed to determine the assets and liabilities accrued over the lifetime of a development	Individual Scanning/tracing Synthetic inquiry In broadest case, mixes quantitative and qualitative factors	Economic impact analysis Environmental impact analysis	Sassone and Schaffer (1978)
Export base models (Chapter 11)/estimates regional changes through a multiplier applied to the development in question	Individual Tracing Synthetic inquiry Quantitative	Economic impact analysis	Tiebout (1962) Isard (1960) Richardson (1969) Sage (1977: Chapter 6)—on input–output analysis
Decision analysis (Chapter 14)/formal aid to compare alternatives by weighing the probabilities of occurrences and the magnitudes of their impacts	Individual Tracing Synthetic inquiry	Impact evaluation Policy analysis	Bross (1965) Howard et al. (1972)—an example Sage (1977: Chapter 7)
Policy capture (Chapter 14)/a technique for uncovering the decision rules by which individuals operate	Group Tracing Empirical inquiry	Impact evaluation Policy analysis	Hammond and Adelman (1976)

[a] KSIM and QSIM are different cross-impact computer programs to perform cross-impact analysis; Dynamo is a program used for system dynamics.

TABLE 5.4. Methodological Concerns in Selection of Techniques

Inferential power	Distinguishes effects of the technology in question from effects due to other causes or to interacting causes; suitable for the assessment design to yield causal understanding
Assumptions	*Explicit* as to assumptions and criteria employed; *robust*, not overly sensitive to analytical assumptions that do not fit the real situation
Resources required	Type of *data required* (necessary precision, accuracy, and level of quantification) is available; *sampling* principles and practicalities consistent with the TA /EIA assignment; *costs* of data retrieval and computational procedures are tolerable; *special skills* available; available *time* to perform the analysis
Objectivity	Emphasis on objective measures
Reproducibility	Insensitive to analysts' biases; reliable in character
Uncertainty	Indicates degree of confidence, likelihood of indicated results
Specificity	Identifies *specific indicators* to be measured; provides for the measurement of *impact magnitude; distinguishes effects* on different geographical or social groups (distributional effects)
Comprehensiveness	*Generalizable* across the range of technologies, different scales, and impact types under consideration; suitable for the entire *time frame* of interest
Comparisons	Suitable for comparing *all the alternatives* of interest; provides a means for reasonable *aggregation of impacts* of different types to allow desirable comparison
Communicability	Suitable for involvement of lay persons in analysis or interpretation; provides a format for communicating results, highlighting key issues

priate, to particular techniques (Chapter 16 presents a measure of the relative credibility of a number of these techniques). The resulting guidelines are simply that. Every one could usefully be violated in the proper circumstances.[1] Experience with the particular assessment situation usefully combines with these abstract principles to form strategies. The social and management aspects of strategies, supplementing the present knowledge considerations, are discussed in Chap-

[1]For the reader's interest, we might mention that we developed rather elaborate numerical scorings for the use of particular techniques (categorized according to inquiring system) for the 10 TA /EIA study components. We also cataloged the five sample TA /EIAs (see Appendix) on the content and process dimensions. On reflection, we have more faith in assessors' good judgment on the use of analytical techniques than in our formal guidelines. Hence we have tried to present factors worthy of consideration, without combining them into an algorithm.

ter 17. The reader who wishes a broader consideration of strategy should turn ahead at this time.

Bounding is a global process. Scanning techniques such as brainstorming and expert opinion are useful to generate broad, divergent perspectives. Decision-oriented assessments should focus on the factors immediately impacting the decision, whereas studies for information can typically be broad-ranging. Qualitative analysis is emphasized at the expense of formal and quantitative work.

Technology description is an empirical inquiry that scans the state of the art and makes use of expertise, within or outside of the project team.

Technology forecasting is a synthetic inquiry tracing the development of technologies and alternatives. Long-term forecasts can make use of opinion measurement, scenarios, and, to a lesser degree, trend extrapolation. Short-term forecasts should more often rely on trend extrapolation, modeling, and opinion measurement.

Social description is an empirical inquiry that scans the information available using expert opinion, demographic data, and social indicators.

Social forecasting is a synthetic inquiry that anticipates future states of society. It is essentially a qualitative process, using primarily opinion measurement and scenarios, with only economic and demographic data being treated quantitatively. Social forecasting is one of the weakest components because of the uncertain accuracy of methods and its generally low credibility.

Impact identification is largely an empirical inquiry depending on the consensus of knowledgeable parties at interest. It scans first-order impacts and traces higher-order impacts usefully drawing on opinion, checklists, relevance trees, and matrices. Identifying impacts is "quick and dirty" in a study with low resources, using opinions and checklists. With high resources, an extensive literature search and greater use of expert opinion is possible.

Impact analysis can use synthetic, a priori, and empirical inquiry in tracing the effects of the technology or project. Different generic classes of impacts are treated differently. Economic impact analysis is quantitative, except in extremely low-budget studies, using such techniques as cost–benefit analysis, export base models, and simulation modeling. Environmental analysis typically is a little more qualitative, but it uses modeling extensively. Social and institutional impact analysis are far more qualitative, making extensive use of opinion measurement and scenarios.

Impact evaluation is typically synthetic and dialectical. Decision analysis and policy capture are possible techniques. Qualitative arguments are quite appropriate here.

Policy analysis uses synthetic, dialectical, and global inquiry to trace through the policy sectors and options potentially involved. Use of techniques in policy analysis is usually quite limited. For existing technologies and short time-frame studies, the policy issues are more specific than for emerging technologies and long time-frame studies. An assessment prepared for the private sector typically has a policy focus narrower than that of a public sector assessment.

Communication of results emphasizes matching knowledge generated in the assessment to the needs and formats of involved parties. The appropriate categories for knowledge are, in this case, those of the users rather than the assessors (see Chapter 16).

In dealing with complexity, qualitative and holistic techniques are preferred by many researchers to quantitative and structured ones. The content and process dimensions of TA/EIA labeled "2" in Table 5.1 are more complex, generally suggesting a possible preference for the less quantitative techniques.

The considerations discussed above provide a broad brush orientation for strategic choice in particular assessments. The idiosyncratic character of each TA/EIA makes highly specific rules less useful than general guidelines. Study strategies used in five assessments are summarized in the Appendix. The relative absence of "techniques" therein is notable.

5.5 CONCLUDING OBSERVATIONS

The first part of this chapter (Sections 5.2 and 5.3) discusses the general problem of bounding an assessment. The second part (Sections 5.4 and 5.5) considers and illustrates strategies for knowledge organization and generation, making use of assessment components, content and process dimensions, and techniques. The following are important observations.

Initial problem definition should question the full range of assessment options before closing any down.

Substantive bounding of the assessment can benefit from a microassessment. Six areas requiring bounding are identified.

Bounding is a continuous process. Although the bounds of a study and its procedural strategy should be set early, their details should evolve over the course of the study.

Interpretive structural modeling is a technique potentially useful in bounding.

Seven content and four process dimensions (Table 5.1) characterize assessments and have significant implications for analytical strategy.

Approaches to knowledge and techniques can be classified as a priori, empirical, synthetic, dialectical, or global.

Some techniques can be worked by an individual; others require a group. Some techniques scan for impacts; others trace cause and effect links.

The main assessment techniques are briefly described in Table 5.3. Methodological concerns in technique selection are raised in Table 5.4.

Complex inquiring systems tend to be most appropriate for bounding, impact evaluation, and policy analysis. Description, forecasting, impact identification, and impact analysis typically make use of simpler inquiring systems.

Small-scale studies, studies of social technologies, and policy-oriented studies tend to be relatively qualitative. Large-scale studies of physical technologies tend to be quantitative and structured.

RECOMMENDED READINGS

Armstrong, J. E., and Harman, W. W. (1977) "Strategies for Conducting Technology Assessments." Department of Engineering–Economic Systems, Stanford University, Stanford, CA. Prepared for Division of Exploratory Research and Systems Analysis, NSF.

An incisive look at study strategy for TA based on experiences on NSF-sponsored TAs.

Arnstein, S. R., and Christakis, A. (1975) *Perspectives on Technology Assessment*. Jerusalem: Science and Technology Publishers.

An interesting treatment by TA practitioners of some of the major problems in doing TA.

Hencley, S. P., and Yates, J. R. (eds). (1974) *Futurism in Education: Methodologies*. Berkeley, CA: McCutchan.

A collection of articles that treats techniques relevant to TA/EIA.

Sage, A. P. (1977) *Methodology for Large Scale Systems*. New York: McGraw-Hill.

Contains thorough treatments of many techniques appropriate for TA/EIA.

Warner, L., and Preston, E. H. (1974) *A Review of Environmental Impact Assessment Methodologies*. Prepared for Office of Research and Development, Environmental Protection Agency.

A good overview of methodological requirements with a discussion of 17 techniques for EIA.

EXERCISES

5-1. Discuss whether an assessment of space colonization should be performed at this time.

5-2. Discuss the typical differences between TAs and EIAs in terms of the content and process dimensions of Table 5.1.

5-3. You are performing an assessment of geothermal energy for the Electric Power Research Institute (the research arm of the electric utility industry). Discuss important considerations for bounding the problem (include the six substantive bounding areas). Set some preliminary bounds for the assessment.

5-4. Bounding (project-suitable)

1. Determine a candidate topic for TA/EIA. [This could well be a brainstorming topic (Chapter 1), a topic used in outlining study steps (Chapter 4), or an ongoing project topic.]
2. As appropriate, form into small groups (three or so members) and address each of the five question areas raised in the first paragraph of Section 5.2. List major topical options and value perspectives possibly pertinent to the study.
3. Suggest a most desirable topic definition and overall approach to the resultant TA/EIA. Explain the choice and note the essential assumptions that your approach accepts. Be realistic, recognizing that every TA/EIA must be performed under resource constraints.
4. If more than one group has performed the exercise, compare and discuss findings.

5-5. Microassessment (project-suitable)

1. Determine a candidate topic for TA/EIA. (This could be a topic studied in a previous exercise or an ongoing project). Arbitrarily specify the purpose of the assessment (see the "microassessment" cameo, especially item 6).
2. As a group (about three to six members), brainstorm possible technical and policy alternatives, and impacts.
3. Assess available data sources. As time and resources permit, perform a literature search to pursue the description of the technology. As appropriate, do likewise for the societal attributes most pertinent and to define the policy arena. Again, as appropriate, gather information to buttress impact speculations.

4. Extend and test descriptions and projected possibilities by quizzing knowledgeable professionals. As the situation warrants, explore the alternative values perspectives by contacting possible parties at interest.

5. Prepare a brief report focused on the key questions addressed by the microassessment. Make appropriate recommendations as to future course of action.

5-6. Interpretive structural modeling (ISM) (project-suitable)

1. Study the description of ISM (Section 5.3.4).

2. Consider a potential study topic, such as one explored in a previous exercise. Identify the pertinent elements in the related technological delivery system (Figure 4.1).

3. Identify an important relationship to explore in relation to the system in question. For instance, you might study whether:
   a. variable *i* influences variable *j*, where the variables are major system decision structures and participants;
   b. variable *i* impacts on variable *j*, where the variables are major economic, environmental, social, and political parameters; or
   c. variable *i* is more important than variable *j*, where variables have such relevant values as energy supply, environmental protection, and provision of employment.

4. Construct a matrix of 0s and 1s based on the answers to the question "Does *i* have the relationship identified to *j*?" for all *i* and *j*. This is a structural model.

5. Discuss the resultant patterning of relations. How might this prove to be useful information in the formulation of an assessment?

5-7. Classify each of the five TA/EIAs summarized in the Appendix in terms of the content and process dimensions listed in Table 5.1. Discuss their use of analytical techniques.

5-8. Modes of inquiry (project-suitable)

1. Select one of the five TA/EIAs described procedurally in the Appendix, or another assessment topic. Consider that TA/EIA in turn, from each of the five perspectives presented in Table 5.2 and Section 5.4.2.2. Describe how an assessment emphasizing each of the perspectives would differ from that performed.

2. Using Table 5.2, classify the analytical techniques used (or to be used) in the TA/EIA by their mode of inquiry. Is there a tendency within the(se) TA/EIAs to use a single mode?

3. Alternatively, reflect on the assessment. Present alternative study strategies that emphasize each of the five modes of inquiry.

APPENDIX: SUMMARIES OF SELECTED ASSESSMENTS

Five sample TA /EIAs are chosen to illustrate the diversity of assessments (Table 5.5). The three EISs range from a minor, local highway project, to a state-size irrigation project, to a national energy program assessment. The two TAs are national in scope, one addressing a social technology and the other, an emerging physical technology.

The summaries focus on the procedural aspects of the studies. They illustrate the diversity of possible strategies for individual assessments; the assessments are not evaluated as "good" or "bad."

Final Environmental Statement for U.S. 59 in Stevens County, Minnesota, prepared by the Federal Highway Administration and State of Minnesota Department of Highways, January, 30, 1974, EIS-MN-74-0180-F.

This EIS of a new 4.5-mile city bypass highway and associated state road development is simple and qualitative in character. Its 27 pages of text include eight pages of excerpts of Minnesota Highway Department specifications to protect the environment. It includes comments received from interested parties and responses by the preparers, along with a number of maps and aerial photographs. Five alternatives are noted. However, only two (differing only in the type of crossing of a railroad track) are examined.

The final EIS follows the Council of Environmental Quality (CEQ) Guideline categories, providing paragraph to page-length comments. A public hearing was held November 1970 with no opposition noted. The draft EIS was distributed for comment on August 1972 to 24 federal, state, and local government units and the Burlington and Northern railroad. The most substantive criticisms were received from regional offices of the EPA and the Department of the Interior. They questioned project effects on river water quality and quantification of the water quality level, the location of the bypass partially in a flood plain, and quantification of effects on wildlife. The preparers responded briefly and qualitatively.

The EIS contains a page on qualitative economic impacts, noting that increased development is the major expected impact. There is no attention to social or institutional impacts and, most importantly, none to the higher-order impacts.

This assessment exemplifies a minimal response to the requirement to file an EIS. It is a common-sense treatment based on available knowledge and opinion without significant use of techniques or data. It is informative on the likely impacts of the proposed highway project, noting, but dismissing without serious examination, alternative routes and the do-nothing alternative. It thus reflects no conflict resolution with respect to project approval or environmental protection. It does mention points requiring careful construction measures to protect a river that is crossed, guard against erosion, and so on. It notes that the major impact of the proposed project is likely to be increased development along the route but does not examine the extent or implications of this.

This study lacks significant technology forecasting, social description and forecasting, impact analysis and evaluation, and policy analysis. It simply says here is a proposed project that everyone wants, and these are the kinds of economic and environmental impacts to be expected.

Final Environmental Statement for the Initial Stage, Garrison Diversion Unit (North Dakota), prepared by the Bureau of Reclamation (Department of the Interior), January 10, 1974, EIS-ND-74-0058-F.

This EIS addresses the initial 250,000-acre segment of a 2,007,000-acre irrigation project.

TABLE 5.5. Five Selected Assessments

	EISs			TAs	
Features	Route 59	Garrison diversion	GESMO	Materials information systems	Geothermal energy
Topic	A 4.5-mile city bypass highway	A large project to divert Missouri River water through North Dakota to the Canadian Red River of the north	Use of recycled plutonium in nuclear power plants	Possible information system arrangements to deal with natural resource availabilities	Potential of U.S. geothermal energy development
Preparer	Federal Highway Administration and Minnesota Department of Highways	Bureau of Reclamation	Nuclear Regulatory Commission	Office of Technology Assessment	The Futures Group
Character	A short, simple, qualitative assessment of a local project	A substantial quantitative assessment of a controversial large-area project	A major, quantitative program assessment on a controversial national policy issue	A substantial, qualitative assessment of a national institutional issue	A substantial, quantitative assessment of a "new physical technology on a national basis
Impact Coverage	Thin	Emphasis on environmental and economic impacts	Emphasis on environmental, economic, health, and institutional issues	Emphasis on institutional and policy analysis	Coverage of the EPISTLE impacts
Methods Used	Common sense	Quantitative environmental cataloging, some models (main cost–benefit analysis cited from other studies)	Cost–benefit analysis, models	Largely expert opinion and literature review	Wide range of "TA techniques" including relevance trees, trend-impact analysis, cross-impact matrices, and models

The diversion of water from the Missouri River, behind the Garrison Dam, for irrigation in North Dakota before flowing to the Canadian Red River of the North has been seriously considered for 100 years. Construction began in 1967. This 800-page final EIS traces back to an 11-page document of January 1971. That was followed by a regional draft environmental statement submitted to CEQ in April 1971. This was determined by CEQ not to meet the requirements of NEPA. In December 1972 suit was filed against continued development without preparation of a final EIS. In January 1973 the Bureau of Reclamation (BuRec) released a 145-page (preliminary) final environmental statement, upgraded to a 246-page draft environmental statement released in April. Public hearings on this draft were held in May 1973 at Minot, North Dakota. Comments were requested or received from over 100 governmental and private parties.

The inputs to the final EIS are diverse. Six Congressional hearings (1957–1965), over 800 public meetings of some form involving BuRec, and over 50 reports each by BuRec and the Bureau of Sport Fisheries and Wildlife have been held or prepared since 1947.

The EIS focuses on the project (already underway) and its environmental (including recreation) and economic impacts. Alternatives to the project are considered only lightly. Social description and forecasting, social and institutional impacts, and policy analyses are essentially ignored. The environmental and economic impact analyses are highly quantitative, using some computer models. The environmental impact study emphasizes water quantity and quality, vegetation, wildlife (including fish), and recreation. The economic impact study emphasizes costs and benefits.

In January 1975 The Institute of Ecology (TIE) released a 100-page critique of the final EIS, prepared by a volunteer review team, in conjunction with a Canadian review of the international impacts. Their review asserts that the analyses in the EIS are generally weak—many assumptions are questionable, impact analyses are one-sided and unsupported, and indirect impacts are not sufficiently considered. The EIS is written by the agency building the project to describe the impacts and measures to mitigate adverse effects—not to decide among alternatives.

TIE claims the water-quality model used was not applied to enough waterways or pollutants. They do not feel that BuRec completed a needed comprehensive survey of existing vegetation units. Quantitative estimates of wildlife effects, prepared by the Bureau of Sport Fisheries and Wildlife were based on 10 sample wetland areas—inadequate according to TIE. Recreation projections did not include objective input from citizens. The cost–benefit conclusions (analyses not provided in the EIS) are critiqued by TIE for being too favorable, ignoring opportunity costs, and not indicating local versus national cost–benefit distributions.

The Garrison diversion project EIS reflects a costly and controversial study on a costly and controversial project. The EIS was not intended to influence the decision to build or not, since building was long underway; however, that decision was reopened under the Carter administration.

Final Generic Environmental Statement on the Use of Recycled Plutonium in Mixed Oxide Fuel in Light Water Cooled Reactors, prepared by the U.S. Nuclear Regulatory Commission (summary volume on Health, Safety, and Environment dated August 1976; NUREG-0002, Vol. 1, is available from the National Technical Information Service, at $6.00 printed copy; there are five additional volumes plus a Supplement on Safeguards).

This program EIS, known as GESMO, considers the impacts of various ways of recycling fuel for nuclear power reactors. It is a major input to the decision by the Nuclear Regulatory Commission (NRC) on whether to use recycled plutonium and/or uranium as fuels and, if so, under what conditions. There are several major documents: the draft and final GESMOs for health, safety, and the environment; and the draft and final special supplemental nuclear safeguards study.

A commercial reprocessing plant for nuclear fuel operated during 1966–1971. Construction on a reprocessing plant at Barnwell, South Carolina began in 1970. The National Resources Defense Council intervened asking that an EIS be prepared on reprocessing in general instead of for that plant alone. In August 1974, the Atomic Energy Commission (AEC) published a draft generic EIS. CEQ considered that EIS incomplete in failing to analyze safeguards to protect the public from theft and misuse of the material. Congress later mandated that the assessment consider safeguards and security-force possibilities. The NRC, successor to the regulatory functions of the AEC, ordered the preparation of the final GESMO along with a draft Safeguard Supplement.

A year was required to produce a draft report of four volumes with 1100 pages at a cost of millions of dollars. NRC staff coordinated an effort that included contract work for portions of the study and expert advisors and workshops to make strategic decisions. The study was eventually framed around five scenarios reflecting *alternative development patterns*: uranium and plutonium recycle [(1) prompt, (2) prompt for uranium and delayed for plutonium, or (3) delayed fuel reprocessing in general], (4) uranium recycle only, and (5) no recycle.

Impacts associated with the various alternatives were based on a forecast through the year 2000, using the Energy Research and Development Administration's low-energy-demand growth projection [WASH-1139(74)]. This forecasts a proportionate energy-supply share to be supplied from nuclear reactors, from which one can estimate waste products, fuel needs, and so on. The nuclear energy projections were input into a series of sophisticated computer models requiring important assumptions. Social description and forecasting are essentially absent from this study.

Throughout, impacts (primarily environmental and economic) are calculated for each of the five alternatives to allow comparison. The econometric work involved workshops of economists to consider assumptions such as the discount rate to use in the cost–benefit analysis. Health effects on nationwide and worldwide bases are modeled. Effects associated with various development phases are separately evaluated—construction, normal operation, attendant transportation and waste-management activities, and possible accidents for model plants.

Institutional impacts are most heavily considered in the Safeguards Supplement, which focuses on the need for a new security force to protect nuclear materials. Over 300 consultants from other federal agencies and special contractors contributed to the draft supplement comparing the effectiveness of private and federal guard forces. A conceptual design of a reference safeguards system using projections of nuclear developments and considering a range of possible threats was undertaken. The study concluded there was no current need for a special federal security agency. Social impacts are not treated in depth, similar to an EIS for individual power plants. Policy analysis treats alternative means for dealing with plutonium and means for mitigating adverse environmental impacts.

In sum, the several elements of the GESMO EIS represent a massive research effort. The studies were run by NRC staff with inputs from private contractors, federal officials, and expert advisors. The results are part of an extensive decision process by the NRC and the Executive toward reaching a decision among the five recycling alternatives.

An Assessment of Information System Capabilities Required to Support U.S. Materials Policy Decisions, prepared by the Office of Technology Assessment, U.S. Congress, December 1976 (OTA-M-40).

This report analyzes information systems and their role in supporting policy making on materials (natural resources other than food) problems, such as shortages. The 249-page report reflects condensation of seven volumes of materials prepared by OTA staff and four outside contractors. The study, one of a coordinated series of OTA studies on materials, required 2 years and about $500,000.

Study strategy was set by OTA staff in response to a request from the House Committee on Science and Technology. The Technology Assessment Board reviewed the study plan. The study's advisory committee was heavily involved in reviewing prepared materials.

This TA was strongly addressed to identification of policy making needs. Literature review, study of pending legislation, and interviews with 25 senior executives and 59 materials specialists in government, industry, and academia led to the conclusion that current information systems were inadequate. The study group determined that objective quantitative criteria were not available to document information needs.

The existing government materials-information systems were then described and considered as bases from which the identified needs could begin to be met. Three approaches (an interagency coordinating group, step-by-step upgrading of existing information systems, and a new integrated system) were analyzed in terms of their technical capabilities and approximate costs (accurate within a factor of 5). The study then explored possible institutional mechanisms to implement the alternatives, finding seven promising enough to warrant careful description and comparison. Qualitative analysis of the possible impacts of these institutional mechanisms covered the governmental, economic, social, international policy making, and legal impact sectors. The impacts were qualitatively evaluated to compare the alternatives. Finally, based on the impact analysis, public policy issues likely to arise in execution of one or more of the institutional arrangements were analyzed.

This TA emphasized policy analysis and institutional issues. The assessors drew heavily on expert judgment and prior studies. They did not find formal analytical techniques relevant to their needs. As has been usual with OTA studies, no policy recommendations were made.

Intended users of the assessment are primarily the Congress, but the report is publicly distributed. The report provides Congress with several options for upgrading materials information systems and examines the relative merits and drawbacks of each.

A Technology Assessment of Geothermal Energy Resource Development, performed by the Futures Group, Glastonbury, CT for the National Science Foundation, April 15, 1975, U.S. Government Printing Office, Stock No. 038-999-992-3, $6.00.

This report deals with some potential futures for geothermal energy in the United States, evaluates these, and makes recommendations to help policy makers capture the desirable aspects of this energy source while avoiding its pitfalls. Most of the report is devoted to analysis of potential impacts.

The six-person core study team included an energy specialist, economist, lawyer, physicist, and two systems analysts. Bechtel Corporation subcontracted the engineering analysis. Total study cost was around $200,000.

The geothermal TA followed a study strategy quite similar to that advocated in this book, incorporating forecasting, impact assessment, and policy analysis. Both conventional and novel analytic techniques were used.

Literature searches reviewed the state of the art of geothermal energy and other potentially competing sources of energy. The components of geothermal technology were specified using *relevance tree* techniques. Experts were interviewed in person, by mail, and by telephone. They were asked to comment on "normal" and "crash" geothermal research and development program *scenarios.* The team added sociopolitical developments suggested in the literature to the two scenarios. They then constructed a model to provide a systematic basis for comparing the consequences of contemplated policies. The model described the rate of growth of geothermal electricity production in the United States based on a simulation of utility company decision making.

The relative significance and probability of impacts resulting from the scenarios were explored through further interviews. Relevance trees indicated how secondary and tertiary impacts might flow from these primary impacts.

The Futures Group is heavily oriented toward the development and use of techniques. Thus this TA places considerably more emphasis on technique than is typical. The study team utilized three approaches to assess the impacts: (1) *cost–benefit analysis;* (2) *trend-impact analysis,* a new forecasting method that permits extrapolations of historical trends to be modified in view of expectations about future events, and (3) a series of "*value matrices,*" in which interest groups were listed on the abscissa, impacts on the ordinate, and each cell contained a number expressing the effect of the impact on the interest group (the overall impact of the technology on the interest group could be determined by summing).

Policies suggested in the interviews or invented by the study team were introduced in an effort to change the evolutionary scenario and its impacts. These policies were tested in the simulation model and by changing factors involved in the cost–benefit, trend-impact, and value-matrix analyses. From the iterative loops involved, a series of potentially desirable policies emerged.

This study considered relatively long-term possibilities, yet it was oriented to present policy options. It was not prepared for a target user, but made available for general policy discussion.

These five summaries illustrate a number of current practices in TA/EIA. The main impact assessment interest lies with technical, economic, and environmental issues. These are the areas that are most developed, where a relatively firm methodology exists, and where the concerns of many parties at interest lie. On the other hand, the treatment of social factors is limited. Policy analysis, despite the fact that it is crucial to TA/EIA, is emphasized unevenly. There is a pronounced tendency to consider only a limited range of options. Small studies (such as highway 59) and policy-oriented studies (e.g., the OTA study) tend to be qualitative. Large studies dealing with physical technologies tend to be quantitative. Studies of social technologies tend to be qualitative, but there are significant exceptions [e.g., Enzer (1974)].

These studies suggest areas where improvement in assessment practice is most needed—social description and forecasting, impact analysis (including values), higher-order impact analysis, and policy analysis. The relative status of methodological development is made clear in Chapters 6–16.

6

Technology Description and Forecasting

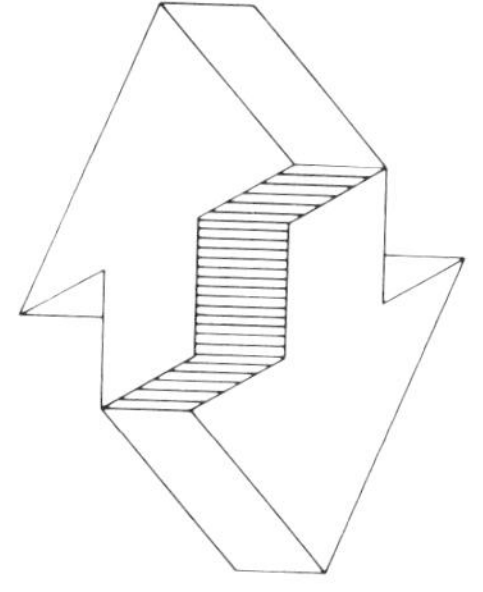

A technology must be carefully described and its future state forecast before impacts can be identified, analyzed, or evaluated. Thus the activities discussed in this chapter precede those presented in Chapters 7–14. The material presented here is designed to assist the reader in preparing a comprehensive description of technology and in projecting it along feasible alternative paths into the future. Clarity dictated that technological and social forecasts be treated separately (in Chapters 6 and 7, respectively). However, they should not be accomplished independently in practice. The opening sections present a brief review of technology description and forecasting in TA/EIA. The interaction of technology and social forecasts is considered next. A discussion of the effects of the level of emergence of technology and subject characteristics on description and forecasting follows. The requirements of a satisfactory description are considered, and an example of such a description is provided. Later in the chapter an overview of techniques, rationale, and propositions relevant to forecasting is presented. Major forecasting techniques are discussed and each is briefly described. Trend extrapolation techniques (in particular, single and compound parameter methods, and envelope and S-shaped curves) are treated. Finally, expert opinion methods, especially survey procedures and the Delphi technique, are presented.

6.1 INTRODUCTION

Practitioners agree that every TA/EIA must include analysis and description of the technology involved. One must have an understanding of the "nuts and bolts" of a technology to perceive the impacts its implementation may bring. For example, the contribution of automobile emissions to air pollution is the result of the type of fuel and power plant it employs. How could impacts of the auto be predicted without an understanding of the engine, its fuel, and the characteristics of its emissions?

Comprehensive description of the state of the art of a technology is necessary but not sufficient for accurate prediction of its future impact. The technology must also be projected along feasible paths into the future. In this projection possible alternatives to the technology must be identified and considered. For instance, the Otto-cycle internal combustion engine that powers today's automobile might well be replaced by external combustion engines using Brayton or Stirling cycles, or even by electric motors. Alternately, the automobile itself could be largely displaced by greatly improved and expanded communications and/or mass-transit systems. The impacts of each alternative would be different from those of the present Otto-cycle technology. Therefore the technology in question must be described and projected, together with realistic alternatives, into the social context of a future time.

Technology and social description and forecasting *cannot, however, be treated as independent of one another*. Thus the techniques of social description and forecasting immediately follow in Chapter 7. Perhaps the most common failure in this area is a technological forecast made without reference to the composition, growth, and change of societal forces that support and control the technology. Although this mistake has been virtually elevated to an art form by the "technology is an uncontrollable force with a life of its own" fraternity, it need not be perpetuated in TA/EIA.

The following section deals with general considerations that form the basis for technology description and forecasting. Subsequent sections deal with technology description and the means for its accomplishment (Section 6.3) and technology forecasting with a few appropriate techniques (Section 6.4). These include monitoring, single and compound parameter extrapolation, envelope and S-curves, Delphi, and survey.

6.2 GENERAL CONSIDERATIONS

The landmark effort of the MITRE Corporation team directed by Martin V. Jones (1971) was mentioned in Chapter 4. A major product of this study is the well-known[1] seven-step methodology (see Table 4.3, p. 56), the second step of which is "Describe the relevant technologies." The emphasis is on a thorough description of both the central technology and supporting and competing technologies (Jones 1971: 1, 43–51). Further, the need to forecast the future state of the art is explicitly noted. J. Coates (1976b) proposed a 10-step approach (Table 4.3) that requires that alternatives to the central technology, called *macroalternatives*, be identified and projected along feasible paths into the future. The technology and its macroalternatives are then compared on the basis of a comprehensive study of their impacts.

Three tasks relate to technology description and projections of alternatives: bounding the assessment domain, data acquisition, and technological forecasting. Data acquisition is considered in Section 6.3, and forecasting of technological alternatives is discussed in Section 6.4. Bounding the assessment has been discussed in Chapter 5. In particular, recall Figure 5.1 (p. 68), which depicts the critical relationship of the bounding process to technology description and projection.

Technology description is often the assessment task for which the most concrete data exist, particularly if the assessment concerns an existing technology. There is sometimes a temptation to overdo the description at the expense of other tasks that involve more uncertainty. Thus some assessments deal with little other than technology description and economic feasibility. Armstrong and Harman (1977: 89) suggest that about 20% of the project resources be spent on technology description and forecasting initially and that an additional 5% be devoted to it during the first iteration of the study.

To summarize, from the earliest methodological studies, the importance of understanding and forecasting the technology being assessed has been recognized. Sufficient information must be gathered to describe the state of the art of the primary technology, supporting

[1]As graduate students, the authors found that the words "well known" were invariably followed by reference to something of which they had never heard. Since graduation they have waited impatiently for an opportunity to perpetuate, in their own fumbling way, this grand old custom.

technologies, and alternative technologies. Further, primary and macro-alternatives must be projected into the future along feasible paths. These projections must be made with an awareness of social forces developed in complementary social projections. Intelligent bounding limits the scope of technological description and forecasting to areas that are consistent with both the technology and the state of society.

6.2.1 Interdependence: Forecasting Technology and Society

No technology is developed, marketed, or adopted outside of a societal context. Society provides the motivation, control, and support for the "technological delivery system" [TDS—see Figure 4.1 (p. 44) and Section 7.2.2]. A technology cannot be described or forecast without reference to society. Society contributes significant public resources to technologies seen as leading to desirable social goals. It has also demonstrated a willingness to withhold support from technologies it deems undesirable (e.g., the supersonic transport) and to control those considered to have dangerous side effects (e.g., automobile emission controls). Both TA and EIA are manifestations of society's search for control and support mechanisms.

There is perhaps no more clear-cut example of the symbiotic relationships between society and a technology than that of the United States and its automobile. The automobile has brought an unparalleled degree of personal mobility to American society, altering that society irreversibly (e.g., city structure, work and transit patterns, mores). Recent changes in societal attitudes toward the environment and safety and economic pressures exerted by foreign oil suppliers have led to a spate of government regulations intended to alter auto technology to suit new societal goals. Forecasts of auto technology that ignore these new societal goals and their possible future manifestations would be of doubtful validity.

It is clear that the integration of technological and social forecasts is a central concern of assessment, but much less clear how it is to be accomplished. In the absence of formal techniques, integration is normally informal and intuitive. Armstrong and Harman (1977: 77) found that assessments of physical technologies project alternatives on the basis of technical criteria and then intuitively supply a consistent social context to complete the description. In contrast, they note that alternatives projected for social technologies are limited by assumed policy options and social goals. Ascertaining both that the social and technological forecasts are mutually consistent and that they represent feasible projections of present states seems a necessary collateral activity in both cases. Forecasts should be iterated, time permitting, until these two requirements are satisfied.

6.2.2 Levels of Emergence and Impact of a Technology

Bright lists seven levels of the emergence and impact of a technology and suggests that forecasters be aware of which of these levels are to be addressed. The levels (Bright 1972: 3-3, 3-4) are displayed in Table 6.1.

The first four levels largely concern characteristics internal to the technology and thus generally have little effect on society, except as they foreshadow impacts that may occur at higher levels of emergence. Levels 1–4 are referred to as the *emergence phase* of a technology. The progress of a technology through the emergence phase is affected by the development of supporting, competing, and alternative technologies.

Levels 5–7 are the *impact phase* of the technology and are the levels of primary interest in TA/EIA. These levels cannot be forecast without attention to the attributes of the future society in which they will occur.

Bright (1972: 4-4) claims that the time involved in moving from scientific hypothesis or speculation through level 6 can be on the order of 10 years for many technologies and is more likely to be 20–25 years for most. The recent history of weapons and nuclear power systems seems to support this. The length of this time holds two important implications for assessors:

1. It is probably unrealistic to consider alternative technologies that have not progressed at least to the latter stages of level 3 (i.e., full-scale prototype) in EIA studies.
2. TA studies should adopt time frames of more than 10 years for assessing technologies that have not moved beyond level 3 (to do otherwise would generally relegate major impacts to times beyond the study bounds).

Unless there are compelling reasons (usually social) to the contrary, these appear to be reasonable rules of thumb. It is the case that the more massive the scale of the technology and the resources required, the more distant in time will be the significant societal impacts.

The preceding discussion has dealt primarily with physical technologies. However, the time necessary to gain a significant base of political support, pass legislation, and develop institutional support for social technologies seems to be of the same order of magnitude. There-

TABLE 6.1. **Levels of Emergence and Impact of a Technology** [a]

Phase	Level	Description	Example
Emergence			
	1	Certain knowledge of nature or scientific understanding will be acquired by . . .	We will know the cause of cancer by 1980
	2	A new technical capability will be demonstrated on a laboratory basis by . . .	We will have a drug to control cancer in mice by 1982.
	3	The new technology will be applied to a fullscale prototype or in a field trial by . . .	We will have experimental success in curing human cancer by 1985
	4	The first operational use or commercial introduction (first sale) of the technology will be by . . .	A drug for cancer control will be introduced commercially by 1987
Impact			
	5	The new technology will be widely adopted by . . .	Cancer drug will be used in 90% of all cancer cases by 1988
	6	Certain social (including economic) consequences will follow the use of the new technology by . . .	Reduction of cancer deaths will add 5 years to average life span by 1990
	7	Future economic, political, social, ecologic, and technical conditions will require creation or introduction of certain new technological capabilities by . . .	Elimination of cancer as a major cause of death will mean that replacement of deteriorating organs with mechanical substitutes will become a major medical requirement by the year 2000

[a] From Bright (1972: 3–3, 3–4).

fore, the implications may be considered to hold for the development of social as well as physical technologies.

6.2.3 Effects of Subject Characteristics

Each technology to be assessed will have different characteristics. Table 5.1 (p. 74) enumerated some of these as content and process dimensions. They will affect the forecast of the technology's progress by producing different levels of political, social, and economic sensitivity, and different relationships to other technologies.

Table 6.2 summarizes how Armstrong and Harman (1977) perceive four of the major assessment dimensions affecting technology description and forecasting. Problem-oriented assessments (see Section 4.3.3) have been relatively rare; technology-oriented assessments are far more common. Project assessments (a third identifiable type—recall Section 4.4.3) usually relate to a single basic technology, although a variety of

TABLE 6.2. Effect of Four Assessment Dimensions on Technology Description and Forecasting[a]

Assessment dimension	Characteristics
Technology-oriented	Projection of a single technology along alternative paths
vs	
Problem-oriented	Comparison of characteristics of several different technologies; projection of several technologies
Physical technology	Technical feasibility limits alternative choices; policy not heavily involved
vs	
Social technology	Political feasibility limits alternative choices; alternatives often closely related to policy options
Existing technology	Feasible alternatives limited by polarization of interest groups; innovative alternatives difficult to introduce
vs	
Emerging technology	Possibilities of innovative alternatives; relatively long time frames necessary and hence high uncertainty in forecasts common
Major intervention	Technological alternatives interact strongly with social projections; policy thrusts likely to be inherent in alternatives
vs	
Minor intervention	Technological alternatives relatively independent of social projections

[a] Based on Armstrong and Harman (1977: 102–112).

alternatives may be addressed. Assessments of physical technologies are usually based on technical and economic feasibility. Nevertheless, institutional and policy factors must be considered eventually, and may modify the initial selection of alternatives. Social assessments, on the other hand, typically focus on political and social feasibility and assign less importance to technical factors.

We consider an existing technology as one which has emerged to at least level 4 of Table 6.1. Existing technologies are more or less well understood. Depending on the level of their development and the time elapsed since their adoption, their first and perhaps even their higher-order impacts may have been recognized. Typically, stakeholder groups have emerged and opinions have become somewhat polarized. For emerging technologies, the lack of general understanding of their nature makes description an uncertain task. If implementation of a technology implies major dislocations or changes in existing social or institutional arrangements, it is termed a *major intervention*. Thus a shift to a hydrogen based energy economy would be a major intervention, whereas construction of even a large number of ocean thermal gradient power-generation stations would not. It is important to note that even minor technological interventions may presage significant societal and institutional changes in the long run. However, by virtue of their scale of introduction or diffusion, society can accommodate such changes gradually without sudden dislocations.

6.3 TECHNOLOGY DESCRIPTION

Technology description addresses the query "What is this technology?" If a colleague asked that question, how would you respond? That would depend, for instance, on the context in which the question is asked, the use to be made of the answer, the characteristics of the technology, and the time available for response. This is no less true of technology description in TA/EIA; . . . it would depend on: whether the assessment is technology- or problem-oriented, how the description will be used in the search for impacts, whether the technology is physical or social, existing or emerging, or representative of a major or a minor intervention, and how much time is available for the description task. The answers to these queries shape the manner in which the description task is carried out.

A sound approach for the assessor is to *begin with a relatively broad coverage of the major aspects of the technology*. A finer descriptive grid can be applied to those aspects that appear most important to the assessment after the major outlines have been traced. This approach provides directions for descriptive activities. It also increases the probability that the information will be in a usable form when the deadline for the task is reached.

A logical starting point for initial coverage is to locate the technology along the dimensions suggested in Table 6.2. Next a tentative decision can be made as to which levels of emergence and impact (Table 6.1) best describe the technology.

6.3.1 A Checklist

We suggest that the MITRE technology description checklist (Table 6.3) provides a useful structure for the next steps of the descriptive task. Some significant modifications may, however, be required in specific instances, particularly for social technologies. A category, titled "agencies and institutions involved," for instance, might logically replace "industries involved."

The basic role of "physical and functional description" is to describe the technology and its major subsystems. Although not specifically called for in the checklist items, an effort should be made to determine what parameters can be used to measure the functional capability of the technology. Such parameters are inherently important of themselves and also because they form the basis for forecasting by trend extrapolation (see Section 6.4.1). Examples of functional parameters include the thrust-specific fuel consumption (pounds of fuel per pound of thrust per hour) for turbojet engines and computing power (number of standard operations per second) for digital computing systems. Such parameters normally depend on more than one subsystem of the technology. For instance, computing power depends on the internal calculating speed of the computer central processor unit (cpu), the time the cpu is idle waiting for input or output, and the memory capacity of the computer (Knight 1973: 378).

The scientific disciplines, professions, and occupations involved in the technology are also included in the "physical and functional description," as are the businesses and industries involved. These considerations not only help to define the technology, but serve as a partial definition of the stakeholders as well. It is desirable to indicate the relative importance of stakeholder groups. This might be expressed, for example, in terms of their numbers and/or centrality to the operation of the technology. The capital investment and/or business activity of the businesses and industries involved with the technology also provide a useful description. The input and output products, design-dimension data, and manufacturing characteristics complete the "physical and functional description."

TABLE 6.3. Technology Description Background Statement for Automotive Emissions[a]

---

1. **Physical and functional description**
   - a. Automotive emissions control technology
     - Modification of present spark-ignition internal combustion engine
       - Positive crankcase ventilation
       - Thermal, catalytic reactors
       - Exhaust recirculation
       - Stratified charge introduced in cylinders, carburetor control, fuel injection
       - Air injection in exhaust
       - Sealed fuel systems
       - Fuel alternatives
     - Development of innovative engines
       - Heat engines (gas turbine, steam, diesel, Wankel, Stirling)
       - Electric
       - Hybrid (electric plus heat or mechanical)
   - b. Scientific disciplines
     - Engineering (automotive, mechanical, highway, systems, city planning, electrical, electronic, chemical, petrochemical, transportation)
     - Fluid dynamics, meteorology
     - Toxicology, pathology
   - c. Industries/businesses involved
     - Primary automotive manufacturing
     - Manufacturing support activities: steel, business services, transport, rubber and plastics, metal stampling, nonferrous metals, machine shops, metal working machinery
     - Wholesale distributors
     - Retail sales and service
     - Fuel refineries, retail outlet stations
     - Automotive parts and repair
     - Automotive export/import
     - Auto rental
     - Auto parking
     - Insurance
     - Motels
     - Highway construction
     - Highway policing/law enforcement
     - Accident-related—medical, legal, mortuary
     - Indirect involvement of power generation stations, refuse burning, home heating, and all industries contributing to the total atmospheric contamination (petroleum refineries, smelters, iron foundries, iron and steel mills, pulp and paper mills, cement plants, fertilizer plants)
   - d. Professions and occupations involved
     - Direct—all involved in the design and production of automobiles and parts, and all of the aspects of distribution, sales, servicing, and utilization
     - Indirect—all nationwide, requiring the automobile in their work or for commuting to their jobs

---

(continued)

TABLE 6.3 *(continued)*

---

e. Products involved

Raw materials in the production of cars and their equipment

Manufactured vehicles, power plants, pollution control packages including catalysts, and fuels

f. Design—dimension data

Maximum allowable automotive emissions by pollutant species, grams per vehicle-mile

2. **Current state of the art**

a. Current state of the assessed technology

Developmental work which has been characterized as lacking orderliness and common direction now undergoing increased attention and focus on all the fronts described in 1(a), above, in efforts to meet stringent emission goals. Principal goal in the domestic and foreign automobile industry is in developing a mass-producible, marketable (not too expensive), reliable, low-polluting power plant.

b. Current state of supporting sciences

Many gaps remain in understanding the toxicity of pollutants and in modeling their dispersion, lifetime, and removal from the atmosphere. The knowledge of public-health damage mechanisms and scope and severity of air pollution is incomplete.

3. **Influencing factors**

a. Technical breakthroughs needed

See 2(a,b), above.

b. Technological factors affecting development

Finding suitable heat-exchanger fluids, high-temperature materials, or catalytic materials which can withstand or remain activated in the engine operating environment. Also, determining combinations of workable materials which are cheap and plentiful.

c. Technological factors affecting application

Finding emission-control devices or power plants which give acceptable performance, mileage, and warranted, maintenance-free service as well as being moderately inexpensive to purchase and operate, providing adequate (urban) range, and being easily refueled.

d. Institutional factors affecting development

Lead time for development and introduction of innovative engines and subsystems must be adequate; otherwise, manufacturers will attempt only to modify the present ICE (internal combustion engine) to update it to meet near-future regulatory requirements.

e. Institutional factors affecting application

Without enforced regulations (use restrictions, taxation, etc.), the public will probably not of their own accord choose to purchase low-emission alternatives which are more expensive and possibly lower in performance, e.g., the poor sales of recently introduced low-lead fuels. Purchase of foreign-produced

---

(continued)

vehicles (especially in the event of power plant breakthroughs abroad) will be strongly influenced by such factors as the price advantage associated with cheaper foreign labor or, conversely, tariffs to deter competition.

4. **Related technologies**

a. Complementary (supporting) technologies
   - Distribution, servicing, and use of automobiles (exclusive of vehicle and part manufacturing)
   - Highways

b. Competitive technologies
   - Urban mass transit
   - Communications
   - Nonautomotive pollution sources
     - Power generation
     - Industrial
     - Waste disposal
     - Natural
     - Jet aircraft /off-the-road vehicles
   - Automotive safety devices

5. **Future state of the art**

Timing—initial operational capability and widespread applications

a. Early 1970s. Present evolutionary development of the ICE, forced by sequential federal requirements for new emission controls, will continue. Overly stringent, near-term regulations may prove to be counterproductive should they divert work from innovative power plants.

b. Mid-1970s. The cleanest possible ICE, e.g., a stratified charge engine with computer control, will be introduced. Also, other innovative power plants will appear in federal demonstration programs.

c. Early 1980s. Pilot production of cars with new power plants will commence. More special-purpose urban vehicles will be utilized by the public (indicative of increased multicar ownership).

d. Mid-1980s. More widespread usage of innovative vehicle types becomes commonplace. The patronage of advanced mass transit and even people-movers in several urban areas improves. Conceptual work on highly advanced concepts with high utilization factor and little or no degradation of the environment will proceed in this time frame.

6. **Uses and applications**

a. Current and prospective

The automobile currently interacts with all aspects of society (employment, shopping, recreation, selection of residence, education, religion, schedule of activities and plans of the family). In the future little change is forecast in automobile usage unless higher costs prohibit utilization by the less affluent.

(continued)

Table 6.3 *(continued)*

More leasing of cars appears to be the trend and more usage of specialized vehicles in the future is foreseen. Possible purchase of more mobile homes could alter community and commuter patterns.

b. Buyers: age groups, incomes, geographic distribution

No strongly discernable trends for change here. More major metropolitan areas will develop throughout the nation, coupled with movement toward the suburbs of those regions and amplification of the attendant problems of achieving necessary and sufficient mobility.

[a] From Jones (1971: 150–152).

"Current state of the art" (item 2, Table 6.3) deals with both the technology being assessed and the state of scientific knowledge relevant to that technology. The present level of operational capacity and gaps or deficiencies that keep the technology from obtaining a higher performance level should be identified. Description of antecedent technologies is appropriate. Current development and research thrusts are also important.

Identification of "influencing factors" (item 3, Table 6.3) begins the process of locating the lines of influence between the technology and the surrounding technical, economic, and social milieu. "Technological breakthroughs needed" and "technological factors affecting development and application" deal with factors that may operate either inside or outside the central technology. The former identify achievements needed to remedy deficiencies described in the "current state of the art." The latter include the technical design criteria against which the success of the system is measured.

"Economic and institutional influencing factors" are considered next. These considerations trace influences between the technology and its environment. Of particular importance are federal, state, or local regulations that affect the application or development of the technology. The development of coal gasification, for instance, has been retarded by the regulation of interstate natural gas prices. This regulation maintains a price differential between natural and manufactured gas that makes coal gasification less economically attractive than it otherwise might be.

Economic or institutional factors can emanate from sponsoring, beneficiary, or third-party organizations, as well as the organization that develops the technology. For instance, the capital investments required to develop the U.S. supersonic transport were too large for any commercial firm, resulting in a need for public aid (and ultimate rejection).

"Related technologies," those that are supporting or competitive to the central technology, should also be identified (item 4, Table 6.3). For example, microprocessor (computer) technology applied to the control of engine operations supports automotive emissions control technology. Mass-transit-system technology, on the other hand, is competitive to the automobile. This step is important not only in the descriptive sense, but also because it provides preliminary indications of technological impact areas. It can also yield clues to areas in which developments may occur that will influence the course of the central technology. In a sense, identification of complementary and supporting technologies more fully describes the lines of influence alluded to under "influencing factors."

The MITRE checklist recommends description of the "future state of the art" (item 5). We feel, however, that this is more appropriate to technological forecasting, which is discussed in Section 6.4.

The "uses and applications" of the technology (item 6, Table 6.3) are also important descriptors. The social importance of most technologies is manifested through their uses, both current and prospective. Potential applications should be carefully examined to identify societal supporting or controlling mechanisms and to determine needs and opportunities that may aid their development. Obviously not all applications of the technology will be on the same level of emergence. Therefore, different time frames may be required for forecasts of each.

One useful way to conceptually organize the information acquired through completing the technology description checklist is to construct the TDS [see Figure 4.1 (p. 44), Section 7.2.2, and Figure 13.3 (p. 341)]. Constructing a TDS indicates the functional relations between technology and its social context. This systems perspective can help focus on those features of the technology most relevant to impact and policy evaluations.

6.3.2 Gathering Descriptive Information

Gathering information for the technology description should begin with a *literature search*. Much of the information about an established technology can be gathered in this manner. Even if consultation with experts is subsequently required, the assessment team needs a basic understanding of the technology to profit from such interaction.

Armstrong and Harman (1977: 24) note that *contact with persons knowledgable with the technology* was the most common source of such data in the TAs they studied. They also found that:

Surprisingly enough, assessments on extensions of existing technologies [e.g., offshore oil operations—Kash et al.(1973) and no-fault insurance—Enzer (1974)] were just as likely to rely upon individual contacts for primary data sources as were futuristics efforts [e.g., geothermal energy—The Futures Group (1975) and life extension—The Futures Group (1976)].

The most common techniques for gathering information from individual contacts are interviews, conferences, workshops, and questionnaires (see Section 6.4.4).

6.4 TECHNOLOGY FORECASTING

6.4.1 Basic Considerations

The TA/EIA process requires the projection of the future state of a technology so that its impact can be assessed. Projections always involve uncertainty that increases with their time span. Forecasting seeks reproducibility by increasing its reliance on what is known and systematizing the estimation of what is not. To do so, it must proceed logically from a basis of explicit data, relationships, and assumptions.

There are at least three rationales that support the assertion that technological progress and development can be forecast (Bright 1972: 3-6). First, examination of the historical growth of technological capability (e.g., speed, power, capacity) reveals a surprisingly ordered pattern of development. Based on this observation, *continuity* of growth seems to be the norm and discontinuities are rare. The rationale of continuity provides the foundation for techniques such as trend extrapolation. Recognition of patterns, such as exponential growth, that characterize the behavior of certain attributes can be used as a basis for extrapolations.

The second rationale is that technological development responds to opportunity and need [for a review of technological innovation, see Kelly et al. (1978)]. It is also sensitive to the allocation of resources and to social control through regulation. By identifying and monitoring such influencing factors, technological progress can often be anticipated.

The third rationale asserts that an understanding of the process of technological innovation aids successful prediction of technological development. This process is reflected in the orderly progression in the levels of emergence (Table 6.1).

There are many techniques for technology forecasting and numerous schemes for categorizing or classifying them [see Table 5.3 (p. 82) and the discussion of its basis for classification]. Another common basis for categorizing forecasts (and hence the techniques used in them) is the distinction between *exploratory* and *normative* forecasts. Exploratory forecasts give a prediction of the future. They provide means for exploring the shape of tomorrow given the state, trends, and premises of today. Normative forecasts, on the other hand, seek to determine courses of action by which to achieve future goals, given conditions existing today. Both types of forecasts may be useful in TA/EIA depending on the needs of the study.

Normative forecasting begins with the identification of future needs and goals. If, as suggested by the second rationale, technological innovation responds to perceived needs, their identification may provide the key to the course of innovation. Bright (1972: 7-1) suggests that the goal of putting a man on the moon by 1970 implicitly included many technical subgoals (e.g., life-support systems, power-storage systems, more heat-resistant materials). These technological goals could well have provided the basis of forecasts of what was to come. However, the goal must be perceived as worthwhile, at least in principle, by a society or by powerful stakeholder groups within that society. Politicians often interpret this criterion as the society's will to achieve the goal. Assessment of the existence of the will to achieve is perhaps the most difficult task of those who would practice normative forecasting. Most normative forecasts are based on relating the goals or needs to the technological or social innovations which are necessary for their achievement.

This section concentrates on three sets of techniques—monitoring, trend extrapolation, and expert opinion methods. Before turning to the particulars of the techniques, we share three critical principles for forecasting based on Ascher's (1978) thoughtful review of both social and technology forecasting:

1. *Methodological sophistication contributes little* to the accuracy of forecasts. Expert judgments and simple extrapolations perform well in comparison with elaborate models [e.g., econometric] or sophisticated trend manipulations.
2. *Basic core assumptions*, not derivable from methodology, are the major determinants of forecast accuracy. When the core assumptions are valid, the choice of method is either obvious or secondary. When they are wrong, methodology can make little difference. For instance, the use of envelope curves [see Section 6.4.3.2] signifies a preconception of future growth in the relevant technology as capability is spurred through cumulative break-

throughs. The use of any form of curve-fitting methodology assumes that a growth pattern will continue.

3. The *time horizon* is the most consistent correlate of accuracy. Shorter forecasts are more accurate, in a nearly linear relationship. Confidence limits should be established for a particular trend; these can be obtained from the spread of experts' opinions as the horizon lengthens.

Principles 1–3, together, imply that a more recent forecast is preferable to an earlier one, even if that were considerably more sophisticated. The negative effects of outdated assumptions (e.g., energy perspectives changed radically in the United States following the 1973 oil embargo) and longer horizons usually outweigh methodological sophistication. This suggests that assessors would often do well to make their own forecasts, even if only limited resources were available, rather than rely on somewhat outdated forecasts. For instance, a quick poll of experts on solar energy should outperform a 5-year-old report on such a fast-changing technology.

Technology forecasts may fail for a number of reasons. Among these are a lack of nerve by the forecasters (for example, watering down opinions in committee), missing converging developments, concentration on specific configurations to the extent of missing known facts, the occurrence of historical accidents, and the effects of ever-present uncertainties. The forecaster who is aware of the underlying principles discussed above and the pitfalls mentioned in this paragraph can effectively use the repertoire of techniques for technological forecasting.

6.4.2 Monitoring

Monitoring is perhaps the most widely practiced forecasting technique. It is based on the assumption that technological change is foreshadowed by changes in the political, technical, economic, ecological, and/or social environments—a view consistent with the discussion in this chapter. Therefore, it should be possible to monitor signals in these environments, analyze them, and forecast the emergence of new technologies. For instance, Bright and Schoeman (1973: 240) note that the closing of the Suez Canal (a political signal) led to alterations in the economics of transporting oil and ultimately to the increase in the size of tankers. One might hypothesize that supertanker accidents (e.g., the Amoco Cadiz off the coast of Brittany in 1978) may lead to social pressures that will, in turn, produce new transportation technologies.

Successful forecasting through monitoring must, of course, involve more than merely gathering data. The forecaster must sift the information being assembled for signals and project their possible implications. Parameters indicative of the emergence of the technological capability must be identified so that monitoring can proceed systematically. Finally, the results of the monitoring activity must be analyzed and synthesized so that forecasts and/or conclusions can be made. An example of a successful industrial monitoring process is displayed in Table 6.4.

TABLE 6.4. Chronology of Major Events Involving Permanent Press Products and Their Impact on Whirlpool[a]

Winter 1963–1964	Rumors in textile industry of new "delayed cure" process for resin applications
April 1964	Confirmation of rumor—Spartanburg, South Carolina
May 1964	First glimpse of process—San Francisco
May–June 1964	Education of Whirlpool personnel about permanent press garments; forecast of doubling of dryer sales in a year
August 1964	"Development conference" arranged for vice-president, laundry, at large fiber manufactory; permanent press garments shown at Whirlpool for first time
September 1964	Textile industry introduces permanent press
January 1965	Permanent press cycles on washers and dryers; first by appliance industry
March 1965	Research project for a "new concept ironer" dropped; monies diverted to other uses
May 1965	Dryer sales reported more than double those of May 1964
November 1965	Introduction of permanent press sheets
February 1966	Volume of dryer increased to approximately 6 ft^3 without change in exterior dimensions; this to accommodate large permanent press sheets and other items
February 1967	Forecast that knitting would grow to challenge weaving for control of garment textiles by mid-1970
May 1967	Work begun on knit cycles for washers and dryers
June 1969	Textile industry forecast that knits would control 30% of trouser market in 1970.
January 1970	Introduction of knit cycles on washers and dryers.
May 1970	"No-iron" concept of textiles postulated
August 1970	Forecast that by 1975, knit structures would constitute 50% of men's-wear market
February 1971	Five textile experts forecast that by 1975 knits would constitute 50% of men's outerwear

[a] From Davis (1973: 614).

6.4.3 Trend Extrapolation

Trend-extrapolation techniques attempt to capture the historical progress of a technology in a mathematical expression or graphical display. Once determined, this information can be used to project or extrapolate performance at a future time, provided that no discontinuities occur.

Trend extrapolation is based on the assertion that today is far more likely to be a time of continuity than not (Martino 1973: 106). Trends should not, however, be extended blindly. The forecaster must, for example, recognize that natural limits (e.g., 100% efficiency) or breakthroughs in competing technologies can cause discontinuities. A real effort must also be made to determine the economic, social, and/or political forces behind the trend as withdrawal or alterations in these factors can also produce discontinuities.

In the following sections two trend extrapolation techniques—single- and compound-parameter extrapolation, and envelope and S-shaped curves—are discussed.

6.4.3.1 Single- and Compound-Parameter Extrapolation

Single and compound parameter approaches seek to extrapolate the level of functional capacity (performance) of a technology. Functional capacity may be represented by a single technical attribute, a parameter, or a combination of such attributes, a compound parameter.

The most fundamental task of the forecaster is to identify parameters that accurately portray the functional capacity of the technology to be forecast. Failure to do so will lead to forecasts that are incorrect or, at best, misleading. Martino (1973: 117–118) identifies five characteristics that an appropriate parameter must possess. It must:

1. Be one that can be measured quantitatively in objective and meaningful terms. Therefore, it must be possible to measure the parameter in terms of the characteristics of the device or its operation (e.g., horsepower or torque).
2. Represent a measure of the actual functional capability. A parameter can fail to do this in two ways. First, it may indicate only a single aspect of that capability. For example, in lighting technology both the amount of light produced (output in lumens) and the amount of energy required to produce it (input in watts) are important. A device could be unsatisfactory either because of its output or required input. Neither alone is, therefore, representative of functional capacity. Second, the parameter must account

for the effects of design trade-off. In the design of a jet engine it is possible to achieve higher thrust levels by sacrificing fuel consumption. The designer can, within limits, trade thrust for fuel consumption, and vice versa. Thus neither alone is representative of the state of the art. The thrust-specific fuel consumption (pounds of fuel /pound of thrust /hour) is one measure of the functional capacity of jet engines.

3. Serve to represent functional capability for a progression of different technical approaches. It must, therefore, be applicable to each succeeding approach and not peculiar to one. With lighting technology, lumens per watt can be used, for instance, as a parameter for devices ranging from the paraffin candle to the gallium–arsenide diode.
4. Be one for which sufficient historical data exist to establish the trend. This is the most restrictive of all limitations on parameter selection.
5. Be selected consistently with respect to the level of emergence and impact (Table 6.1) of each successive technical approach. Suppose, for instance, the forecaster is investigating the progress of commercial air transportation. It would not do to compare the functional capacity of the DC-3 at the time of wide airline adoption (level 5) with that of the jet transport at the time Boeing demonstrated the first full-scale prototype (level 3).

When an appropriate parameter has been selected and sufficient historical data gathered, the forecaster is ready to examine past behavior to determine whether a trend exists. This is most easily done by constructing a graphical representation. In such a graph, time t is chosen as the independent variable and plotted on the abcissa (horizontal axis), and the parameter P is the dependent variable plotted on the ordinate (vertical axis).

A surprising number of technological systems display an exponential growth of functional capability. If this is true of the parameter in question, it will appear as a straight line when plotted on semilogarithmic graph paper (the vertical axis is $\ln P$, and the horizontal axis is t).

As an example of the process, consider Figure 6.1 (Martino 1973: 112) which represents the top speed of U.S. combat aircraft. The straight line superimposed on the graph represents the trend in military aircraft speed. The graphical format is semilogarithmic; thus the linear variation implies speed has shown an exponential growth over the time period 1900–1970. If an equation is needed, a regression fit

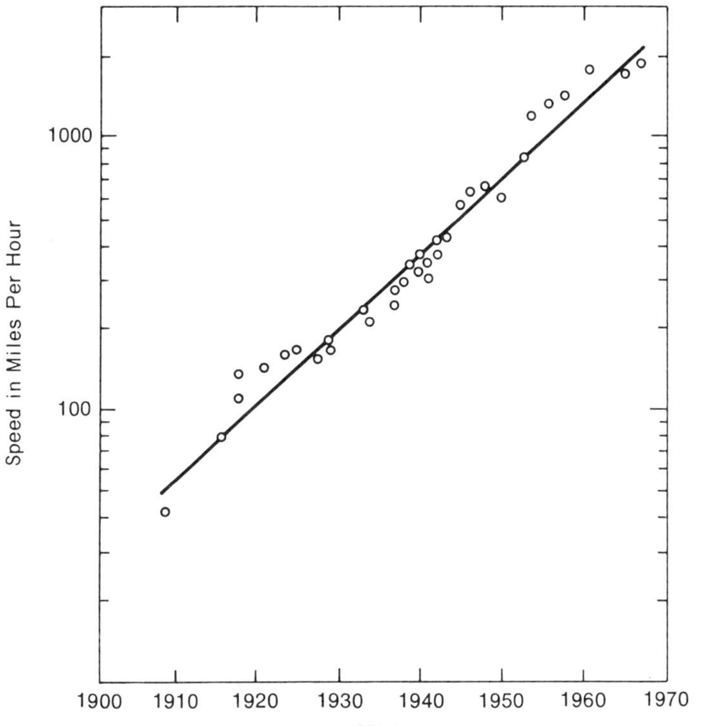

FIGURE 6.1. Top speed of U.S. combat aircraft.
Source: Martino (1973: 113).

can be obtained.[2] Alternatively, the line can be drawn in by eye (estimated) and the slope and intercept determined graphically to give the equation. In this case the equation is

$$\ln P = -118.30 + 0.064\,t,$$

where t is the calendar year and P the speed in miles per hour. If an equation is not required, a simple graphical extension of the line

[2]Least-squares linear regression is appropriate here. The best-fitting straight line is defined as the line that makes the sum of the squared deviations around it minimal. It is referred to as a regression line. Mathematically, for the line $P = (a + bt)$, P is the dependent variable; t is the independent; a is the intercept, the value of P when $t = 0$; and b is the slope of a line relating values of P to values of t. Then b can be computed as

$$\sum_{i=1}^{n}(t_i-\bar{t})(P_i-\bar{P}) \Big/ \sum_{i=1}^{n}(t_i-\bar{t})^2,$$

where $\bar{t}$ and $\bar{P}$ are the mean values and the summation is over all the n observations.

through the time period of the forecast will do quite well (provided no discontinuities are foreseen).

Whether the extrapolation is done mathematically or graphically, the forecaster should be convinced that factors that could cause discontinuities will not come into play within the time period of the forecast. Figure 6.2 represents the growth of a technology, crude oil tankers, for which a political factor, the first Middle East crisis, caused a discontinuity. Closing of the Suez Canal during the crisis presented a need that was fulfilled by the technology through the design and construction of larger-capacity tankers. Note that the dashed line in Figure 6.2 that is passed through the peaks of the capacity variation (this is called an *envelope curve;* see the next section) takes a distinct jump (discontinuity) as it passes through the year of the crisis. Simmonds (1973: 227) suggests that a natural limit, water depth, will restrict tanker dead weight to approximately one million tons (10^6 tons). If political, economic, social, and ecological factors support increases in tanker size, it is reasonable to suppose that technology will be altered to overcome this limit. Depending on the technological approach, this may or may not produce yet another discontinuity.

Both of the examples given thus far use single parameters to repre-

FIGURE 6.2. Oil tanker size.
Source: Simmonds (1973: 227).

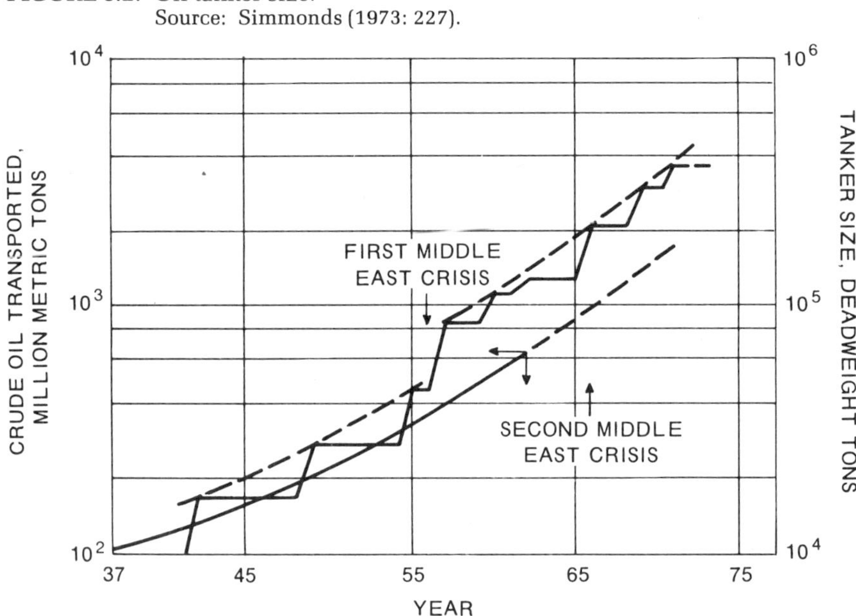

sent the level of functional capacity of the technology. That is to say, each parameter measures only one attribute of the technology (e.g., speed, cargo capacity). Single parameters, however, often fail to represent a measure of the actual functional capability of the technology (the second of the five characteristics required of parameters). When this occurs, a parameter that includes more than one attribute of the technology (a compound parameter) must be developed. Three compound parameters—thrust-specific fuel consumption, computing power, and lumens per watt—were mentioned earlier in this chapter.

The productivity of transport aircraft is another variable that requires a compound parameter. Both capacity and speed are important aspects of that productivity. Therefore, the compound parameter "seat miles per hour," a product of speed and capacity, is sometimes used. Note that the product, rather than the quotient, is chosen because productivity increases when either speed or capacity is raised. Construction of compound parameters is critically dependent on an accurate perception of the functioning of the technology. Compound parameters are forecast in the same fashion as single parameters.

6.4.3.2 Envelope and S-Shaped Curves

An envelope curve is simply a smooth curve forming a boundary, usually the upper, for the variation of a performance parameter. The dashed lines tangent to the peaks of the curve in Figure 6.2 represent the envelope curve for tanker size. The performance of many technological systems grows in an exponential fashion. This growth results from the increase of the performance of a number of different technical approaches that follow one another sequentially. In Figure 6.1, for instance, the technology of the modern jet-powered combat aircraft superseded that of the propeller-driven metal monoplane, which, in turn, replaced the biplane constructed of wood and fabric. If the aircraft speeds for each succeeding technological approach were pursued, they would display an S-shaped growth curve characteristic of the development of a single technology.

In the start-up phase of their development, technologies typically encounter numerous developmental problems that must be overcome before rapid strides in performance can be achieved. The rapid development of the jet aircraft was, for example, delayed until problems in high-speed aerodynamics could be solved and materials developed to withstand the high temperatures and centrifugal forces produced in the turbine section of jet engines. Once developmental problems are solved, performance typically increases rapidly, fulfilling the promise that gave rise to the technology. Rapid advance is followed by a slowing of the rate of increase as the last advantage is wrung from the technology. Typically, during this latter phase, scientists and engineers are faced with increasingly complex problems whose solution yields only marginal increases in performance until physical or economic limits of the approach are reached.

Intrinsic (time-independent) and extrinsic (time-dependent) constraints can operate to slow the rates of change of parameters near the top of an S-shaped curve. In cases involving the *diffusion* of a technology within a fixed population, as in the case of television sets in American households, market saturation can limit the use of technology. Another example of a peaking effect is the substitution of one technology or material for another, as in the case of the substitution of steamships for sailing ships. In these cases there is an upper limit—100% of American households or ships—which cannot be exceeded.

The function commonly used to describe saturation is the S-shaped or logistic curve sketched in Figure 6.3.[3] Bright (1978) claims that data should be accumulated, as available, to a level of 20% substitution to forecast accurately the time course of the substitution.

In cases of innovation, where no intrinsic upper limit exists, state-of-the-art constraints on a technological system operate to slow the development of important key parameters in each technological configuration. For example, in Figure 6.4 the biplane's time of development was from approximately 1910–1930, at which point its top speed

[3]The mathematical expression of the logistic function is $F(t) = B/[1 + A \exp(-kt)]$, where B is the upper bound (e.g., 100% saturation) and $B/(1 + A)$ is the lower bound.

FIGURE 6.3. The time form of the S-shaped curve.

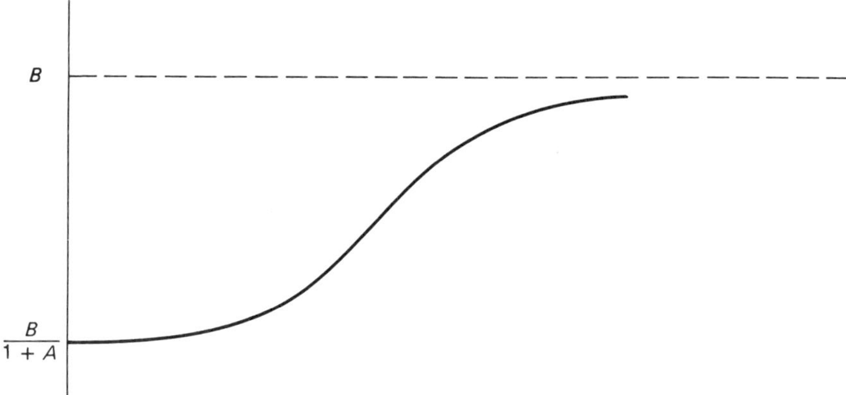

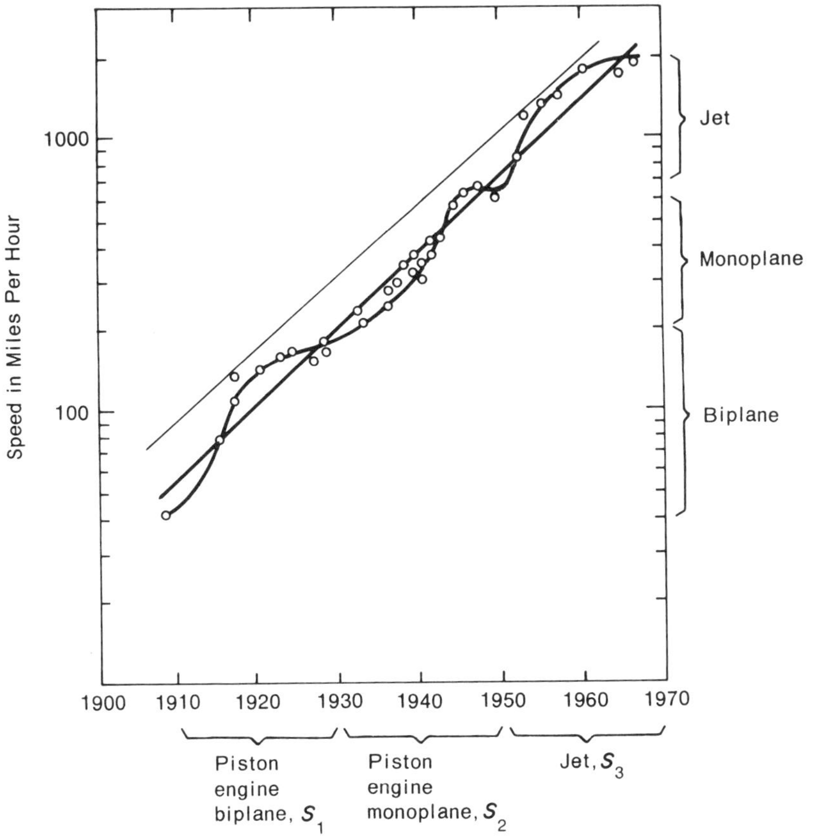

FIGURE 6.4. Figure 6.1 redrawn to show development of the performance of various types of combat aircraft, illustrating the application of the S-shaped curve.

could not be increased significantly. It was replaced by the monoplane, which around 1950 gave way to the jet. Each technology's development may be roughly represented by an S-shaped curve. The jet, which replaced the monoplane for high-speed applications, had a significantly higher speed. Thus when one looks at parameters embodied in successive configurations, one often sees development of the technologies peaking at higher and higher values. Ayres (1969) refers to considering the development of a parameter in successive technological configurations as aggregating. The light line tangent to the upper bound of the trends of the individual technologies is called the "envelope" curve, as mentioned above.

Since all forecasting, technological or otherwise, involves uncertainty, a forecast represented by a single curve does not accurately portray the confidence limits that the user can reasonably place upon the forecast. Therefore, it is common practice to use two envelope curves—one representing the upper, and the other the lower, bound—to display this uncertainty. By adopting this representation, the forecaster bounds the region within which the actual development is projected to occur. The range of performance bounded by the envelope curves naturally will broaden as the time frame of the forecast increases, reflecting the growing level of uncertainty.

As a final caveat, Bright (1972: 6-6) warns that the forecaster should not assume that when the S-shaped curve of a particular technology no longer supports the envelope curve, it will immediately fall from use. Many such technologies remain active for a variety of reasons including custom, economics, reliability, personal preference, and special operational considerations. Remember, although the car has largely displaced the horse as a means of transportation, there is still a lot of horse manure to clean up after many holiday parades.

6.4.4 Expert Opinion Methods

Expert opinions are currently more mundane than in the ancient days of the biblical prophets and the Greek oracles. However, "asking the person who knows" can produce a quick sense of the prospects in a particular subject area. Expert opinion can also be used to generate a more formal and credible assessment.

Two types of expertise can be identified as potentially useful in TA/EIA (Mitchell et al. 1975). The first belongs to persons with an extensive special knowledge about a topic (e.g., researchers on laser fusion to discuss prospects for breakthroughs). The second rests in representatives of a subpopulation whose attitudes or actions influence the forecast topic (e.g., attitude of community residents toward a proposed highway modification). A critical concern is identification of "the expert." Expertise may be "certified" by a variety of means—educational degrees, professional memberships, peer recognition, and even self-proclamation. Mitchell et al. (1975) offer a useful way to clarify the nature of expertise by asking "expertise relative to what?" Two dimensions serve to assess expertise for a given TA/EIA application: (1) the *topical object*, with special concern for theoretical foundations and scope (experts on barge traffic may not be helpful on surface transportation generally) and (2) the *sponsors* and end users of the study; "expertise" is not independent of values (the Corps of Engineers and the Sierra Club would likely concur that certain expertise was relevant some of the time, but not always).

Expert opinion methods involve four critical assumptions, drawing on the preceding considerations:

1. The forecast topic can be delineated to make possible the identification of pertinent experts.
2. The nature of the topic is such that these experts can predict the probable future course of developments.[4]
3. A representative sample of experts can be tapped to participate in the study.
4. These experts will be seen as credible by both the study sponsors and study users.

Expert opinion methods range widely in elaborateness, costliness, time requirements, and expertise required to execute them. We consider four types of expert opinion forecast: (1) genius forecast, (2) survey forecast, (3) panel forecast, and (4) Delphi.

Genius forecast (Cetron and Monahan 1968: 146) consists in finding a person whose cognitive powers cover the area of interest, and asking your questions. The answers constitute the forecast. Although its reliability is poor, it can provide a quick, perceptive, and unambiguous view of an area. Genius forecasting avoids problems involved in achieving consensus. Informal contact with special experts can be especially helpful in describing a technology under assessment and understanding the prospects for future development.

Surveys imply polling a group of experts about their opinions [for survey methodology, see Warwick and Lininger (1975)]. Surveys are useful when a group of appropriate respondents can be identified and when interaction among the respondents is not considered necessary (see the cameo entitled "Surveys"). Surveys may be used in many ways in TA/EIA—for example, to forecast technological and social changes or to understand how affected persons will respond to a development or a particular policy.

Panels involve interaction among experts. Typically a group is convened, given a time limit and a budget, and asked to report on a particular topic, which may be a particular issue within a TA/EIA, or the assessment itself. "Technology: Process of Assessment and Choice," one of the original reports dealing with TA, was written by a panel appointed by the U.S. National Academy of Sciences (1969) at the behest of the House Committee on Science and Astronautics. A more substantive inquiry, "Assessing Biomedical Technologies," was per-

[4]Wise (1976), in reviewing 1556 technological predictions made during 1890–1940, concludes that the accuracy of predictions appears at best weakly related to general technical expertise, and unrelated to specific expertise.

SURVEYS

Surveys are a prime source of information in TA/EIAs. They can be used to: *describe* behaviors (what people do) or attributes (what people are), *predict* future conditions (what people believe), and/or *evaluate* particular conditions (what people prefer). To accomplish the intended uses, one must assure that the survey produces the kind and amount of data needed for the assessment. This demands considerable sophistication.

Surveys may be formal (carefully structured) or informal. Three important forms are face-to-face interviews, telephone interviews, and mail questionnaires. The logistics of conducting a survey vary dramatically with the form, the information sought, and the selectivity used in choosing a group of "experts." Some of the issues to be considered in preparing a survey are as follows:

1. Obtaining a representative sample.
   a. Can the population be defined?
   b. Can selected respondents be located?
   c. Can nonresponse bias be avoided?
   d. Are findings generalizable?
2. Constructing an appropriate questionnaire.
   a. What length is feasible?
   b. How complex can questions be?
   c. Can open-ended questions be used?

formed by another panel appointed by the U.S. National Academy of Sciences (1975). Panels are suitable when the subject under consideration requires discussion. This process is more interactive and costly than a survey. Panel studies are appropriate for identifying important problems, for preliminary microassessments, and for delicate policy analysis and evaluation. Dominant personalities and the bandwagon effect may affect a panel's deliberations.

Our final expert opinion forecast technique is the *Delphi* method, named after the oracle of old. Delphi [cf. Dalkey and Helmer's original paper (1963) and Linstone and Turoff's in-depth book (1975)] is a technique that interactively iterates the responses of surveyed experts, thereby combining some of the advantages of surveys and panels (see the cameo entitled "Delphi"). Delphi has been a popular technique in technological and social forecasting. It can serve to generate systematic

3. Obtaining accurate answers.
   a. Can social desirability pressures be minimized?
   b. Does the interviewer distort responses in any way?
4. Fulfilling administrative requirements.
   a. Can trained personnel be obtained?
   b. Is sufficient time available?
   c. Can costs be kept reasonable?
   d. Are provisions made for distribution, follow-ups, coding, and computer analysis (as appropriate)?

Adequate survey procedures are not easily developed. Survey research entails a delicate communication process in which slight differences in wording (clarity and tone) of questions and attitudes of questioners can dramatically alter responses. Survey results must stand up to criteria of validity (control of systematic errors in measuring what is intended), reliability (control of random errors), and sampling (representativeness and size for statistical inference). Assessors are rarely suitably trained to be interviewers in formal surveys. Even for informal surveying (e.g., phoning selected experts), they should seek guidance from professionals in preparing the survey.

thought about future courses of events (e.g., technological breakthroughs) that are difficult to treat by other means. However, its *credibility among assessment users is not high* (see Table 16.2).

The Delphi technique has been a center of controversy (Sackman 1974; Goldschmidt 1975). Some difficulties are the following:

1. It capitalizes on group suggestion to pressure toward consensus, yet it is unclear whether such consensus yields accurate forecasts.
2. The director's control in structuring the process (i.e., selecting items and verbal responses) may suppress other valid perspectives of the issue and yield biased results.
3. Lack of item clarity or common interpretation of scales and feedback responses may lead to invalid results.
4. Participants may become demoralized by the demanding nature of the process (particularly if not compensated for their time).

DELPHI

Delphi is an expert opinion survey with three special characteristics—anonymity among participants, statistical treatment of responses, and iterative polling with feedback. The procedure is as follows (Sackman 1974: 7–8):

1. Typically a structured, formal, paper-and-pencil questionnaire is administered by mail, in person (e.g., at a conference), or through an interactive, on-line computer console. Participants do not meet face-to-face.
2. The questionnaire items may be generated by the study director, participants, or both.
3. The questionnaire is administered for two or more rounds; participants respond to scaled objective items, and sometimes to open-ended responses as well.
4. Each iteration is accompanied by statistical feedback on each item, which usually involves a measure of central tendency (e.g., the median), plus some measure of dispersion (e.g., the upper and lower quartile values—those between which one-half of the responses lie); sometimes the entire frequency distribution of responses is provided.
5. Respondents in the upper and lower quartiles may be asked to justify their responses; selected verbal feedback may be provided on each iteration (individual responses are kept anonymous).
6. Iteration continues to some point of diminishing returns that is determined by the director; convergence of opinion is sought, but not forced.

There are many variants on the Delphi procedure; for instance, Mitchell et al. (1975) describe a "mini-Delphi" simplified for use by small groups gathered at a conference.

Delphi can be used to forecast particular technological or social events. It can be used to determine policy preferences of a particular group. Conditions that make Delphi an attractive alternative to panels or surveys include the following (Linstone and Turoff 1975: 4).

1. The problem is not amenable to analytical techniques, but its solution could benefit from subjective group judgment.
2. The individuals who could potentially contribute most to the problem solution represent groups with extremely diverse expertise and experience who lack a history of effective communication with each other.
3. More individuals are needed than can effectively interact in a face-to-face meeting.
4. Time and cost make frequent group meetings infeasible.
5. The efficiency of a face-to-face meeting can be increased by a supplemental group communications process.
6. Disagreements among individuals are so severe or politically unpalatable that communications must be refereed and/or anonymity assured.
7. Bandwagon effects or the domination by a single individual or group of individuals must be avoided to assure the validity of the results.

Figure 6.5 summarizes a single item from a three-round Delphi (Roper and Brophy 1977). On the first-round, each participant was asked to estimate the time of occurrence for the events on a "clean" questionnaire. For the second round, respondents received statistical feedback as shown (the peak of the "house" is the median first round response; the ends are the lower and upper quartiles), plus a "√" indicating that respondent's first round reply. Participants whose responses lay outside the interquartile range were asked to either adjust their estimates or supply a rationale (see Figure 6.5). For the third round, participants received the latest group estimates and the rationales for individual disagreements. In the third and final round, all participants were encouraged to supply statements supporting or contradicting the group position. Results were displayed in several ways: (1) as illustrated in Figure 6.5, (2) with items displayed in predicted chronological order of occurrence, and (3) in order of probability of occurrence, calculated as

$$1 - \frac{\text{number of ``never'' responses}}{\text{total number of responses}}.$$

Users of Delphi should carefully consider the cautions noted in the text.

Statement of event	Estimated year in which event will first attain a better than 50% chance of occurrence												
	Now	1978	1979	1980	1981	1982	1983	1984	1985	1986	1987	1988	Later	Never
1. All practicing engineers are required by law to be registered nationally or in one or more states.														

WRITTEN RATIONALE FOR AGREEMENT OR NONAGREEMENT*

ROUND II: RATIONALE FOR NONAGREEMENT WITH GROUP MEDIAN DATE

RATIONALE FOR EARLIER OCCURRENCE

1. Engineering associations are talking about this and polling their memberships now.
2. Internal interests and external pressures (conservation groups, court action on liability, etc.) may get at least one state to move quite soon.
3. This will be a political decision. Political expediency will be served by passing such a law in 1984 (an election year). Recent information indicates that enabling legislation has been introduced into the legislatures of at least three states.
4. With all the government regulations facing industry, I think some states will require this in the next four to five years.

RATIONALE FOR LATER OCCURRENCE

1. This program has been discussed and proposed by many (starting in the 1940s). Related to the entire registration process including recertification, and so on.

2. Don't think it will progress to the point of law in the next eleven years. Present federal trends seem to indicate *less* regulation by laws.
3. My NEVER based on definition of "practicing engineers." Assume this means all holders of jobs which sometimes require elements of engineering. If you mean individuals with full-time responsibility exclusively in engineering, then 1987 or sooner.

ROUND III: COMMENTS SUPPORTING RATIONALE OR GROUP MEDIAN DATE

MEDIAN OCCURRENCE

1. This has been a long time coming, but it is inevitable. When the general public becomes aware of the impact engineers have on the environment, health, and product safety they will demand assurance of professional competency.
2. The federalization of licensing for engineers is likely to parallel that of lawyers, rather than physicians. I strongly disagree with the assertion that "Present federal trends seem to indicate *less* regulation by laws."

LATER OCCURRENCE (INCLUDES "NEVER")

1. The government moves ever so slowly. While there is constant pressure by professional engineering groups to speed registration, unless there is some obvious reason (e.g., a tragedy) I find it hard to believe that it will happen soon. I also believe that universities should stress registration.
2. I believe that "*never*" is proper as we are referring to "occasional" engineers. It is my understanding that the great majority of engineering degree holders are occasional engineers (e.g., management responsibilities). In this capacity their work normally has no direct relationship to the law or the public for whom the law is written.
3. Some registration may be required for plant or process engineers; however, it is unlikely that product engineer registration will be required.

FIGURE 6.5. Summary of a three-round Delphi forecast.
Source: Roper and Brophy (1977).

5. Not exploring disagreements may also cause dissenting participants to withdraw, biasing the results.
6. Delphi practice has tended to be shoddy with respect to the principles of good survey practice.

To close on a general cautionary note, there are a number of points of concern that are common to expert opinion methods in greater or lesser degree. Whereas sociologists do not attempt to build bridges very often, engineers (and other assessors) seem too willing to use expert opinion techniques with naïveté and indifference to professional standards. Whether using Delphi, survey, panels, or even genius forecasting, assessors should beware of crude questionnaire design, poor planning for statistical interpretation, casual definition of "experts," short-cut approaches that are not validated and lack reliability measures, and illusions of precision.

6.5 CONCLUDING OBSERVATIONS

The present and future impacts of a technology cannot be identified or analyzed unless the assessor understands how the technology works now and how it will work in the future. Thus both descriptive and forecasting activities are necessary.

The descriptive activity is perhaps best begun by specifying the level of emergence of the technology (see Table 6.1) and the major characteristics or dimensions of the assessment topic (four are analyzed in Table 6.2). Within this general framework, a comprehensive description can be pursued using a format such as that suggested by the MITRE Technology Description Background Statement (Table 6.3).

Description should do more than merely trace the physical and operational characteristics of the central technology. It should locate the lines of influence between that technology and the surrounding technical, economic, and social milieu; that is, it should characterize the TDS involved. This can be accomplished by careful examination of uses and applications of the technology, supporting and competing technologies, and influencing factors.

The progress of technology cannot be forecast confidently without considering the future state of the society within which it must operate. The assessor must ascertain that technology and social forecasts are mutually consistent and represent feasible projections. Forecasts should proceed logically from a basis of explicit data, relationships, and assumptions and should be reproducible by other forecasters starting with the same information. Technology development proceeds in a surprisingly ordered pattern unless acted on by exogenous factors. It also responds to opportunity and need. These two factors, coupled with the assumption that understanding the process of technological innovation will allow successful forecast of its progress, form the rationale underlying technological forecasting.

Three general principles are offered for forecasting:

methodological sophistication contributes little;

the core assumptions are critical; and

short time horizon correlates strongly with increased accuracy.

Taken together, these imply that forecasts should be current, even if simple. Three techniques are elaborated in the chapter:

1. Monitoring is the most commonly used forecasting method; it depends on insightful interpretation of gathered information.
2. Trend extrapolation has many variants—all emphasize continuity of relationships and depend on choice of parameters. Single and compound parameter extrapolations can sometimes be aided by assuming exponential or logistic (S-shaped) functional relationships. Envelope curves aggregate over continuing technological developments.
3. Expert opinion methods include genius forecasts, surveys, interactive panels, and Delphi. Some general cautions are to
   a. design questionnaires with professional care;
   b. plan for statistical interpretation;
   c. take care in identifying experts and obtain a representative sample;
   d. use validated approaches where possible; and
   e. avoid the illusion of precision.

RECOMMENDED READINGS

Armstrong, J. E., and Harman, W. W. (1977) "Strategies for Conducting Technology Assessments," report to the Division of Exploratory Research and Systems Analysis Division, National Science Foundation, Washington, DC: Department of Engineering–Economic Systems, Stanford University, Stanford, CA, November 1977.

A good overview of assessment techniques, including technology description and forecasting. Includes perceptive insights as to the conduct of assessments as well.

Ascher, W. (1978) *Forecasting: An Appraisal for Policy-Makers and Planners*. Baltimore: Johns Hopkins University Press.

An insightful review of the promises and perils of technology forecasting and social forecasting.

Bright, J. R. (1978) *Practical Technology Forecasting: Concepts and Exercises*. Austin, TX: The Industrial Management Center.

An excellent introduction to techniques of forecasting. A number of well-thought-out exercises are included.

Bright, J. R., and Shoeman, M. E. F. (eds.) (1973) *A Guide to Practical Technological Forecasting*. Englewood Cliffs, NJ: Prentice-Hall.

An excellent series of articles covering the spectrum of technology forecasting with emphasis on actual applications.

Linstone, H. A., and Turoff, M. (eds.) (1975) *The Delphi Method: Techniques and Applications*. Reading, MA: Addison-Wesley, Advanced Books Program.

Everything you ever wanted to know about Delphi and were afraid to ask. Also includes cross-impact analysis and KSIM (see Chapter 8 of this text).

Warwick, D. P., and Lininger, C. A. (1975) *The Sample Survey: Theory and Practice*. New York: McGraw-Hill.

A solid "how-to" book on surveying.

EXERCISES

6-1. 1. Select several technologies with which you are acquainted and classify their level of emergence and impact (Table 6.1).

2. Research a specific technology (e.g., railroads, aircraft) and trace its course of development through the seven levels of emergence and impact given in Table 6.1.

6-2. Obtain a completed technology assessment [e.g., *A Technology Assessment of Coal Slurry Pipelines*, Office of Technology Assessment (1978)] and

1. Classify the level of emergence and impact of the central technology.
2. Locate the technology along the topic dimensions described in Table 6.2.
3. Compare the technology description presented in the assessment with that suggested by the MITRE format (Table 6.3). Discuss any differences that you find.

6-3. (Project-suitable) Select a technology with which you are acquainted and prepare a brief description in outline form. If more than one technology is to be described (i.e., in your class), choose ones with different levels of emergence and emphasize different assessment topic dimensions.

6-4. Discuss modifications that you feel would be necessary to the MITRE technology description format when the assessment topic is a social technology.

6-5. Listed in the accompanying table are historical data for the installed electrical capacity of U.S. utilities and industries for 1902–1970.

1. Plot the variation of generating capacity with time. Try graph paper with arithmetic scales first and then semilogarithmic paper.
2. Form a curve to represent the data for 1910–1930 and another for 1946–1970. If computer software is available, a least-squares fit can be used to determine the equations of these lines; if not, determine them graphically.
3. Are the two portions of your curve from part 2 exponential? Can you account for the discontinuity by the variation of events in the social, economic, or political realms?

Installed Generating Capacity in Electric Utility and Industrial Generating Plants (in Megawatts)[a]

Year	*Capacity*	*Year*	*Capacity*	*Year*	*Capacity*
1970	360,327	1950	82,850	1930	41,153
1968	310,181	1948	69,615	1928	36,782
1966	266,816	1946	63,066	1926	32,936
1964	240,471	1944	62,066	1924	25,923
1962	209,576	1942	57,237	1922	21,317
1960[b]	186,534	1940	50,962	1920	19,439
1958	160,651	1938	46,873	1917	15,494
1956	137,342	1936	43,582	1912	10,980
1954	118,878	1934	42,545	1907	6,809
1952	97,312	1932	42,849	1902	2,987

[a] From U.S. Department of Commerce (1975).

[b] Denotes first year for which Alaska and Hawaii are included.

6-6. (Project-suitable) Using brainstorming (Chapter 1) or some such technique, construct a list of three or four events that you see as critical to the development of some technology (e.g., the technology most central to your business, educational technologies, etc.). Construct and administer a small-sample, three-round Delphi poll. Focus on the prospects for occurrence of the critical events. Use class members, colleagues, or some available group to minimize the time necessary for the process.

6-7. (Project-suitable—parts 1 or 2)

1. Select a physical technology. Discuss how you would forecast its development for the next 25 years using the methods and techniques introduced so far. Where would your data come from? What sort of accuracy do you expect?
2. Do the same for a social technology.
3. Discuss the differences you have found, if any, between forecasting a physical technology and a social technology.

7
Social Description and Forecasting

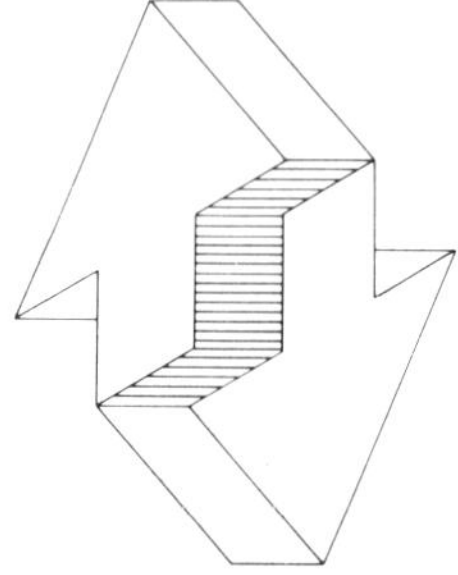

This chapter is intended to help the assessor describe the present and forecast the future state of society. Special emphasis is placed on the interaction of social and technological factors. A model of the technological delivery system (TDS) is introduced as one means of delineating the social–technological interface. Social indicators, measures of social system performance, are also introduced and their role in description and forecasting is described. A strategy for performing societal description in TA/EIA is presented in conjunction with the TDS model and social indicators. General approaches to the description of future societal states are considered next. Four general principles, continuity, self-consistency, stakeholder group divisions, and cause–effect linkages that support social forecasting are discussed, and general characteristics of social forecasting are probed. Finally, two widely applicable methods of forecasting, scenarios and trend-like approaches are discussed.

7.1 INTRODUCTION

The interdependence of technology and social descriptions and forecasts was emphasized in Chapter 6. The technological environment is only one portion of the larger social context. This consists of the complex system of interacting technical, legal, environmental, economic, political, and social processes, and their institutional guises. The technological organization must draw upon the larger society for factors such as capital, natural resources, specialized manpower, and tools. It is subject to direction by government and private policies based on societal norms. Its operation is constrained by competition for the factors of production. Thus, together with forecasting the technology in question, the assessor must predict the future societal state in which it will exist. Only in this manner can the resources that will be expended by society to support and control the technology be gauged.

This chapter presents some methods by which the assessor can approach both social description and forecasting. General considerations, including a model for the technological delivery system (TDS), are first presented to provide orientation for the tasks. Next a format for societal description is discussed. Social indicators and the understanding of values are singled out for attention as important both to social description and forecasting. Basic principles underlying the forecast of alternative societal states are then considered and, finally, two techniques that hold promise for integrating social considerations are presented.

This chapter owes its conceptual roots to Chapter 2—Technology and Society. It also has a particularly intimate connection with Chapter 12 (on social impact analysis). Because such analysis requires societal description and a logical sequence of task steps has evolved to accomplish it, Chapter 12 contains material that relates closely to this chapter.

In the treatment of values, this chapter again has close bonds to Chapters 2 and 12. Our procedure is to introduce the interdependence between technology and values in Chapter 2, emphasize the description of values and how they may change over time here, and concentrate on how particular technologies may impact values in Chapter 12.

7.2 GENERAL CONSIDERATIONS

7.2.1 Social Context—Critical for TA/EIA

Assessment studies have generally focused on the description and forecast of the technology and have paid little attention to social context. The limited attention displayed tends to be static. Yet, when one is concerned with the impacts of a technological development 50 years in the future, it should be obvious that societal context will not be what it is today. Braudel (1976) suggests three temporal perspectives for dealing with social context: (1) *events*—focuses on a time frame of years (e.g., the 1973 oil embargo), (2) *institutions*—focuses on time periods of decades (e.g., "Project Independence" for U.S. energy self-sufficiency), and (3) *enduring patterns*—changes over the course of centuries. Forecasting social context requires merging all of these. There are suggestions that certain enduring patterns of Western industrial society are changing, such as reductions in consumption of luxury goods (Harman 1976). If this is correct, it implies a need for critical attention in TA/EIA to future societal context.

Whereas relegating societal forecast to the background may suffice for EIA and short-term TA studies, it can produce erroneous or misleading results for longer-range assessments. The tendency to slight social forecast can certainly be understood. Technology description and forecasting, for example, are simple and deterministic by comparison. Armstrong and Harman (1977: 78) summarize the state of the art too well when they write that

> The general consensus with regard to projection of alternative societal contexts (and there is probably more consensus here than on some other aspects of TA) appears to be that (A) it is desirable and (B) nobody really knows how to do it . . . at least within the typical constraints of time and resources.

There are, however, approaches and techniques useful for this task. The discussion of some of these takes up the remainder of this chapter.

7.2.2 The Technological Delivery System

Technological activity and the larger societal context of which it is a part interact through a complex and only partially understood system of relationships. As with most such systems, its operation can be better understood through the use of a simplified model. The model is intended to parallel reality but, by eliminating some of the higher-order complexities of the real system, to limit analysis to the primary interrelationships. Such a model has been proposed by Wenk and Kuehn (1977). A schematic representation is displayed in Figure 4.1, to which the reader should refer at this time (p. 44).

The model conveys the notion of a technological delivery system (TDS). Each TDS is specialized to deliver a specific product, whether aircraft or education. The TDS is assumed to be composed of all the institutions and individuals necessary to develop and control the ensemble of technical, legal, economic, political, and social processes required for the functioning of the system. (Most of these institutions, of course, interact with more than a single TDS.)

The components of the TDS are institutions, each of which is designed, organized, and focused to perform a specific task. Connecting links between institutions represent channels of communication. These links indicate the flow of information upon which the institutions act within their constraints. The model is composed of four basic elements:

1. Inputs to the system, including capital, natural resources, manpower, tools, knowledge from basic and applied research, and human values.
2. Institutional and organizational groups, both public and private, that play roles in the operation of the TDS or the modification and control of its output.
3. System processes by which the institutional actors interact with each other through information linkages, market, political, legal, and social processes.
4. System outcomes including both direct (intended) and indirect (unintended) effects on the social and physical environments.

To understand better the TDS and its behavior, it is useful to consider its operation first in a static (i.e., time-independent) sense. Basic and applied research organizations in universities, industry, private think tanks, and government develop knowledge and capabilities that provide a *push* for the delivery of new technological outputs. Consumers provide *pull* through their demand for goods and services. The push and pull are coupled through the management of the technological organization that senses demand and capability, gauges external constraints, and assembles and organizes the factors of production.

External constraints to the operation of the technological organization can be of a social, technical, economic, and/or environmental nature. Social constraints include cultural or traditional factors, such as resistance to the development and use of birth-control pills. They can also be institutional (e.g., union opposition to mechanization). Technical, economic, and environmental constraints include, for example, simple lack of technical capability, competition for scarce factors of production, and air-quality standards.

Government institutions select and prioritize the value preferences of both the general public and individual stakeholder groups. Policies and programs then formalize these preferences. The performance of the technological organization and its output are strongly influenced by government through regulation, subsidy, research and development

(R&D) programs, and so on. The direction and output of the system has become increasingly a shared public/private responsibility in recent years. It is important to note that impediments to the delivery of desired outcomes can develop from within the government sectors of the TDS. Thus factors such as conflicting value preferences, constraint of information flow, inadequate or inaccurate information exchange, and bureaucratic inertia can constrain the delivery of goods and services. Figure 13.3 (p. 341) illustrates a specific TDS, solar energy for private housing, that is particularly sensitive to such factors.

7.3 STATE OF SOCIETY DESCRIPTION

7.3.1 Elements

The critical interrelationship between technology and society has been emphasized thus far in this chapter. The TDS produces outputs that impact and alter society and is, in turn, controlled, shaped, and altered by society. A technology introduced into different social structures would, in general, produce different impacts and different impacted parties as well as different approaches to control of the technology. Both technology and society must thus be described before impact identification and analysis can proceed.

Adequate description of the whole of society would be an impossible task. Fortunately, it is not required in TA/EIA. What is needed is a delineation of those elements of the social context that are affected and affect the operation of the particular TDS. The TDS model [Figure 4.1 (p. 44)] and the technology description (Section 6.3) provide benchmarks from which to begin.

Technology description [Table 6.3 (p. 106)] has already indicated important components of the TDS such as the factors of production (e.g., capital, natural resources, specialized manpower, tools,), scientific disciplines, industries/businesses, professions and occupations, products, and supporting technologies. Government institutions and policies have been at least partially identified under the categories of institutional factors affecting development and application of the technology. Finally, the delineation of uses and applications has indicated some of the impacts and impacted parties that proceed from the outcomes of TDS operation.

Not all important elements of the social context, however, have been dealt with by the technology description. There are also certain overarching concerns (threshold attributes) of the larger society that are of supreme importance to the state of society description. For example,

Jones (1971: 54) identifies four such concerns, expressed in the form of assumptions about the United States:

1. There will be no major *war* involving the United States in the period studied.
2. There will be no internal conflicts sufficiently serious to undermine or overthrow our present democratic system.
3. There will be no major shifts in the balance of *power between government and private decision making,* or between federal, state, and local governments.
4. Although the *value system* will change, most citizens will still accept and operate in terms of the so-called Puritan ethic, which presumably prizes work and personal achievement.

These four major societal attributes presumably enjoy a "go/no-go" status, and, as Jones puts it, if they are violated, "all bets are off." These exemplify core social assumptions, so critical to any forecasting effort (Ascher 1978).

Second to the major threshold attributes in scale are those related to *national conditions*. These characteristics are germane to all assessments to some degree or another. Examples of national conditions include population size, adjusted gross national product (GNP) growth rate, proportion of federal spending in defense and civilian programs, and the shift in industrial structure from manufacturing to service. Many of these conditions can be described in terms of social indicator data (Section 7.3.2).

Underlying the macrolevel elements are the finer-scale elements that are *specific* to each assessment. Those elements of the social microstructure that are central to a particular study should be identified and defined; and, where appropriate, measures should be chosen to scale them. Depending on the focus of the TA/EIA, these may relate to international, national, regional, or local conditions. As indicated in Chapter 12, appropriate methods vary—national social description may be well served by quantitative social indicators; local description may demand qualitative, primary data collection.

The MITRE study (Jones 1971) listed six major categories for use in describing the state of society: values and goals, demography, environment, economics, social factors, and institutions. Suggestions on completing these categorial descriptions appear in different sections of the book.[1]

[1]The object is to keep specific areas of expertise as units (e.g., economics), rather than to treat them repeatedly within TA/EIA components (e.g., economic state of society description, economic impact identification, economic impact evaluation).

HOW TO PERFORM SOCIETAL DESCRIPTION

The following sequence of steps is meant to be suggestive, not definitive. It suggests a way to organize the state of society description task in a TA /EIA.

1. Describe the technology [see Section 6.3 and the checklist provided in Table 6.3 (p. 106)]. This provides a place to begin the societal description and a first cut at some of the necessary information as well.
2. Sketch the elements and interrelationships of the TDS (see Section 7.2.2). This provides a systematic model by which to organize the social-context information.
3. Identify the major threshold attributes and core social assumptions (see this section). These will be critical for the social forecast to follow.
4. Describe the pertinent *general* social conditions (e.g., national scope) and *topic-specific* social features (e.g., local factors). For both, acquire necessary information in the six major categories:
   a. values and goals (see Section 7.3.3);
   b. demography (see Section 7.3.2 and Section 12.3);
   c. environment (see Chapter 10);
   d. economics (see Chapter 11);
   e. social factors (see Section 7.3.2 and Section 12.3); and
   f. institutions (see Chapter 13).
5. As impacts are identified and analyzed, revise and expand the social description.

The cameo entitled "How to Perform Societal Description" suggests a strategy to accomplish the state of society description.

Having set forth the general strategy, we turn now to two specific areas of importance for social description and forecasting.

7.3.2 Social Indicators

Social indicators are aggregate measures of various phenomena that collectively indicate the state of a society or some subset of that society. Raymond Bauer (1966: 1) has defined social indicators as "statistics, statistical series, and all other forms of evidence—that enable us to assess where we stand and are going with respect to our values and goals, and to evaluate specific programs and determine their impact."

SUGGESTED INDICATOR GROUPS

Abt Associates (1975) have delineated the following social indicator groups:

mobility indicators, including birthplace, length of residence, and region of residence variables;

ethnic-composition indicators, including native language and parentage indicators;

family-structure indicators, including parent presence and sibling structure indicators;

educational-attainment indicators, including indicators of public vs private school populations and years of schooling completed;

economic-structure indicators, including age of labor force, unemployment, income-source indicators, and occupational profile data;

poverty indicators, including the percent and nature of community families below the poverty level;

income-structure indicators, consisting of indicators of the dollar income of community members; and

general demographic indicators, including sex, total population, and housing-count data.

Since the publication of Bauer's seminal work, significant progress has been made to begin to accumulate such measures of social well being (Knezo 1973). A recent U.N. Statistical Office survey (1976) documented no fewer than 29 countries with social-trends books published or in preparation. The U.S. Office of Management and Budget's (1974) *Social Indicators* provides time-series information on health, public safety, education, employment, income, housing, leisure and recreation, and population.

The cameo entitled "Suggested Indicator Groups" is further suggestive of useful social indicator information for use in TA/EIA. Olsen et al. (1977) present another detailed listing of social-life factors and indicators suitable for community-level analysis. They also offer provisional (debatable, but interesting) specifications of favorable indicator levels (e.g., the closer to 100 persons per square mile, the more favorable the county population density).

Sheldon and Land (1972) identify three functional types of indicator:

(1) problem-oriented or direct policy indicators which are intended for direct use in policy and program decisions;
(2) descriptive indicators intended primarily to describe the state of society and the changes taking place within it; and
(3) analytic indicators that serve as components of explicitly conceptual and causal models of the social systems or some particular segment thereof.

Each of these types of indicator has a different potential application. However, all are time-series data at some level of aggregation and reflect both objective conditions of society and subjective perceptions of life experiences.

Current social theory is not really adequate to make use of Sheldon and Land's type 3 indicators. The present social indicators are thus empirical entities gathered at different levels of aggregation. The indicators identified in this section are of this nature—descriptive indicators (type 2 in Seldon and Land's classification). They do not clarify the dynamics of interaction between social elements. They offer quantitative counterparts to the well-established economic indicators and help focus on social outputs (e.g., personal mobility) instead of dollar inputs (e.g., transportation expenditures). Forecasting of social indicator trends is feasible because they are basically descriptive of social states, neither tightly related to particular causes of those states nor localized to particular contexts.

Hence, insofar as available social indicators are not localized, they are more useful in general TA/EIA social description than in topic-specific description. In no sense are available indicators tightly related to project outcomes on a cause–effect basis (Sheldon and Land's type 1). This is their most telling shortcoming for TA/EIA use.

On the positive side, social indicators are generally widely applicable and can be used with very little additional information about the societal context. Further, they are explicit and relatively easily interpreted. Finally, they can be used with the well-developed and documented techniques found in technological forecasting.

A closely allied notion concerns measuring social welfare in terms of the "quality of life." An interesting volume by the U.S. Environmental Protection Agency (1973b) highlights quality-of-life considerations. Quality of life provides a basic goal against which policies can be assessed via objectives (e.g., health and recreational opportunity) and their respective social indicators (e.g., morbidity rates and accessibilities to recreational activities). This provides a rationale, if not a tight metric, to incorporate social considerations into the policy processes. Furthermore, quality-of-life measures are suitable for consideration of the individual's status, in contrast to most social indicators, which are highly aggregate in orientation. A checklist of such indicators is provided in Section 12.6.

7.3.3 Values and Value Change

In this section we are concerned with human values as a part of the social context for technological developments. (In Chapter 12 we focus on how such developments can alter human values.) This section addresses, in turn, the following questions:

What are values?

Can they affect technology?

Do they change?

Can value change be usefully anticipated?

"Value" is a word that is both often and loosely used. There are two foci around which its meanings cluster. One refers to the values held by persons, consisting of *conceptions of desirable states of affairs* (Williams 1967). The other centers on *properties possessed by things*, states of affairs, and /or processes that refer to capacities to produce the desirable states of affairs. For instance, we may value the neutron bomb if we perceive it serving to preserve our freedom and security.

The issue of whether values affect technological development has been considered in Chapter 2. The answer is "yes." Values are the bases for goals, in turn the bases for policy actions. They provide criteria for judging the worth of policy outcomes (e.g., the effects of building of a dam). As discussed in Chapter 2, technological development is considered highly sensitive to life-style and consumer preferences. Were social values to shift toward a simpler, conserving life-style, dramatic changes could result in manufacturing practices, markets, and imposed regulations in such sectors as automobiles, clothing, and housing.

The foregoing discussion implies a sense that values can change and that such changes may be critical to description of future societal states. As Toffler (1969: 2) points out,

> In the last 300 years, however, the rate of value change appears to have speeded up—to the point at which major shifts in the value system of a society become apparent *within* the span of a single lifetime and within even shorter periods.
>
> This acceleration of value change is one of the most dramatic developments in the entire cultural history of the human race. It shatters the

presumed identity between one generation and the next. It makes untenable the assumption that the values of future generations will resemble our own, and also makes it impossible to predict future values by simple straight-line projection.

Both Williams (1967) and Rescher (1969) have formed typologies of value changes. Rescher's "Modes of Value Change" are illustrated in a cameo.

The cameo classifies Rescher's value changes as upgrading or downgrading. The point is that there are many diverse ways in which acceptance of a given value can occur. Table 7.1 provides a compilation of directions of value change for major American values (Williams 1967). This sort of a presentation may form the basis of a trend analysis of values.

Value changes may be local and isolated, as, for example, in the

TABLE 7.1. Changes in American Values[a]

Value-belief complexes	Directions[b] of change (approximate period)	
	1900–1945	1945–1966
Activity	Indeterminate	–
Work	–	–
Achievement	–	+ (post-Sputnik I)
Success	+	+
Material comfort	+	+
Humanitarianism (domestic)	+	+
Humanitarianism (war)	+	–
"Absolute" moral orientation	–	Indeterminate
Practicality	+	–
Efficiency	+	–
Science and secular rationality	+	+
Progress	+	–
Freedom	Indeterminate	–
Equality	+	+
Democracy	+	Indeterminate
Conformity (to social pressure)	+	+
Individual personality	+	Indeterminate
Nationalism	+	– to +
Racism—group superiority	–	–
Totals		
Increase	13	8
No change or indeterminate	2	3
Decrease	4	8

[a] From Williams (1967).
[b] Minus sign indicates decrease; plus sign, increase.

MODES OF VALUE CHANGE [from Rescher (1969:89)*]

Modes of upgrading		Modes of downgrading
Value acquisition	—1—	Value abandonment
Increased redistribution	—2—	Decreased redistribution
Rescaling upward	—3—	Rescaling downwards
Widening redeployment	—4—	Narrowing redeployment
Value emphasis	—5—	Value deemphasis
Restandardization by a raising of standards	—6—	Restandardization by a lowering of standards
Retargeting by adding implementation targets or by giving higher priority to existing ones	—7—	Retargeting by dropping implementation targets or by giving lower priority to existing ones

*In *Values and the Future: The Impact of Technological Change on American Values*, edited by Kurt Baier and Nicholas Rescher.

discussion of impacts on life-style; or they may pervade an entire society. In general, the mass media have broken down some value differences based on economic and social class, and on geographic location. An individual's values may even change on relocation. Generally, pervasive value changes are long range and indirect. Different subgroups in a society may undergo value changes in different directions at the same time.

Important properties of value changes are the *type* of value change, the *pervasiveness* of the change, the *rate* at which the change takes place, the *direction* of the change, and the presence of *related value trends*. These must be considered in attempting to forecast values.

The two most likely predictive techniques are obtaining the views of experts and extrapolation of historical value trends (techniques discussed in Chapter 6). The latter is exemplified, on a national scale, in the work of Williams (1967) cited in Table 7.1.

One central consideration in applying these techniques is the fact that values often conflict with one another, both in the abstract, and in the competing demands their realization and pursuit make on man's finite resources. Thus when a change occurs in the rationale that constitutes the operative framework within which a value is pursued, we may expect a series of stresses on our scale of values leading to a rescaling or reordering. But how is one to predict the character of this value response? Two key factors to consider in predicting this change are cost and benefit.

Maintaining a value exacts a cost manifested as an investment of various resources. The extent of the requisite investment will be affected by changes in the environment; "cleanliness" comes cheaper in modern cities than in medieval ones, and the achievement of "privacy" costs more in urban than rural environments. The maintenance of a value will obviously be influenced by its cost. When this becomes very low, we may tend to depreciate the value as such. When it becomes high, we may either depreciate the value in question as such (the "fox and the grapes" reaction)—or rather more commonly—simply settle for lower standards for its attainment. (Think of "peace and quiet" in this era of sonic booms and ambulance sirens.) The benefits of values arise from both the need felt for maintaining a value and the tangible benefits resulting from its maintenance.

We note two different approaches to value-change prediction. First, the Stanford Research Institute (Elgin and Mitchell 1977) attempted to profile the American people in terms of their levels of attainment on Abraham Maslow's needs hierarchy (five levels—survival, security, belongingness, esteem, and self-actualization). They then projected this to the year 1990 based on subjective estimates of underlying demographic and value trends. The altered needs profile was then used to predict changes in American societal values. Second, Martin Ernst in a TA of electronic funds transfer by Arthur D. Little, Inc. (1975) generated purposely "far-out" scenarios. These helped to expand the TA team's vision of the future and enabled them to think about value systems different from our present ones.

7.4 FORECASTING THE STATE OF SOCIETY

7.4.1 General Principles

Because of the interactive nature of technology and society, forecasts of both the future state of the technology and the society within which it is embedded ideally should be done simultaneously. Henschel (1976) discusses very perceptively the differences between physical and social forecasts. In current state of the art, however, the assessor will generally need to iterate between the results of two semiautonomous forecasts to assure that both represent feasible and mutually compatible future states.

Armstrong and Harman (1977: 76–77) list four common-sense principles on which the forecast of societal state rests. The first is that social systems tend to exhibit *continuity*. Even during periods of extreme disruption (e.g., wars and revolutions), most of the elements of a social system continue to function without rapid or discontinuous change.

The second principle is that social systems tend to exhibit a *self-consistency* in their internal structure. Thus societies have strong cohesive forces that assure that different elements of the society cannot long pursue radically different courses.

A third principle states that *stakeholder group divisions* within the larger society produce tension and conflict that are the sources of change. One useful method to produce state of society forecasts is, therefore, to examine areas of conflict and tension to determine stakeholder needs and desires. This principle parallels the technological forecasting rationale that maintains that technological development responds to need.

The final principle is that social systems exhibit characteristics that suggest the operation of *cause and effect* linkages. For example, we know (or think we know) that money supply actions by the Federal Reserve Board affect both inflation rates and employment level. This principle underlies the supposition that cross-impact techniques such as trend-impact analysis (to be described shortly) are appropriate vehicles for social forecasting.

Ascher (1978: 205–207) observes that social and political forecasting generally refers to actions and attitudes relating to the "deference values" (e.g., respect, affection, and power) rather than to material well-being values (e.g., wealth and skill). Five distinguishing characteristics affect the conduct of such forecasting:

1. The topics of social forecasting are generally highly alterable through human volition because few constraints are imposed by limited material resources.
2. There is seldom a consensus on the preferred direction of changes. This provides a potential for more radical changes in trends than in technological or economic measures.
3. Social attitudes are less cumulative than material growth patterns. The predominance of a value does not necessarily imply its future acceptance.
4. Single, discrete events are often very central to social forecasts (e.g., the outbreak of a war, an election outcome).
5. The meaning of one condition or event may depend on a whole constellation of other conditions (e.g., a Democratic presidential victory in 1988 does not mean much unless one knows what that party will represent).

TREND-IMPACT ANALYSIS
[Based on Gordon and Stover (1976)]

Trend-impact analysis combines trend extrapolation (Section 6.4.3) with a cross-effect matrix formulation (Section 9.3) to build and then modify a portrait of societal change.

The first step in this approach is to select long-term societal trends that are germane to the TDS or individual institutional actors within the TDS. These trends might represent changes in societal norms and value preferences, governmental regulations, institutional structures, or demands for the goods and services produced [see Figure 4.1 (p. 44)].

Next, the trends are extrapolated. A computer program facilitates this step.

The historical data (trends) are next augmented by considering possible future high-impact events. Four types of event are noteworthy:

1. bounding conditions (e.g., a maximum credible jet plane speed is reached);
2. a changing bias (e.g., different life-styles reduce the demand for airplane travel at high speed);
3. an unprecedented event (e.g., legislation outlaws aircraft with certain characteristics); and
4. a surprise (e.g., a breakthrough in technology greatly reduces the costs of achieving certain aircraft speeds).

Ascher's observations caution the would-be social forecaster against the simplistic use of trend-like approaches.

We consider two general analytical approaches to forecasting future societal states. The first is to adapt the techniques of technology forecasting. In this approach these techniques are applied to the projection of social indicators or other important social parameters. Thus the advantages inherent in well-established techniques accrue to the forecaster. There are, however, disadvantages as well. Social indicator projections typically develop only the skeletal outlines of a future state. A general social theory to interrelate them, by explaining cause–effect

A list of such events is prepared, using such techniques as literature search, Delphi, or expert opinion survey. Each selected event is judged according to its

1. probability of occurrence as a function of time;
2. impact on the trends under study, including:
   a. lag between occurrence and trend response,
   b. time between occurrence and maximum impact on the trend,
   c. magnitude of that largest impact,
   d. time from occurrence until the impact reaches a final or steady-state level, and
   e. magnitude of that steady-state impact.

The Futures Group approach plots revised trends in light of the occurrence of the events. This yields improved trend forecasts and potential insight into the effectiveness of particular policies or actions that might act to change the trends.

If one lacked the resources or inclination for such a quantitative trend impact analysis, a qualitative approach could still be helpful. For instance, Rosove (1973) described a reverse sort of interaction in which one considers how certain trends will impact critical areas (which can include events or other trends). This qualitative approach relies on a contextual map that crosses the trends on one axis against the impacted areas on the other in a matrix format. Each cell contains information on the relevant relationships, including "subtrends." This information can be qualitatively assessed on a cross-impact basis (see Section 9.3.1) to incorporate judgments of the form "if this trend were to reach such a level, then the probability and/or magnitude of this event would be X."

relationships, does not exist. Thus many aspects of the forecast must be uncertain and unclear. Further, the task of selecting appropriate social indicators to forecast is not as well understood as the choice of the analogous technical system performance parameters. Obtaining sufficient data to define indicators and establish trends may not always be possible. Finally, there are the cautions raised by Ascher.

Trend-impact analysis represents a combination of trend-extrapolation (Section 6.4.3) and cross-effect matrix (Section 9.3) approaches. We discuss it here as an example of technology forecasting techniques (see Chapter 6) with a useful potential in social forecasting. See the cameo entitled "Trend-Impact Analysis."

The second approach to forecasting future social states is the mechanism of the scenario. This is discussed in the following section.

7.4.2 Scenarios

A scenario is a descriptive sketch or chronological outline of a possible future state of society. It attempts to produce a holistic view of the pertinent social context as it relates to the technology being assessed. It seeks to define the major contextual elements and to depict their relationship to one another. Scenarios are usually presented in story form, making them easily read and understood.

Two sorts of scenario can be differentiated. One describes a possible future state of society, a "snapshot." The other presents a "future history," that is, a plausible chain of events that lead from the present to a particular future state.

Scenarios can have a range of possible uses. In general, they portray *possible* future societal states; they are not projections of what is expected to actually happen. Keeping this in mind, scenarios can serve to

- alert users to critical *policy* issues, through their contextual richness;
- suggest areas to *monitor*, as playing potentially important roles in affecting the emergence of future states of society;
- yield *insight* through the scenario construction process regarding interrelationships and critical factors; and
- sensitize people to the gravity of a development by bringing the potential *impacts* to life through vivid portrayals.

Scenarios are most useful when contrasted with other scenarios. Forecasts usually present a set of alternative scenarios, generally between two and six. Three is particularly popular, allowing for two extremes and a "most likely" in-between. A base case, or surprise-free projection, is often constructed; interesting variations are then counterposed to it. Sometimes the set of scenarios is limited to relatively reasonable alternatives (e.g., depiction of differences for the year 1990 resulting from U.S. emphasis on coal as an energy source vs nuclear power). In other forecasts, scenarios may be highly improbable but useful vehicles to span the reasonable probabilities through illustration of extreme cases. For instance, a year 2050 energy-focused scenario set might range from a glorious fusion energy breakthrough case to an economic collapse case caused by an inadequate energy supply.

To be effective, scenarios must focus on significant differences. They must delicately balance contextual richness against relevance to the

TDS under scrutiny. Explicit consideration of topic definition, core assumptions, data bases, and interpretations are important and should be reported along with presentations of the scenarios (Mitchell et al. 1975). Even so, it is often difficult for a user to gauge the implications of changes in key variables or major assumptions for the overall forecast. As Ascher (1978: 207) notes:

> The multiple nature of the events and conditions of a scenario makes its appraisal difficult, since the scenario forecast can be partially right and partially wrong, and usually there are no explicit indications of which elements are to be regarded as more important and by how much.

Construction of scenarios is a blending of insightful forecasting with storytelling. The underlying forecast should derive from the best available trend extrapolations and expert consensus, whether the aim is to generate probable or far-out scenarios. Storytelling makes it compelling—consider Orwell's book-length scenario *1984*. The cameo entitled "Scenario Construction" suggests a systematic procedure for scenario preparation. Chapter 20 offers some mild examples in three scenarios of the future for TA/EIA.

In summary, we believe that to be usable a social projection needs to employ a combination of techniques. For example, scenario construction might be preceded by the selection and forecast of social indicators relevant to the technology under study.

7.5 CONCLUDING OBSERVATIONS

The technological environment is only one portion of the larger societal context. This societal context influences the technological sector and is influenced and altered by it in turn. To describe and forecast one without the other would be of doubtful validity. The TDS has been suggested as a useful vehicle for visualizing many aspects of the interaction of technology and society. A strategy for societal description is offered that combines the TDS with social indicater and qualitative information suitable for most TA/EIAs.

An iterative process should be followed to produce forecasts of the technology and the future societal state that are both feasible and consistent with each other.

If social indicators are to be used for the forecasting process, their similarity to technological system performance parameters can be capitalized on to employ the technology forecasting techniques described in Chapter 6. Unfortunately, the absence of an underlying social theory makes it difficult to confidently interrelate the various ele-

SCENARIO CONSTRUCTION

The following steps when combined with the artistry of the storyteller (no recipe provided herein) can yield effective scenarios.

1. Define the paramount issue of concern and the appropriate dimension or dimensions to represent it.
2. Identify the potential user audience for the scenarios, the main uses, and the points of particular interest to the audience.
3. Specify the suitable geographic scope, time horizon, and range of alternatives (implying the number and range of scenarios also).
4. Identify the TDS central to the scenarios.
5. Create a checklist of relevant factors to be scripted into the scenarios. The TDS should provide the focus; scanning the components of EPISTLE (see Section 4.5.6) should yield the factors.
6. Compile and organize the information base with the best available forecasts of the relevant factors. As appropriate, draw on trend extrapolation, monitoring, relevance trees, expert opinions, and plain intuition.
7. Sketch out the baseline (surprise-free) scenario and then the alternatives. These may build on key policy choices, probable future states of society, particular technological breakthroughs, anticipated levels of the key factors, and so on.
8. The scenarios themselves may simply present the outcome of the interacting forces as future state portraits, or chronologies of the significant events and developments building to the future states may be detailed.
9. Evaluate and interpret the implications of the scenarios, to be presented along with them.
10. Circulate review drafts, and then revise the scenarios accordingly.

ments of the skeletal vision of the future society such as forecasts portray. Complementary holistic portraits of possible future societal states can be constructed through the mechanism of the scenario. Scenarios are usually constructed in sets representing a surprise-free projection of the future and several variations. It is important to realize that no matter how compelling the logic and prose of such scenarios, they do not represent forecasts of the future, but rather forecasts of *possible* futures.

Clearly, the forecast of societal states is far more complex and uncertain than the forecast of technology. However, its importance demands the attempt. Here common sense, insight, and intuition are still the assessor's best guides.

RECOMMENDED READINGS

Henschel, R. J. (1976) *On the Future of Social Prediction.* Indianapolis: Bobbs-Merrill Company.

A thought-provoking analysis of the problems and potential of social forecasting.

Jones, M. V. (1071) "A Technology Assessment Methodology: Some Basic Propositions." Report MTR 6009, Vol. 1 (NTIS #PB 202778–01). Washington, DC: The MITRE Corporation.

A detailed review of the MITRE checklist for describing the state of society and numerous examples of its application.

Mitchell, A., Dodge, B. H., Kruzic, P. C., Miller, D. C., Schwartz, P., and Suta, B. E. (1975) *Handbook of Forecasting Techniques.* Report 75–7 to U.S. Army Corps of Engineers Institute of Water Resources. Fort Belvoir, VA: Stanford Research Institute.

An excellent discussion of scenario construction including history, strengths, weaknesses, and an example. Also, excellent characterizations of some dozen forecasting techniques.

Wenk, E., Jr., and Kuehn, T. J. (1977) "Interinstitutional Networks in Technological Delivery Systems," in *Science and Technology Policy: Perspectives and Developments* (J. Haberer, ed.). Lexington, MA: Lexington Books, 53–175.

A detailed discussion of the TDS concept.

EXERCISES

7-1. Gasohol is a mixture of 10% ethyl alcohol and 90% gasoline. Supposing that the alcohol were to be produced from grain and farm waste products such as corn stalks, discuss the technological delivery system that would be involved.

7-2. (Project-suitable; change the topic to that of the project assessment)

1. Describe the current state of society in regard to the legalization and commercial production of marijuana.
2. Develop a scenario of a surprise-free future for the United States 15 years from now. Then construct a radical variation scenario.

7-3. Discuss any differences you perceive between the projection of parameters measuring technological system performance and social indicators.

7-4. Describe situations, if any, that you can visualize for which a technology might be confidently forecast without recourse to a state of society projection. If you find none, explain why the state of society projection is necessary.

7-5. Market-analysis studies are a limited form of technological–social forecast. Discuss a product that has failed (e.g., the Crosley car, an early subcompact produced in the late 1940s and early 1950s) and, if possible, relate those failures to changes in the state of society or errors in its description.

7-6. (This exercise draws on the expert opinion methods described in Section 6.4.4) You are the social scientist helping to prepare an EIS of a new dam situated near a small city of 10,000 people. The dam will produce hydroelectric power and new recreational opportunities for this secluded, traditional community.

1. Describe how you would obtain expert opinions on the future economic implications of the dam for the region. Note the form of survey you would use, and highlight how you would address the four issues listed in the cameo entitled "Surveys" in Chapter 6.
2. Describe how you would determine local attitudes toward the dam construction. Proceed as in step 1.

7-7. (Project-suitable) Describe the TDS for the topic of a TA or EIA. Sketch out the most critical actors and linkages for the formulation of policy actions for this TDS [use Figure 13.3 (p. 341) as a model].

8

Impact Identification and Policy Considerations

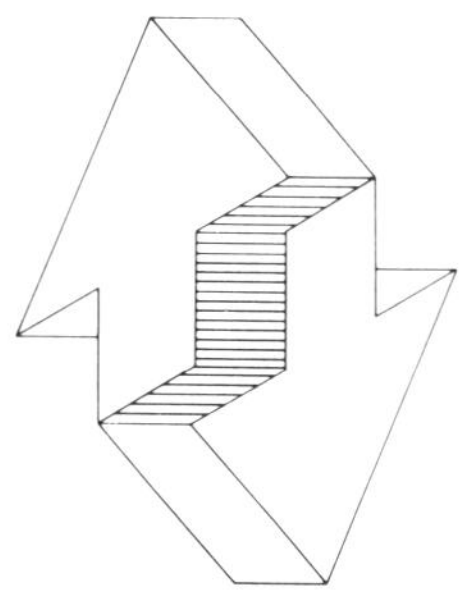

This chapter assists the reader in the selection and application of strategies and techniques for the identification of impacts in TA/EIA. Impacts consist of primary and higher-order effects resulting from activities of a technological delivery system (TDS). As such, this discussion provides a necessary prelude to the more detailed analysis of those impacts considered in Chapters 9–13. The chapter describes elements of candidate strategies for impact identification. It then suggests criteria for selecting from among them ones appropriate to a particular assessment task. A number of techniques of impact identification are next briefly described and their relative suitability considered with regard to differing assessment characteristics. These techniques are checklists, matrices, relevance trees, and morphological analysis. Finally, general considerations and criteria for the identification of policy options are raised.

8.1 INTRODUCTION

Impact identification is an important feature of an assessment. It can grow only from a thorough description of the technology in question and an equally thorough description of the society within which the technological delivery system (TDS) operates (Chapters 6 and 7). Further, impact identification is symbiotically related to impact analysis (Chapters 9–13) and evaluation (Chapter 14) and is a basis for policy analysis (Chapter 15).

Impact identification, impact analysis, impact evaluation, and policy analysis cannot be executed as sequential steps. They must proceed together iteratively, based on knowledge of the TDS. A preliminary "quick and dirty" microassessment can be invaluable for providing an initial context within which to commence the search for impacts—a search that will necessarily be refined as the assessment unfolds.

Strategies and techniques for impact identification must be well chosen so as to discover not only the immediate, or first-order, impacts, but also to trace the delayed and unexpected impacts that flow from them.

The difficulty of impact identification is further compounded because the decision-making activities of the TDS are decentralized and semiautonomous. This makes the decision-making process difficult to predict, since decisions are made under conditions of uncertainty by individuals who may not interact with one another directly (Wenk and Kuehn 1977).

Nearly all impact identification exercises, regardless of the technique employed, produce more impacts than can be dealt with in the analysis phase, given the usual constraints of available time and resources. Thus the assessment team must select the "important" impacts to be analyzed. The determination of which impacts are most important is of course "a judgment call." That judgment must be based on a sound understanding of the TDS, the sponsor, and the interests of potential users of the TA/EIA, as well as estimates of the probability, timing, and extent of the impacts themselves.

Policy considerations should be included early in the TA/EIA activity. We treat identification of policy concerns together with that of impacts because policy is the means for modifying impacts, and because impact identification suggests the types of policy options available. If policy concerns are left until after treatment of impacts is complete, as is too often the case, policy analysis appears as an "add-on" that is poorly related to the impacts. As impacts are identified, team members should consider who may make relevant policy and the range of possible actions (but they should not unduly constrain the impact analysis to fit perceived policy courses). This lays the foundation for subsequent analysis in the TA/EIA.

This chapter is constructed so that the elements of the impact identification task appear in more or less the same order that they would be encountered in a typical assessment. Impact identification strategies and suggested criteria for selecting an appropriate strategy are considered first. Specific techniques for identifying impacts are treated next, followed by a discussion of technique selection. Finally, selection of an appropriate range of impacts and considerations of policy are developed in the closing sections of the chapter.

8.2 IMPACT IDENTIFICATION STRATEGIES

What if? . . . What then? . . . Who would win? . . . Who would lose?

Impact identification is largely a process involving the systematic application of imagination and intuition. This does not mean that wild guesses will do! Impacts that can neither be supported by compelling rationale nor substantiated by causal relationships destroy the credibility of an assessment as a decision making aid. However, it does mean that there is no "sure-fire" algorithm for impact identification.

Strategy lays down the grand structure within which the imagination and intuition of the assessment team are to function; specific techniques provide the formal schemes for their systematic application. To gain perspective on the strategies for impact identification, we first consider three key dimensions: (1) reductionist vs holistic perspectives, (2) scanning vs tracing approaches, and (3) internal assessment team vs external input to the identification process.

8.2.1 Some Dimensions

Perhaps the most fundamental decision faced by an assessment team is the question of whether to view the impact field as a whole or to subdivide it into constituent parts. The decision to follow one approach or the other will imply different strategies and techniques for the identification task.

The most commonly employed approach subdivides the complex impact field into smaller, more easily examined sectors. Such strategies are referred to as *reductionist*. Often the sectors are chosen so as to coincide closely with the disciplinary areas represented on the assessment team. The rationale for this arrangement is that team members are most comfortable and productive when dealing with material in their own area of expertise. One such division, used in Chapters 10–13, is suggested by our acronym EPISTLE, which divides impacts into:

Environmental,
Psychological,
Institutional/political,[1]
Social,
Technological,
Legal, and
Economic.

Other variants have been used in assessments. Dee et al. (1972) utilize ecology, environmental pollution, aesthetics, and human-interest sectors; an Arthur D. Little, Inc. transportation study (1971) suggests noise, air quality, water quality, soil erosion, ecologic, economic, and sociopolitical sectors. The Multiagency Task Force (1972), seeking to establish guidelines for multiobjective water-resource planning, has advanced sectors dealing comprehensively with biological, physical, cultural, and historical resources, and pollution factors. The "granddaddy of them all" in terms of completeness, however, is a breakdown prepared by Leopold et al. (1971) that divides environmental impacts, biological conditions, cultural factors, and ecological relationships into 88 separate subdivisions (Table 8.2).

Division of tasks according to disciplinary sectors has disadvantages as well as advantages. It can lead to a fragmented, multidisciplinary assessment, rather than interdisciplinary study, unless strong pressures to integrate are imposed on the study team.

Another approach involves the division of impacts by *impacted parties*. This arrangement means that the impacted individuals, groups, and institutions are the focus. For example, the TA on electronic funds transfer (Arthur D. Little, Inc. 1975) divided the impacts by parties at interest into four general categories of business (depository financial institutions; nondepository financial institutions; consumer oriented, nonfinancial businesses; and nonconsumer oriented, nonfinancial businesses), three categories of individual, and four roles of government for categorizing governmental bodies and agencies. Obviously, a categorization of potential impacts by the affected parties would have to be customized for each assessment.

It seems clear that no one scheme for subdivision is adequate for every assessment task or sponsor requirement. For instance, in some cases it may be fruitful to consider major societal values or problems

[1]Military and/or strategic considerations could also warrant explicit consideration.

and their intersection with the development in question (e.g., how would the development affect U.S. energy supply?) It is also apparent that initial sector decisions will frequently need revision as the identification process proceeds and the dimensions of the problem unfold. The need to integrate impacts across sector boundaries should be anticipated.

Reductionist strategies are, of course, not without difficulties. Significant impacts may fall through the cracks between sectors. Furthermore, sooner or later, impacts from the various sectors must be reassembled into a continuum, cross-linkages identified, and higher-order impacts traced. The linkages between primary and higher-order impacts may be distorted by the sector perspective from which they have been viewed.

Difficulties with reductionist strategies have led to the suggestion that *holistic* strategies be employed to view the impact field as a whole. Holistic impact identification typically consists of allowing the field of impacts to emerge as the study develops in an informal, ad hoc manner without the use of prestructured categories.

Two major problems attend the use of holistic impact identification. First, the emergence of major issues may be very slow unless stimulated either by structured group consideration of impact significance (e.g., formal or informal modeling) or the pressure of rigidly enforced decision deadlines. Such failure to focus can lead to cost and time overruns and/or failure to identify significant impacts. Second, the unstructured approach implied by these strategies can adversely affect the morale and efficiency of team members accustomed to highly structured approaches or uncomfortable with uncertainty.

A second strategic dimension in addressing impacts is the distinction between *scanning* and *tracing* techniques discussed in Section 5.4.3. Scanning techniques are characterized by direct, single step identification of significant impacts. Tracing techniques, on the other hand, generally rely on construction of structural relationships between the various elements of the impact field.

The primary emphasis of scanning methods is on schemes for searching the impact field so as to minimize the probability that significant impacts will be overlooked. Although these methods rely strongly on the assessor's intuition, insight, and background knowledge, they usually provide a framework within which to exercise those qualities (e.g., a checklist or matrix to systemize identification). These methods, however, stop short of structuring the interrelationships that have been identified.

Tracing methods emphasize development of the structure within the impact field, and between technology and policy sectors and the impact field. This structure may be expressed as a formal model or as a "chain" of causes and impacts that are related logically and/or chronologically. Clearly, once initial elements have been identified, the structural continuity of tracing methods greatly facilitates the search for additional impacts.

There is no reason why a mixture of scanning and tracing techniques cannot be utilized, provided that time and resources permit. Scanning techniques might be applied, for example, to locate impacts and identify interrelationships among them. A tracing technique might then be used to develop a structure connecting the related impacts and to extend the impact search. The application of more than one identification method will, in general, enrich the resulting impact set.

The utilization of human resources by the assessment team is the final strategic dimension we will consider. (The important dimension of quantification is considered under impact analysis.) An assessment team may decide to *internalize the identification process* or elect to *utilize outside resource persons* to play a role in the task. The choice depends on such factors as time, resources, opportunity, and, to some extent, the nature of the technology being assessed.

It is generally wise to supplement team perspectives of the impact field with those of individuals intimately concerned with the technology or its potential effects. Contact with such groups can be accomplished in a number of ways, including workshops, conferences, individual consultation, and surveys. Frequent consultation is more productive than a single exchange of information.

In most instances, stakeholder and expert groups will be readily identifiable. Where identification is not immediate, sources such as the technical literature, newspaper records, and comments on EISs may provide leads.

There are some problems in using resource people. Individuals may articulate personal views as well as those of the group they represent. Therefore, contact with several representatives of a group is desirable as a means for cross-checking perspectives. In many instances, particularly where futuristic systems are involved, it may be difficult for individuals to anticipate the impacts that the technology may produce. Workshops and conferences designed to move people "up the learning curve" are valuable here. Finally, some stakeholders who will be strongly impacted, such as consumers, lack sufficient group identity and cohesiveness to state their perspective strongly and articulately. Their composition may be so diffuse that truly representative individuals may not be identifiable.

The assessment team may, on the other hand, choose to rely on its internal expertise to generate impacts. This decision may be motivated by time and resource constraints. In specific instances, internalization may be necessary to ensure the independence of the assessment team and guarantee the credibility of the study.

8.2.2 Criteria for Constructing a Strategy

The strategy for a specific assessment can be assembled from the dimensions discussed in Section 8.2.1. Adoption of a reductionist or holistic view of the totality of impacts, the use of scanning or tracing techniques, and the utilization of external or internal personnel strategy all depend on the characteristics of the technological development under assessment, the impacts, and the team members; sponsor and study user requirements; and resource and study time constraints.

The technology description (Chapter 6) must either precede or parallel impact identification. The technology's characteristics will influence the selection of method, quantification, and resource personnel. For example, if the technology to be assessed is a relatively well-defined hardware system (e.g., alternative automotive propulsion systems), tracing methods can probably be utilized to advantage. Scanning methods could then be less intensively applied to provide cross-checks for completeness. A complex social technology (e.g., a welfare program), on the other hand, can be expected to produce a wide variety of diffuse "out of context" impacts. Here the emphasis could be placed on scanning techniques with lesser attention to tracing methods.

The characteristics of the impacts themselves will shape strategy. In particular, the balance between impacts treated quantitatively and qualitatively deserves comment. We caution that the identification of impacts not be skewed toward those that are quantifiable. Assessors should avoid the tendency to denigrate qualitative impacts. There may too often be an analogy between assessors emphasizing the quantifiable and the drunk who looks for lost keys beneath a street lamp: not because they are there, but because the light is better.

Perhaps no single item will affect the selection of strategy elements more pervasively than the characteristics of the study team. The attitudes of individual members, their willingness to participate in the "give and take" of the group dynamic, their ability to work outside as well as within their disciplines, and especially their capacity to deal with uncertainty are all important determinants. If team members are unable or unwilling to venture outside their areas of disciplinary expertise, for example, the use of a reductionist view is probably mandatory.

This then favors emphasis on tracing methods to integrate across sector boundaries.

Criteria for EISs and, hence, sponsor requirements are relatively specific. For many assessments, however, frequent contact with the sponsor will be necessary to define major portions of the study requirements. If carefully examined, these requirements will strongly influence most elements of the strategy. The assessor should determine, as far as possible, what information the sponsor and potential study users need to know.

Time and resource constraints are not peculiar to TA/EIA studies. The iterative nature of TA/EIA and the degree of uncertainty that usually surrounds an assessment, however, place especially severe demands on the team to budget time and resources carefully. Elaborate impact identification procedures (e.g., modeling and extensive quantification) may not be worth their cost. The most perceptive identification of impacts is a detriment, for example, if it consumes the time and resources necessary for their thorough analysis. In particular, policy relevance should be kept in mind (see Section 8.5).

There is often an urge to compile an exhaustive list of impacts. This urge must be tempered by available time and resources. Generally a few significant impacts, well developed and analyzed, and an indication of potential "sleepers" will be of more use in decision making than a more extensive, but less thoroughly considered, set of impacts.

8.3 IMPACT IDENTIFICATION TECHNIQUES

Techniques for impact identification are of two basic types: scanning or tracing (recall Section 8.2). The techniques described in the following paragraphs are representative of these types. We emphasize techniques that are relatively simple and are potentially useful for impact identification in TA/EIA.

8.3.1 Scanning Techniques

The simplest scanning technique is the *checklist*, which is merely a listing of potential impacts. The checklist is employed as a guide for the assessor to ensure a more or less exhaustive impact search. A checklist may frequently be nothing more than a very detailed listing of the sectors chosen for a reductionist strategy. Tables 8.1 and 8.2 [from Leopold et al. (1971)] list 100 possible actions on the environment and 88 potential impacts—a truly monumental compilation.

TABLE 8.1. Possible Actions [a]

A. Modification of regime
- a. Exotic flora or fauna introduction
- b. biological controls
- c. Modification of habitat
- d. Alteration of ground cover
- e. Alteration of groundwater hydrology
- f. Alteration of drainage
- g. River control and flow modification
- h. Canalization
- i. Irrigation
- j. Weather modification
- k. Burning
- l. Surface of paving
- m. Noise and vibration

B. Land transformation and construction
- a. Urbanization
- b. Industrial sites and buildings
- c. Airports
- d. Highways and bridges
- e. Roads and trails
- f. Railroads
- g. Cables and lifts
- h. Transmission lines, pipelines, and corridors
- i. Barriers, including fencing
- j. Channel dredging and straightening
- k. Channel revetments
- l. Canals
- m. Dams and impoundments
- n. Pies, seawalls, marinas, and sea terminals
- o. Offshore structures
- p. Recreational structures
- q. Cut and fill
- r. Tunnels and underground structures

C. Resource extraction
- a. Blasting and drilling
- b. Surface excavation
- c. Subsurface excavation and retorting
- d. Well drilling and fluid removal
- e. Dredging
- f. Clear cutting and other lumbering
- g. Commercial fishing and hunting

D. Processing
- a. Farming
- b. Ranching and grazing
- c. Feed lots
- d. Dairying
- e. Energy generation
- f. Mineral processing
- g. Metallurgical industry
- h. Chemical industry
- i. Textile industry
- j. Automobile and aircraft
- k. Oil refining
- l. Food
- m. Lumbering
- n. Pulp and paper
- o. Product storage

E. Land alteration
- a. Erosion control and terracing
- b. Mine sealing and waste control
- c. Strip mining rehabilitation
- d. Landscaping
- e. Harbor dredging
- f. Marsh fill and drainage

F. Resource renewal
- a. Reforestation
- b. Wildlife stocking and management
- c. Ground water recharge
- d. Fertilization application
- e. Waste recycling

G. Changes in traffic
- a. Railway
- b. Automobile
- c. Trucking
- d. Shipping
- e. Aircraft
- f. River and canal traffic
- g. Pleasure boating
- h. Trails
- i. Cables and lifts
- j. Communication
- k. Pipeline

(continued)

TABLE 8.1 *(continued)*

H. Waste emplacement and treatment
  a. Ocean dumping
  b. Landfill
  c. Emplacement of tailings, spoil, and overburden
  d. Underground storage
  e. Junk disposal
  f. Oil well flooding
  g. Deep well emplacement
  h. Cooling water discharge
  i. Municipal waste discharge including spray irrigation
  j. Liquid effluent discharge
  k. Stabilization and oxidation ponds
  l. Septic tanks, commercial and domestic
  m. Stack and exhaust emission
  n. Spent lubricants

I. Chemical treatment
  a. Fertilization
  b. Chemical deicing of highways, etc.
  c. Chemical stabilization of soil
  d. Weed control
  e. Insect control (pesticides)

J. Accidents
  a. Explosions
  b. Spills and leaks
  c. Operational failure

Others
  a.
  b.

[a] From Leopold et al. (1971).

TABLE 8.2. Possible Environmental Impacts [a]

A. Physical environment
1. Earth
  a. Mineral resources
  b. Construction material
  c. Soils
  d. Landform
  e. Force fields and background radiation
  f. Unique physical features

2. Water
  a. Surface
  b. Ocean
  c. Underground
  d. Quality
  e. Temperature
  f. Recharge
  g. Snow, ice, and permafrost

3. Atmosphere
  a. Quality (gases, particulates)
  b. Climate (micro, macro)
  c. Temperature

4. Processes
  a. Floods
  b. Erosion
  c. Deposition (sedimentation, precipitation)
  d. Solution
  e. Sorption (ion exchange, complexing)
  f. Compaction and settling
  g. Stability (slides, slumps)
  h. Stress–strain (earthquake)
  i. Air movements

B. Biological conditions
1. Flora
  a. Trees
  b. Shrubs
  c. Grass
  d. Crops
  e. Microflora
  f. Aquatic plants
  g. Endangered species
  h. Barriers
  i. Corridors

(continued)

TABLE 8.2 *(continued)*

2. Fauna
   a. Birds
   b. Land animals including reptiles
   c. Fish and shellfish
   d. Benthic organisms
   e. Insects
   f. Microfauna
   g. Endangered species
   h. Barriers
   i. Corridors

C. Cultural factors

1. Land use
   a. Wilderness and open spaces
   b. Wetlands
   c. Forestry
   d. Grazing
   e. Agriculture
   f. Residential
   g. Commercial
   h. Industrial
   i. Mining and quarrying

2. Recreation
   a. Hunting
   b. Fishing
   c. Boating
   d. Swimming
   e. Camping and hiking
   f. Picnicking
   g. Resorts

3. Aesthetics and human interest
   a. Scenic views and vistas
   b. Wilderness qualities
   c. Open space qualities
   d. Landscape design
   e. Unique physical features
   f. Parks and reserves
   g. Monuments
   h. Rare and unique species or ecosystems
   i. Historical or archaeological sites and objects
   j. Presence of misfits

4. Cultural status
   a. Cultural patterns (lifestyle)
   b. Health and safety
   c. Employment
   d. Population density

5. Man-made facilities and activities
   a. Structures
   b. Transportation network (movement, access)
   c. Utility networks
   d. Waste disposal
   e. Barriers
   f. Corridors

D. Ecological relationships such as:
   a. Salinization of water resources
   b. Eutrophication
   c. Disease–insect vectors
   d. Food chains
   e. Salinization of surficial material
   f. Brush encroachment
   g. Other

Others
   a.
   b.

[a] From Leopold et al.)1971).

A checklist suitable for a specific study can be constructed "from scratch" or borrowed in whole or part from numerous lists appearing in existing assessments.[2] Brainstorming (Chapter 1), Delphi technique, panels, and surveys (Chapter 6) have all been used to generate elements for the construction of checklists.

A *matrix* formulation is similar to a checklist, except that two or more dimensions are required for its representation. These added dimensions provide increased capacity to represent interdependence of

Footnote 2 on next page.

impacts and their causal relationship to the technology being assessed. For example, if the Leopold lists (Tables 8.1 and 8.2) are arranged vertically and horizontally on a page, open cells are created at the intersections of the various rows and columns (see Figure 8.1a for an example). The interrelationship between each action and each impact can be displayed within each cell. Also, by arraying the same categories of impacts on both axes of the matrix, one can search out interaction effects (i.e., higher-order impacts) (see Figure 8.1b for an example).

Several different representations can be used within cells to indicate a relationship. For example, Sage (1977) suggests that a diagonal mark ⧄ be used to indicate a mild interaction and an "X" ⊠ to indicate a more severe one. As an alternative to symbolic representation, brief phrases could be placed in the cell to describe characteristics of the interaction. Alternatively, subjectively or analytically quantified information could be included in each cell to indicate the magnitude, importance, and/or direction of the interrelationship (e.g., −10, very detrimental, to +10, very beneficial). Figure 8.1a shows a matrix representing causal relationships between selected phosphate mining activities and environmental impacts. In this representation, a number from 1 to 10 is placed on the left-hand side of the cell to indicate the severity of the impact. The number on the right-hand side selected from the 1−10 scale portrays the impact importance. Clearly, the validity and utility of this scaling needs to be established before it is used. Such a numerical approach displays more information than do symbols. Care should be exercised in using and interpreting such information, however, as the numerical form of presentation may disguise the imprecision and subjectivity inherent in its determination.

Other sorts of checklist can be utilized in the two-dimensional matrix formulation. One is a vertical stakeholder listing coupled with a horizontal display of technology-related activities. Such a matrix can be used to identify the distribution of costs and benefits of a technology.

Matrices with more than two dimensions are possible, but graphical

---

[2]Numerous checklists and matrix representations are available, especially from EIA studies. These representations are often specific to a particular assessment task. For example, for water-resource studies, see Dee et al. (1972, 1973), Central New York Regional Planning and Development Board (1972), Multiagency Task Force (1972), and Tulsa District, U.S. Army Corps of Engineers (1972); for highway and transportation studies, see Adkins and Burke (1971), Institute of Ecology (1971), Kranskopf and Bunde (1972), Arthur D. Little, Inc. (1971), McHarg (1968), Walton and Lewis (1971); for energy and power plant considerations, see Western Systems Coordinating Council (1971), U.S. Atomic Energy Commission (1972), Tamblyn and Cederborg (1975), Jopling (1974); for development, see Moore et al. (1973), Sorensen (1971), Stover (1972), Wirth and Associates (1972), Nathans and Associates (1972), and Burchell and Listokin (1975).

Possible environmental impacts (from Table 8.2) \ Possible actions (from Table 8.1)	B.d. Highways and bridges	C.c. Surface excavation	D.f. Mineral processing	H.c. Emplacement of tailings
A.2.d. Water quality		2/2	1/1	2/2
A.4.b. Erosion	2/2	1/1		2/2
B.1.f. Aquatic plants		2/2		2/3
C.3.h. Rare and unique species	2/5	2/4	5/10	

(a)

Possible environmental impacts (from Table 8.2) \ Possible actions (from Table 8.1)	A.2.d. Water quality	A.4.b. Erosion	B.1.f. Aquatic plants	C.3.h. Rare and unique species
A.2.d. Water quality				
A.4.b. Erosion				
B.1.f. Aquatic plants				
C.3.h. Rare and unique species				

(b)

FIGURE 8.1. Sample environmental impact matrices concerning a proposed phosphate mining lease: (a) partial impact, (b) partial cross impact. Note: The number on the left of a cell represents the severity of the impact; that on the right indicates the impact importance. These are scaled from 1 (least severe or important) to 10 (most severe or important).

representation becomes increasingly difficult as the number of dimensions increases. Usually matrices of more than two dimensions are displayed as two or more two-dimensional matrices sharing a common axis. Subdivision of matrix row or column entries into more detailed matrices is possible as well. In fact, if the reader will allow a bit of license, larger matrices have smaller matrices that feed upon their vagaries, smaller matrices have smaller matrices, and thus on to infinity.

Information displayed in matrix cells can be generated in various ways. It can be produced by the assessment team from personal knowledge or literature research. Alternatively, the entries can be obtained through interaction with experts and/or stakeholder groups. Whatever means are used, the time, effort, and resources required to "fill in" a matrix increase dramatically with the detail of the checklists employed.

As noted earlier, matrices can also be used to indicate the existence of higher order impacts and causal relationships (see also Section 9.3). Thus matrices provide a starting point for understanding the tracing techniques presented in the next section.

8.3.2 Tracing Techniques

The *relevance tree* is the simplest and most commonly employed tracing method. Relevance trees graphically represent the interrelationships (or linkages) between various members of some set of elements. Trees can be considered as graphical outlines that proceed more or less linearly from the general to the specific. Relevance trees are closely related to the decision trees utilized in decision analysis and share many characteristics with those representations (see Section 14.4.2). Perhaps the most common examples of trees are family trees and the organizational chart. In the latter, the various levels represent a hierarchy of authorities; in the former, family lineage is displayed, more or less chronologically.

Several different trees may be utilized for impact identification. The *impact tree* is used to portray the causal relationships between some activity of the TDS and first-order impacts, and from first-order to second-, third-, and higher-order impacts. The definition of higher-order impacts as "impact of impacts" is thus explicitly displayed by the tree structure.

To construct a tree, the assessor must ask the question "What if?" at each node. At the first level, for example: "What if an activity were pursued? . . . What would be affected?" Answers to these questions can be supplied from any or all of several sources, including personal knowledge of the development process involved, literature research, analogy to similar activities, or consultation with experts or stakeholder groups.

A tree representing impacts from the excavation activities of surface coal mining has been partially displayed in Figure 8.2. The branches originating at the node of an impact tree are not necessarily an exhaustive or mutually exclusive set. (Impact trees differ from decision trees in this characteristic.) At the node on the first level of this tree the question was posed, "What would happen if surface excavation were begun?" As a result of this activity, for example, the basic ground form would be disturbed, coal removed, people employed, and any archeological sites or objects in the excavation area would be disturbed. Next, "What if the basic land form were disturbed?" This could lead to alteration of scenic views, the loss of land for agricultural production, and changes in drainage patterns. Changes in drainage patterns could, in turn, lead to erosion, acid drainage from materials uncovered, and so forth. Thus in the tree arrangement each impact is causally related to those with which it is connected.

Impact trees can be developed a step further by assigning subjective probabilities of occurrence to each impact. Such estimations could also be made of the timing and severity of each impact. The sum of the probabilities estimated for the events emanating from a particular node need not be unity since the impact list on the next level may not be exhaustive or mutually exclusive. In addition, the same impact may have more than one cause, thereby intertwining the branches.

Other types of relevance tree can also be useful in impact identification. For example, a tree representing activities that are necessary to support the basic technology could be constructed. A tree depicting the technologies necessary to support sea farming (Landis 1971) is shown in Figure 8.3. Each branch of this tree could be examined in turn to determine impacts arising from it. Another variation of the relevance tree would be a stakeholder tree. This would involve, for instance, a chaining of development activities together with affected parties.

The advantage of the tree representation is that it simultaneously displays both the *linkages* between activities and impacts and the *scope* of the impact field. Further, the structured framework of the method tends to make the search for impacts self-propagating. Disadvantages include the fact that early omissions may bypass sections of the field, leaving some impacts undetected. In addition, the process requires a fairly extensive knowledge of the structure of the impact field, a knowledge that may take time and resources to develop.

In addition to relevance trees, tracing methods include conceptual and mathematical models (Chapter 9) and input–output analysis

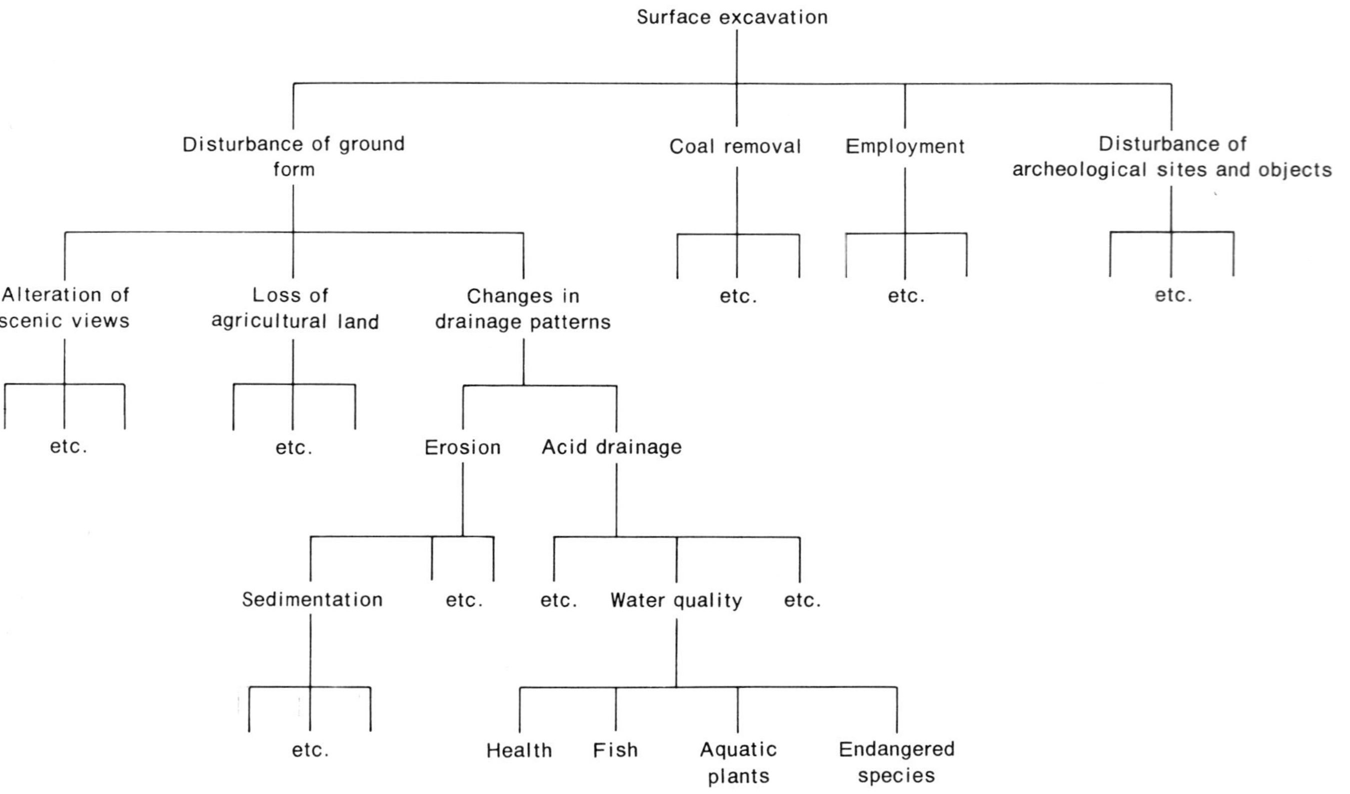

FIGURE 8.2. Hypothetical impact tree for surface coal mining.

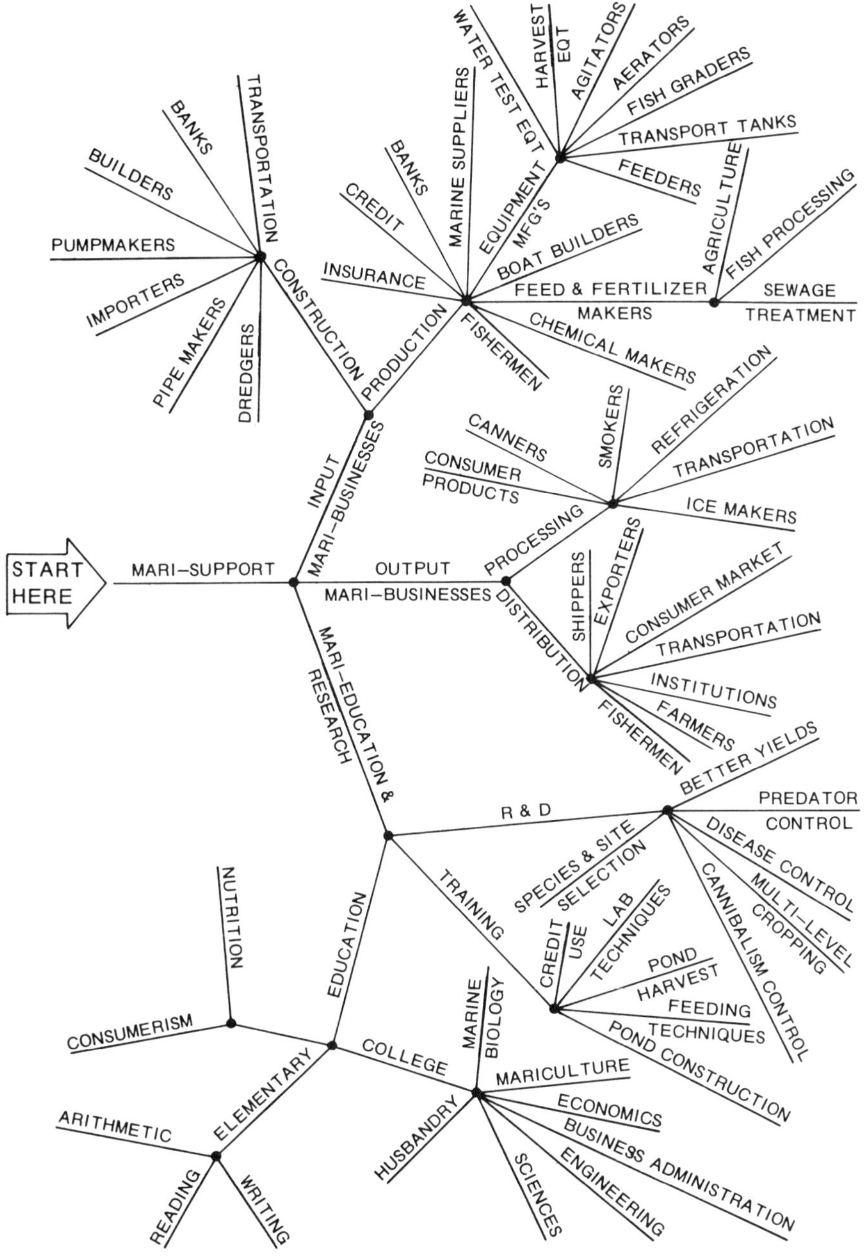

FIGURE 8.3. Relevance tree displaying mari(marine)-culture supporting technologies.
Source: Landis (1971).

(Chapter 11). The interested reader can translate these from the context in which they are presented to impact identification. Table 8.3 lists some additional impact tracing (and scanning) techniques.

Some methods combine characteristics of both scanning and tracing methods; morphological analysis is one. Although morphological analysis is designed to provide an exhaustive search (using scanning techniques), it nevertheless assumes a field of inquiry that is composed of structurally related elements (as do tracing techniques). The cameo entitled "Morphological Analysis" explains this technique [see also Ignatovitch (1974)].

8.3.3 Criteria for Selecting Techniques

The strategic considerations of Section 8.2 form a basis for the choice of techniques. To facilitate the selection of specific techniques, general characteristics of various techniques are summarized in Table 8.3, where we emphasize general methods of potential applicability for impact identification in TA/EIA studies. Table 8.3 allows the compari-

TABLE 8.3. Characteristics of Selected Impact Identification Techniques [a]

METHOD	Tracing/scanning method	Time requirements	Resource requirements	Use of external resource people	Degree of input quantification	Degree of output quantification	Chapter where discussed
Modeling	T	H	H	*	H	H	9
Surveys	S	*	*	Required	L	*	6
Delphi	S	M–H	*	Required	L	L	6
Literature search	S/T	*	L	L	L	*	
Historical analogies	T	L	L	L	L	L	7
Scenarios	T	L	L	M	L	L	7
Matrix approaches	S	M	M	*	L	L	8, 9
Checklists	S	L	L	L–M	L	L	8
Morphological analysis	S/T	M	M	*	*	*	8
Relevance trees	T	*	M	M	L	L	8, 14
Brainstorming	S	L	L	M	L	L	1

[a] Legend: S, scanning method; T, tracing method; L, low; M, medium; H, high; *, variable—depending on scope or intent of application, or expertise and composition of team.

son of the character of the technique, time and resource requirements, its use of human resources, and its information characteristics for a number of techniques discussed in this book. Techniques specific to particular disciplines, which may be utilized for impact identification, are not listed in the table.

We recommend that, if at all possible, more than one approach should be chosen initially. This gives the assessor a degree of flexibility should problems arise. It also provides the opportunity to compare techniques as the study develops. In general, different techniques will highlight differing types of impacts. Thus multiple-technique approaches hold the promise of richer and more complete identification.

8.4 SELECTIVITY

Assessment time and resources are seldom sufficient to allow more than a handful of impacts to be pursued thoroughly. Furthermore, an extensive impact list can be overwhelming or confusing to a decision maker, particularly if it mixes impacts of varying significance. Therefore, the crucial problem in impact identification is frequently not the compilation of impacts, but rather the determination of which are most significant.

Before attempting to single out significant impacts, the assessor should be certain that some order or systemization has been imposed on the impact set. If one or more of the techniques described in Section 8.3 has been used in the identification process, that order is assured. If, however, the holistic approach has been followed and significant impacts have simply been allowed to emerge during the process, some ordering principle may need to be applied post hoc to provide structure.

To aid in determining the major impact areas to be examined, the assessors should consider such factors as sponsor and study user requirements, stakeholder perspectives, the characteristics of the impacts and technology, possible policy options, and time and resource constraints on analysis. Berg (1975a) has suggested criteria for bounding, four of which may be applied to selecting impacts for extensive consideration. These criteria are centrality, resource limitations, cognitive limitations, and political factors.

Centrality reflects the assessor's judgment as to which impacts are most critical to the sponsor, stakeholders, and policy makers. Centrality is frequently difficult to judge from the perspective of the assessment team alone, even when a thorough consideration of the sponsor's requirements has been made. Contact with experts and representatives of

MORPHOLOGICAL ANALYSIS

The morphological approach requires that the field that is to be searched satisfy a number of conditions (Ignatovich 1974). One condition is that sufficient "pegs of knowledge" (i.e., concepts or principles) exist within the field to allow the analysis to be developed. More restrictive for most applications, however, is the requirement that the problem must be clearly defined and the essential parameters exhaustively identified.

Perhaps the best way to understand morphological analysis is to examine an application. Therefore, the five steps of morphological analysis will be considered using Zwicky's enumeration of possible types of jet engines (1962, 1969). Zwicky's original formulation in 1943 identified the possibility of the V-2 rocket-propulsion system then under development by Germany, a system discounted by Churchill's scientific advisor because it utilized liquid rather than solid fuel propellent.

Step 1: *Concisely define the problem.* What is the totality of all pure medium jet engines containing single sample elements and activated by chemical energy?

Step 2: *Localize and analyze all parameters.* Zwicky identified the following parameters and alternative discrete values of their elements:

Parameter		*Elements*	
P_1	Source of chemically active mass	P_1^1	Intrinsic, P_1^2 extrinsic
P_2	Location of thrust generation	P_2^1	Internal, P_2^2 external

the sponsor, stakeholder, and policy making groups may be useful in this regard.

A graphical approach to judging centrality consists of representing impacts in a three-dimensional array. The coordinates of each impact are specified by a set of three numbers representing subjective judgments of their severity, timing, and probability. Depending on the assessment context, certain regions of the array will generally be more central to the study than others. Impacts that are judged as severe, near

P_3	Source of thrust augmentation	P_3^1	Intrinsic, P_3^2 extrinsic, P_3^3 zero
P_4	Location of thrust augmentation	P_4^1	Internal, P_4^2 external
P_5	Jet	P_5^1	Positive, P_5^2 negative
P_6	Possible thermal cycle	P_6^1	Adiabatic, P_6^2 isothermal, . . . , P_6^4
P_7	Medium	P_7^1	Vacuum, P_7^2 air, P_7^3 water, P_7^4 earth
P_8	Motion	P_8^1	Translatory, P_8^2 rotary, . . . , P_8^4
P_9	State of propellent	P_9^1	Gaseous, P_9^2 liquid, P_9^3 solid
P_{10}	Type of operation	P_{10}^1	Continuous, P_{10}^2 intermittent
P_{11}	Igniting of propellent	P_{11}^1	Self, P_{11}^2 nonself

Step 3: *Construct the "morphological box" (i.e., multidimensional matrix containing all potential solutions).* There are 36,864 possible solutions for the problem posed (two alternative values for $P_1 \times$ two alternative values for $P_2 \times \cdot \cdot \cdot$).

Step 4: *Evaluate all possible solutions contained in the box with respect to the purposes to be achieved.* The unreasonable solutions are eliminated from the potential solutions (e.g., combinations containing both P_1^2 and P_7^1, extrinsic source of fuel when operating in a vacuum), leaving 26,880 possibilities.

Step 5: *Select the optimal solutions and apply practically, provided the necessary means are available.* The technical feasibility of individual solutions and sets of solutions is evaluated.

Morphological analysis could, at least in theory, be applied to search the impact field in an exhaustive fashion. The chief limitation to this application would seem to be that of obtaining a clear problem definition and exhaustive identification of parameters.

term, and highly probable, for example, may be judged as most central.

Each impact that passes the centrality test should be considered to determine the difficulty of analysis and availability of required information. In this fashion *resource requirements* could be estimated and compared to limitations. Impacts requiring resources beyond those available would necessarily be excluded from in-depth analysis. However, they should be flagged for policy making consideration and possible future study.

Impacts that satisfy centrality and resource-limitation criteria should be considered in the light of *cognitive limitations* that may exist. As impact chains or cause and effect linkages are pushed farther and farther from their origins, uncertainties compound. These uncertainties are the products of information and understanding limitations, as well as methodological imperfections. The analysis of impacts stretched beyond these limitations can flaw the credibility of the assessment.

The final criterion, *political factors,* addresses two different considerations. First, if the TA/EIA study "is intended to affect policy, to alter institutions, to redistribute impacts, it has political content" (Berg 1975b). In striving to make political content more effective, it seems clear that certain impacts will be more significant than others. Second, given the realities of resource constraints, some impacts must of necessity be excluded. Which are to be included, and which neglected?

It is impossible to overemphasize the importance of exercising selectivity in the preparation of the final impact set. Although this will be a task attended always by ambiguity and uncertainty, the criteria described in this section can provide guidance in its execution.

8.5 POLICY CONSIDERATIONS

The focus of assessment should be policy. Assessment results will generally be useful to sponsors only insofar as they delineate effective policy alternatives.

Unfortunately, policy consideration has historically been among the weakest links of the assessment chain. There have been notable exceptions, however, such as the TA on offshore oil and gas by Kash et al. (1973). There are, of course, reasons for this general failing. Not the least of these is the complexity of the task. This complexity is compounded, however, by a tendency to begin policy analysis only after the rest of the study has been completed. Sometimes this activity has been initiated only when funding is running out and team members have left the TA/EIA for other tasks. Effective policy analysis should be an activity that is integral to the entire assessment effort, and sufficient time and resources should be provided for its completion. We recommend, therefore, that policy consideration commence no later than the point at which significant impacts are selected for further analysis. From this point on, policy analysis should proceed hand-in-hand with impact analysis and evaluation.

We ask the reader now to turn to Section 15.4.3 and, especially, to Figure 15.1 (p. 389). As discussed there, the policy system comprises the set of potential actions substantively affecting the subject of the assessment and the actors having an interest in these actions. This policy system should be identified as the impacts are being identified. Figure 15.1 graphically displays the interaction between developing knowledge of impacts and that relating to policy. As the study of impacts develops in Chapters 9–14, it is well to keep in mind the flow of events described in Figure 15.1 so that policy considerations will be intimately joined to the study of impacts.

In identifying the policy system, some of the techniques listed in Table 8.3 may prove effective. Modeling, excluding qualitative models of social decision making processes, is probably not very useful for this task. Input from persons involved in decision making on the subject of the assessment is a very effective starting point. Assessors can formulate the policy system by working outward from the initial suggestions of policy makers until nothing new and significant emerges.

8.6 CONCLUDING OBSERVATIONS

In impact identification the assessor should not only seek to identify all impacts that would proceed from a technological activity, but also select those that would be most critical. Impact identification should proceed iteratively with policy analysis activities as each contributes to sharpening the perspective of the impact field.

The identification task is largely systematic application of imagination and intuition for which no sure algorithm can be set down. Therefore, strategy and techniques must be carefully chosen so as to enhance team imagination and intuition without imposing debilitating constraints.

Differences in sponsor requirements, the technology being assessed, and team characteristics mitigate against a single appropriate strategy for all assessments. However, strategic considerations involve:

- type of perspective—reductionist or holistic;
- selection of identification techniques for scanning and/or tracing the impact field; and
- choice of resource people to employ in impact identification—only those on the project team, or persons external to the team as well.

Successful strategy must blend these decisions with the following study characteristics and constraints:

- characteristics of the technology;
- characteristics of the impacts;
- characteristics of the team members;

sponsor and potential study user requirements;

resource constraints; and

study time constraints.

There is a vast array of techniques for impact identification; however, most can be classified as being one of two basic types: scanning or tracing.

Scanning techniques are characterized by direct, single-step impact identification.

Tracing techniques rely on schemes for structuring relationships among impacts.

Regardless of the strategy and search techniques employed, however, more impacts will generally be identified than study time and resource constraints allow to be thoroughly analyzed. Therefore, the assessor must be selective as to which are chosen for analysis. The assessor may wish to augment team perspectives with those of outside experts, stakeholders, and/or policy makers in making this selection. During this selection, each impact should be judged on the basis of the following criteria:

centrality (significance in the context of the study task);

resource limitations;

cognitive limitations (information and understanding limitations and methodological imperfections); and

political factors (internal and external value considerations).

The selection of a subset of impacts from those that have been identified is an activity having profound implications for the integrity of the entire assessment. Although it will be a task attended always by ambiguity and uncertainty, the suggested criteria can guide its performance.

Assessment results will generally be useful to sponsors only insofar as they delineate effective policy alternatives. Unfortunately, policy analysis has historically been among the weakest links of the assessment chain. Many of the problems associated with this task can be mitigated, however, by beginning the consideration of policy early in the assessment. We suggest that these considerations begin during the selection of impacts for further study and proceed concurrently with the remainder of the assessment. We also suggest that most assessment teams will profit in this regard from the perspectives of individuals who are directly involved in the policy-formulation process.

RECOMMENDED READINGS

Berg, M. R., Chen, K., and Zissis, G. J. (1975) "Methodologies in Perspective," in *Perspectives in Technology Assessment* (Sherry R. Arnstein and Alexander N. Christakis, eds.). Jerusalem: Science and Technology Publishers, pp. 21–44.

Offers good descriptions and comparisons of basic methods of impact identification.

Ignatovich, F. R. (1974) "Morphological Analysis," in *Futurism in Education: Methodologies* (Stephen P. Hencley and James R. Yates, eds.). Berkeley: McCutchan, pp. 211–234.

Complete discussion including implicit and explicit assumptions. Includes Zwicky's original jet-engine analysis.

McGrath, J. H. (1974) "Relevance Trees," in *Futurism in Education: Methodologies* (Stephen P. Hencley and James R. Yates, eds.). Berkeley: McCutchan, pp. 71–96.

Clear and fairly comprehensive discussion of relevance trees.

Warner, M. L., and Preston, E. H. (1974) *A Review of Environmental Impact Assessment Methodologies*. Washington, D.C.: U.S. Government Printing Office.

Includes a detailed discussion of impact identification and various approaches to it.

EXERCISES

8-1. A major freeway is to be constructed through the center of your campus or near your place of business.

1. Develop a checklist to include the potential environmental, biological, cultural, and ecological *impacts* that you feel are appropriate.
2. Develop a checklist composed of *activities* associated with the construction phase of the freeway project.
3. Develop a checklist composed of the *stakeholder groups* associated with the construction phase of the freeway.

8-2. Construct an impact matrix that utilizes the checklist from Exercise 8-1 (1) as the vertical axis and that of 8-1 (2) as the horizontal axis. Use checks (√) to indicate impacts and activities that are related.[3]

8-3. Use the appropriate two checklists from Exercise 8-1 to construct a matrix that can be used to identify those groups which would win or lose during the construction phase. Use a "+" to indicate a "winner" and a "−" to indicate a "loser."[3]

[3]If these exercises are performed by an individual, we suggest that the dimensions of the matrix be no greater than 6 × 6. Larger matrices can be handled by small groups by partitioning them and assigning the components to individuals or subgroups.

8-4. Apply the criteria suggested for selectivity to the impacts identified in Exercises 8-2 and 8-3.

8-5. Pick a branch of the hypothetical impact tree for surface mining shown in Figure 8.2 and carry it out to three additional levels.

8-6. Select one activity from the checklist constructed in Exercise 8-1. Use the matrix constructed in Exercise 8-2 as a starting point to construct a three-level tree for impacts proceeding from that activity. Estimate probability, timing, and severity for each element of the tree.

8-7. Construct a tree representing the technological activities necessary to support a doubling in coal production in the United States within the next 10 years.

8-8. Construct a support tree for the activities necessary to execute the impact identification task of a technology assessment.

8-9. (Project-suitable) Develop and defend an overall impact identification strategy for a technology or project assessment, including choice of techniques.

9

Impact Analysis

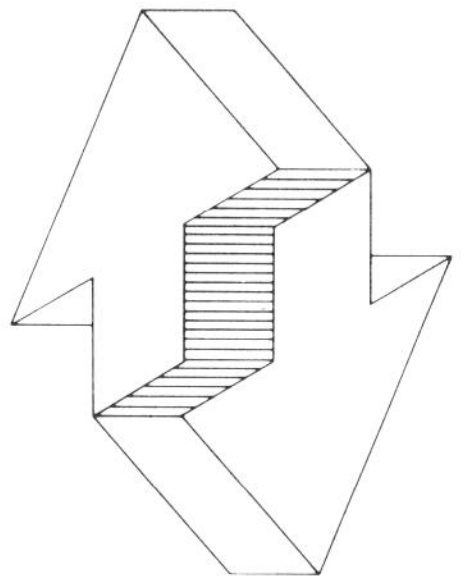

This chapter develops the groundwork for impact analysis—that phase of a TA/EIA devoted to estimating the likelihood and magnitude of the effects of a development. It provides the foundation for the specific analyses of environmental, economic, social, and other impact sectors (Chapters 10–13). This chapter focuses on the use of models. Initial sections provide a perspective from which to select and apply models in impact analysis. Advantages and dangers inherent in the use of quantified models are emphasized. The remaining sections address particular modeling approaches. Cross-impact analysis and simulation, as exemplified in the case of the KSIM model, are detailed. The chapter concludes with a discussion of sensitivity analysis.

9.1 INTRODUCTION

Analysis links the identification of significant impacts with their evaluation and the formulation of policy options to deal with them. Analysis adds form and substance to the consideration of the impacts already identified. Impact analysis addresses questions regarding points such as the probability, timing, severity, and diffusion of each impact; who will be affected and how; their probable response; and how significant the higher-order impacts will be.

This chapter logically follows impact identification (Chapter 8). It develops generic approaches to impact analysis that set the tone for dealing with specific types of impacts, in later chapters—environmental (Chapter 10), psychological (Chapter 12), institutional/political (Chapter 13), social (Chapter 12), technological (Chapter 13), and economic (Chapter 11). It anticipates impact evaluation (Chapter 14) and policy analysis (Chapter 15).

This chapter emphasizes modeling techniques that can be used in a variety of impact analysis contexts. The generic techniques include cross-effect matrices and simulation models, as well as other modeling approaches. The sensitivity of such analyses to changes in parameter values is also considered.

Section 9.2 consists of a discussion of the intellectual basis of modeling and the characteristics of models. Ayres' (1966) classification of models provides a framework within which different approaches can be understood. Problems and advantages in the use of models, as well as trends in their use are discussed to complete the perspective.

The consideration of specific models begins in Section 9.3. Cross-effect methods are discussed first. Cross-impact matrices are considered in special detail and a U.S. Army Corps of Engineers' example is used as a vehicle to illustrate their use.

Section 9.4 is devoted to the consideration of simulation models. A cross-impact simulation model, KSIM, is discussed in some depth. Next, we introduce a number of approaches that, although not presently heavily used, may have potential in TA/EIA. The chapter concludes with the consideration of the sensitivity of models to changes in parameters, an important determinant of a model's utility.

9.2 GENERAL CONSIDERATIONS

In a general sense, all impact analyses proceed from some explicit or assumed model. J. Coates (1976a: 144) defines a model as "any systematic interrelationship of elements and components into a system which is intended to parallel in structure, form or function some real world system." Impacts result from the interaction of a technological development with its societal context—that is, the activities of the technological delivery system (TDS). Impacts are thus understood in terms of a framework, or model, such as the TDS.

9.2.1 Model Characteristics

Models can be distinguished from one another on the basis of various characteristics. Ayres' (1966) hierarchy of models provides a useful basis for categorization.

1. Analytic model. The patterns of events can be predicted and explained in terms of more fundamental "laws" with wide applicability (e.g., Maxwell's electromagnetic theory, Einstein's special and general theories of relativity).
2. Empirical–phenomenological model. The pattern is adequately predicted by a mathematical formulation with empirically fitted parameters (e.g., Landau's two-fluid quantum model of superfluids).
3. Quasi-model [more than (4) and less than (2)]. Qualitative operational predictions can be tested (e.g., Darwin's theory of survival of the fittest).
4. Metaphor or analogy. The outlines of coherent patterns are perceived (e.g., the mechanical analogy of an electrical circuit).
5. Conjecture. Probable positive correlation between some pairs of observations (e.g., weather and sun spots).

Ayres' hierarchy is ordered (from analytic models to conjecture) in the direction of decreasing precision, explicitness, and mathematical elegance. It also corresponds roughly to decreasing demands on the modeler's time and resources and, in practice, decreasing quantification. This order does *not* necessarily correspond to decreasing usefulness in the impact-analysis task.

Analytic or empirical–phenomenological models can be constructed only for those impact sectors with relatively well-developed theories. Such models are often highly quantified. Frequently referred to as *formal models*, these approaches can require computer facilities. System dynamics and certain economic and environmental models are representative of quantified formal models.[1]

[1]Examples include "World 2" (Forrester 1971), "World 3" (Meadows et al. 1972) and Strategy for Survival (Mesarovic and Pestel 1974); National Socioeconomic Model (Low 1974), Wharton Annual Industry Forecasting Model (Preston 1972), and Chase Econometrics Industry Forecasting System (Evans 1969); Strategic Environmental Assessment System (U.S. EPA 1974b), and Materials–Process–Product Model (Ayres et al. 1974).

Approaches based on metaphor, analogy, and conjecture are characterized as *informal models*. It is important to note that policy makers, faced with exceedingly complex and interrelated issues for which no well-developed explanatory theory exists, rely almost exclusively on informal models.

Danger and promise are both inherent in the use of formal modeling. Several investigators have considered both these aspects of modeling (Pavitt 1972; J. Coates 1976a). The major dangers include the following:

The construction of computer models can be extremely expensive in both time and resources. This is especially true if extensive software and/or data-base construction are required. Even after the model is completed, considerable expert consultation may be required to validate it.

The model, even after validation, may not be credible to the sponsor for a number of reasons. There may be a failure of model maker and policy maker to communicate. This can be a particular problem with a new model whose complexity makes its operation and output difficult to explain. This should be anticipated in planning the model and especially in selecting formats for its output. Further, policy makers experienced in the use of informal models may naturally distrust the ability of formal models to represent the nuances of complex issues.

There may be a tendency to attempt quantification of factors "which are either for practical or intrinsic reasons difficult to quantify, or which are not likely to be accepted in quantitative terms by decision makers and the lay public" (Coates 1976a).

The assessor may fail to make sufficiently clear the assumptions on which the model is based, or those assumptions may be impractical or naive from the vantage of the policy maker.

The model may be used as a substitute for critical thought and its output accepted without question by virtue of the sophistication of the technique and hardware used to produce it.

Provided that these dangers are recognized and avoided, quantified modeling offers a number of potential advantages:

Assumptions about the present and future states of society are explicit.

The investigation of alternative policies and their implications is systematic.

Known facts and the predictable consequences of policies are presented so that assessor and policy maker can concentrate their expertise on factors where technical and physical judgment are essential (Pavitt 1972).

A foundation for periodic evaluation and review is established.

Considering these lists of potential dangers and benefits, the use of formal models can either strengthen a TA/EIA or weaken it. Hence models must be selected carefully to produce appropriate output within the available resources.

9.2.2 Models in Impact Analysis

Howard (1968) has suggested a three-dimensional formulation of the problem space that is useful in visualizing the role of models in the analysis task. The three dimensions chosen for this description (Figure 9.1) are (1) *degree of uncertainty* (deterministic → probabilistic), (2) *complexity* (few variables → many), and (3) the *time factor* (static → dynamic). J. Coates (1976a) has observed that TA/EIAs are primarily concerned with corner number eight of the problem space—many variables, dynamic, and probabilistic. This is due to limitations of theory

FIGURE 9.1. The problem space.
Source: Howard (1968).

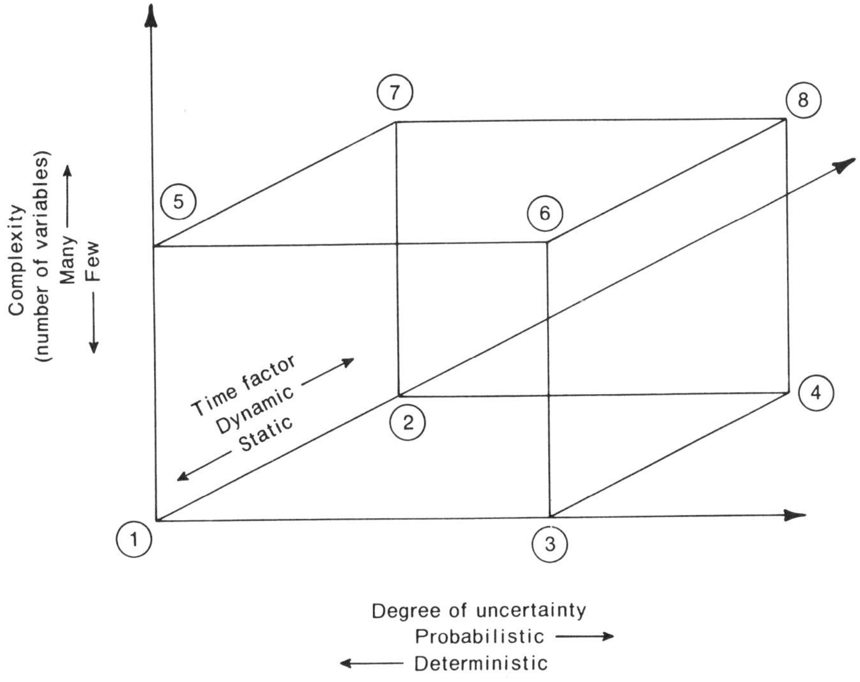

and data, inherent uncertainties, the number of variables, and the dynamic character of assessment subjects.

Historically, neither EISs nor TAs have placed heavy reliance on sophisticated quantified modeling techniques. EISs have been dominated by the need to gather requisite information under constraints of time, money, and manpower. The opportunity to develop formal models has consequently been severely limited. Maps, overlays, checklists, and surveys have been frequently utilized; matrices, networks, and specialized models (e.g., noise-level and water-quality models), less frequently employed. Generally, assessors have adapted the basic modeling elements to meet the topical needs of their study within time and resource constraints. TA practitioners typically deal with issues that are of longer range and broader scope than those of EIAs, yet they have tended to be equally pragmatic.

The use of formal models in EIA appears to be on the increase. The major EISs prepared in 1974 and 1975 included more reliance on quantitative models than earlier efforts. Examples include noise-level and water-quality modeling approaches (J. Coates 1976a).

Armstrong and Harman (1977: 50), after reviewing a number of TA studies, noted that

> Computer models were by far the single most popular quantitative technique of impact assessment, being used in some fashion in 10 of 24 TA's surveyed (5 of the primary 10). About half of these were input–output models; another quarter were in some other way generally economic, and the remainder were for simulation of physical systems.

They concluded that such quantitative models "were best used in an ancillary role" in the impact analysis, and were "especially important for helping to define economic and physical elements of technological alternatives." Rossini et al. (1978) found considerable sentiment among assessors against the use of formal models and techniques as major components of wide-scope TAs. Objections that have been raised to the use of models include their high cost in study resources; the lack of hard data; the required, but often unjustified, degree of quantification; poor results obtained in areas of high uncertainty; and a general feeling that model output lacks validity.

Different levels of model (on Ayres' hierarchy) are suitable for the analysis of different sorts of impact. Generally speaking, economic, technological, and some environmental impacts are amenable to formal modeling approaches. Thus models that are sophisticated, fine grained, and highly quantified are relatively common in such areas. Social, institutional, political, and legal impacts, on the other hand, are rarely amenable to analytic approaches and only slightly more accessible through empirical–phenomonological models. Impact analysis must, of course, deal with all sectors of the impact field. Therefore, for many assessment topics, it is unlikely that formal modeling alone will suffice or even that it will be very useful. Either "models" from more than one level of the hierarchy, or a single informal model will generally be required.

There is a temptation to consider a model solely in terms of its product, the output, and to forget that it also reflects a *process*. Often this process of constructing the model can be more valuable than the model output. Model construction, generally a team rather than an individual effort, demands the focused study of the various impacts, delineation of impact field structure, and careful attention to the assumptions underlying the modeling approach. The exchange of viewpoint, knowledge, and perspective demanded by model development can yield large dividends in understanding the impact field. It can also be extremely valuable in the integration of individual knowledge and expertise of team members (see Chapter 17). The best models for use as process vehicles have relatively simple structures that are unobtrusive to the user.

Models appropriate for the analysis of impacts cannot be chosen arbitrarily. The assessor must consider the compatability of the model characteristics with those of the assessment subject and of the impacts selected for analysis. Sponsor requirements and data availability must also be considered. J. Coates (1976a) observed that the avoidance of formal models can be "profitable particularly in the preliminary assessment of a subject which has not been addressed before." Further, the time, resource, and expertise requirements of constructing computer-based analytic or empirical–phenomonological models will often preclude their use in the average TA/EIA activity. Incremental alterations of existing models are more suitable. Even in this latter case, data requirements may rule out use.

We now turn to specific modeling techniques that may prove useful in certain TA/EIA contexts.

9.3 CROSS-EFFECT MATRICES

9.3.1 Overview

Cross-effect matrix is a generic term referring to a broad family of techniques that share the notion of an orderly cross-comparison of important factors. Cross-matrices are extremely versatile and can be used in many assessment tasks. The application of matrices in impact identification was discussed in Section 8.3. Depending upon the need, matrix approaches can incorporate trends, activities, events, goals, impacts, policies, and so forth.

As discussed in Section 8.3, a cross-effect matrix is constructed by arraying one list of factors vertically and a second list horizontally. The cells display data portraying the interaction between each row and column entry. Such data can be expressed in various forms including quantitative estimates of importance, conditional probabilities, or quantitative estimates of magnitude. A matrix utilizing the same list of factors for both vertical and horizontal dimensions is referred to as a *cross-impact matrix*. When entries in cross-impact matrix cells quantitatively describe the degree to which a column impact intensifies a row impact, the matrix can form the basis for a simulation model such as KSIM (Section 9.4.2).

In Ayres' hierarchy, cross-effect matrices are structurally empirical–phenomenological models. However, a large part of the information filling the matrix cells is generated by means of informal models or expert opinion. In practice, therefore, matrix models are more nearly quasi-models than empirical–phenomenological ones. (Their combination of structural model and data classifies them as a synthetic inquiring approach in the terminology of Section 5.4.2.) In terms of the three-dimensional representation of the problem space (Figure 9.1), cross-effect matrix approaches lie on leg 3–6, generally nearer to 3 than 6. That is, they can be probabilistic, are generally static, and usually deal with a fairly limited number of variables.

An array of potential cross-effect formulations that could be used in TA/EIA is presented in Table 9.1 to display the variety that is possible. Note that diagonal cells represent potential cross-impact matrix formulations (e.g., technology × technology, society × society, etc.). A few of the more important formulations are discussed briefly in the following paragraphs.

Technology × Technology. This cross-impact matrix can be used to display the manner in which developments in one technology affect the progress of another. For example, development of an improved dielectric material would greatly improve storage battery characteristics, thus giving impetus to the development of the electrically powered automobile.

Technology × Society. This matrix portrays the first-order impacts of activities of the TDS on the state of society. The cell entries can be used to indicate effects of technological activities on the probability, magnitude, or diffusion of societal effects. This matrix is useful in social-impact assessment.

TABLE 9.1. An Array of Potential Matrix Formulations in TA/EIA

Major TA/EIA dimension (impactee) \ Major TA/EIA dimension (impactor)	Technology	Society	Impact	Policy	Time
Technology	One technology depends on another	State of society influences development and implementation of technologies	Impacts may suggest alternatives in the technology being implemented	Policies may influence development and implementation of technologies	Passage of time influences state of development and implementation of technologies
Society	Implementation of a technology can alter the state of society	Changes in societal state may breed other changes in its state	Impacts may cause changes in the state of society	Changes in policy may lead to changes in state of society	Passage of time influences society and its perceptions (social forecasting)
Impact	Implementation of a technology can alter environmental—social state	Changes in society may impact environment, political factors, etc.	Impacts can interact with each other and lead to higher-order impacts	Changes in policy can lead to various environmental, social, etc. impacts	Passage of time can lead to alterations in intensity, type, or diffusion of impacts
Policy	Technologies implemented may alter both current and future policies	Changes in society may cause changes in its policies	Impacts can lead to alterations in current or future policies	Policies can affect or alter other policies	Passage of time can lead to alterations in policy positions

Policy × Impacts. This matrix could be used in policy analysis to display the impacts that might be produced by adoption of policy options available to the decision maker. The information provided in the cells can be used to describe the effect of policy on the diffusion, severity, or probability of impacts. For example, an air-pollution control board might use it to investigate the impacts of various policy alternatives for enforcing clean air standards on the economy, institutional structure, and environment of its state.

Impacts × Impacts. This matrix is the most widely used of the various arrangements suggested in Table 9.1. When people speak of "the" cross-impact matrix, this is generally the formulation to which they refer. The impact × impact matrix portrays the manner in which impacts from some TDS activity interact to beget other impacts. It is extremely valuable in the analysis of higher-order effects.

There is a wide variety of cross-effect matrices and their applications. Specific matrices are discussed throughout this book as topically appropriate (e.g., impact identification in Chapter 8; input–output in Chapter 11). The following section concentrates on cross-impact analysis, the matrix application that has received the most attention in TA/EIA.

9.3.2 Cross-Impact Analysis

Cross-impact analysis had its origin in "Futures," a simulation game developed by Gordon and Helmer in the 1960s (Gordon and Hayward 1968). The term "cross-impact" has come to represent a family of different analytical techniques, especially those dealing with the way in which (1) impacts interact to produce higher-order impacts and (2) the occurrence of certain events affects the likelihood of other events. The former consists of tracing impact chains through successive impact × impact matrices. The latter technique, perhaps better termed cross-event analysis, has been widely studied.[2] This framework can be used either qualitatively or quantitatively to explore the change in likelihood of one event, given the occurrence of another. The following discussion presents a step-by-step approach that provides a means to account for the probabilistic nature of future events and their impact upon the likelihood of related events.

The recommended procedure involves two stages: (1) construction of the cross-impact matrix and (2) simulation using the matrix to determine a revised set of mutually consistent event probabilities.

[2]Compare the works of Gordon and Hayward (1968), Gordon (1969), Enzer (1970, 1971, 1972), Turoff (1972), and Duval et al., (1975). For a thorough development of the cross-impact analysis methodology, see Sage (1977: 176).

To construct the matrix, the assessor can follow these seven steps:

1. *Identify* a list of the events critical to the assessment topic. Create a matrix by arraying this list vertically and horizontally (see Table 9.2). This list can be obtained by a variety of means, including opinion measurement, literature searches, and brainstorming. It should be carefully chosen since the omission of an important event can produce misleading results. It is helpful, but not necessary, to list the events in the expected order of occurrence.
2. Estimate the probability of occurrence of each event; these initial estimates provide the *marginal probabilities*.[3] This may be accomplished by a survey of expert opinion. These marginal probabilities are entered in a convenient location on the cross-impact matrix, typically within parentheses.
3. Estimate the conditional probability (i.e., the probability of event i given that event j has occurred) for each cell above the diagonal.
4. Compute the *acceptable range of conditional probability* for each cell above the matrix diagonal. Using the marginal probabilities estimated in step 2, the conditional probability of any pair of events can be shown to fall within the limits of a well-defined range of values.[4] Probabilities outside the acceptable range can be retained if a solid rationale for them exists, but the corresponding entries below the diagonal should be estimated rather than computed.
5. For each estimate in step 3 falling within the range computed in step 4, Bayes' rule[5] can be used to compute the appropriate condi-

[3]Our probability notation is as follows:

$P(i)$ = the probability that event i will occur, or the marginal probability of i;
$P(i|j)$ = the conditional probability that event i will occur, given that event j occurs;
$P(i|\bar{j})$ = the conditional probability that event i will occur, given that event j does not occur;
$P(i \cap j)$ = the probability that both i and j will occur (the intersection);
$P(i \cup j)$ = the probability that i or j or both will occur (the union).

[4]Sage (1977: 178–181) shows through use of the laws of conditional probability and the probability of compound events that

$$P(i) \leqslant P(i|j) \leqslant [P(j)]^{-1}P(i), \text{ if } j \text{ is enhancing;}$$
$$1 + [P(j)]^{-1}[P(i) - 1] \leqslant P(i|j) \leqslant P(i), \text{ if } j \text{ is inhibiting.}$$

Thus in Table 9.2, $.5 \leqslant P(4|1) \leqslant .6$, rather than .4 as estimated there.

[5]Bayes' rule states that events i and j must be related by the expression

$$P(i|j) = [P(j|i)/P(j)]\,P(i).$$

If the marginal probabilities $P(1)$ and $P(3)$ and the conditional probability $P(3|1)$ are estimated, the conditional probability $P(1|3)$ is fixed. Using data for these events given in Table 9.2, Bayes' rule gives $P(1|3) = .74$, rather than .8 as estimated there.

TABLE 9.2. Cross-Impact Matrix Displaying the Probabilities of Interrelated Societal Changes [a]

If this event were to occur ↓	... then the "new" probability of occurrence of these events would be			
	1. Lower population	2. More leisure	3. More urban	4. More household
1. Lower population (.8)[b]		.99	.65	.4
2. More leisure (.9)	.85		.65	.6
3. More urban (.7)	.8	.9		.7
4. More household (.6)	.8	.9	.8	

[a] From Mitchell et al. (1975: 135).

[b] Values in parentheses ( ) are the initial, or marginal, probabilities. Values in the matrix are referred to as conditional probabilities. For example, the marginal probability that event 2 will occur is $P(2) = .9$. The probability that event 2 will occur if event 1 occurs (conditional probability) is $P(2|1) = .99$.

tional probability below the diagonal. Corresponding entries below the diagonal for estimates outside the range should be estimated rather than computed.

6. Examine entries below the diagonal computed via Bayes' rule to ensure they are reasonable in the light of a priori information that may be available. If they are not, they can be altered.
7. Compute the entries for the nonoccurence matrix.[6] Remember, the nonoccurrence of an event can affect the probabilities of other events just as surely as can its occurrence. The nonoccurrence probabilities corresponding to entries in the occurrence matrix

[6]From the conditional probability law,

$$P(i)|\bar{j}) = \frac{P(i) - P(j)\,P(i|j)}{1 - P(j)}$$

From Table 9.2, $P(1) = .8$, $P(2) = .9$, and $P(2|1) = .99$, so $P(2|\bar{1}) = .54$.

TABLE 9.3. Cross-Impact Matrix Displaying Statistically Consistent Probabilities

If this event were to occur ↓	. . . then the "new" probability of occurence of these events would be			
	1. Lower population	2. More leisure	3. More urban	4. More household
1. Lower population (.8)		.99	.65	.50
2. More leisure (.9)	.83		.67	.60
3. More urban (.7)	.74	.86		.70
4. More household (.6)	.67	.90	.82	

falling outside the acceptable range (step 3) may be less than zero or greater than one. In such instances they should be set equal to zero or one, respectively.

As implied by the seven steps, estimation of cross-impact probabilities is not a simple task. Accurate estimates of conditional probabilities are hard to come by. Compilation of a sensible and consistent matrixful of them is very difficult and never certain. Table 9.3 illustrates data from Table 9.2 adjusted to satisfy Bayes' rule and the ranges computed in step 3.[7] Note that this does not imply that the conditional probabilities are more nearly correct than those of Table 9.2. Rather, if the marginal probabilities presented were correct and mutually consistent, the conditional probabilities would necessarily be those of Table 9.2.

The next stage in cross-impact analysis is to simulate the effects of the conditional interrelationships. A computer can be used to calculate a final marginal probability of each event given the interrelationships depicted by the conditional probabilities in the occurrence and nonoc-

[7]Marginal probabilities and conditional probabilities from the cells of Table 9.2 above the diagonal were taken as the estimates for this computation. Conditional probabilities in Table 9.2 that fell outside the range indicated by the expression of footnote 4, were altered by the minimum amount necessary to move them within the acceptable range.

currence matrix cells. This "Monte Carlo simulation" proceeds as follows:

1. One event is selected randomly from the list (say, event 2 in Table 9.3).
2. A random number between 0 and 1 is generated and compared to the marginal probability of the event to determine if it occurs (suppose the number is .8, since $P(2)=.9>.8$, event 2 is assumed to occur).
3. The marginal probability of each *remaining* event is replaced by its conditional probability of occurrence or nonoccurrence depending on whether the event is assumed to occur or not (here we read across row two of Table 9.3 and find $P(1)=.83$, $P(3)=.67$, and $P(4) = .60$).
4. A second event is selected randomly from those remaining (events 1, 3, and 4) and the process in steps 1–3 is repeated using the probability of the event as altered in step 3.
5. The process described in the preceding steps is repeated until all events have been selected. All marginal probabilities are then returned to their initial values and the game is "replayed."
6. Each time the game is "played" the events that *occur* are noted. The tabulation of occurrences divided by the number of games is taken as the final (marginal) probability for each event. The game is typically replayed 1000 times. The initial marginal probabilities listed on the matrix are then replaced by the final values. These can then be used as desired in the TA/EIA.[8]

Many variants on this cross-impact scheme are possible. For instance, in the example for the Corps of Engineers depicted in Tables 9.2 and 9.3 (Mitchell et al. 1975), the assessors could also use:

1. A society × technology cross-effect matrix relating the four societal events (Table 9.2) to four key Corps' missions (e.g., a −1, 0, +1, +2 scale could indicate the magnitude of impact of the societal change on each of the four mission activities).
2. Analysis of the implications of particular events or policies by appropriately changing matrix parameters. For instance, the initial marginal probability of lower population growth rate (event 1

[8]Final marginal probabilities based on a 1000-run simulation varied little from the initial estimates for the example. Using the conditional probabilities of Table 9.3, we obtained $P(1) = .80 \pm .01$, $P(2) = .89 \pm .01$, $P(3) = .70 \pm .02$, and $P(4) = .60 \pm .01$. For those of Table 9.2, the results are $P(1) = .80 \pm .01$, $P(2) = .90 \pm .01$, $P(3) = .69 \pm .01$, and $P(4) = .57 \pm .01$.

in Table 9.2) could be set to 1.0. The simulation could then be replayed.[9]

Sage (1977: 191–193) presents an interesting alternative to the approach developed here. This approach involves the use of a process such as ISM (Section 5.3.4) to develop the explicit structure between the events comprising the matrix dimensions. This structure, typically presented graphically as a relevance tree, is studied closely to determine which probabilities must be estimated and which can be computed or estimated from others.

The most obvious limitations to cross-effect matrix modeling are the sensitivity of results to the completeness of the lists used to formulate the matrix and the uncertainty of the estimates used to fill the matrix cells. Conventional wisdom applied to their generation should be examined carefully, for as the old saying goes, often "it ain't what you don't know that hurts you, but what you know is true that ain't."

The cross-impact approach as developed in this section deals only with the occurrence of paired events. For example, we consider the conditional probability of event 2 given that event 1 occurs. We compute such probabilities as that for the joint occurrence of events 2 and 1, or for the occurrence of either event 1 or 2, or both.[10] The estimation of the effect of the occurrence of two (or more) events on a third is theoretically possible within the framework developed here. However, the complexity increases rapidly, and the conditional probabilities that must be estimated are even more difficult conceptually.

In closing, cross-impact analysis provides a logical framework within which to accommodate objective and subjective estimates of future events on an equal basis. Thus it can serve to compare alternative views of the interrelationship of impacts, activities, trends, and events, and their implications for the future.

9.4 SIMULATION MODELS

9.4.1 Overview

Simulation models portray the dynamic (time-dependent) behavior of a system as it varies over time. For instance, a simulation model of

[9]A 1000-run simulation using the conditional probabilities of Table 9.2 for $P(1) = 1.00$, yielded $P(2) = .93 \pm .01$, $P(3) = .66 \pm .02$, and $P(4) = .53 \pm .01$.

[10]The probability that both events i and j will occur is given by

$$P(i \cap j) = P(j)\ P(i|j) = P(i)\ P(j|i).$$

The probability that either event i or j or both will occur is given by

$$P(1 \cup j) = P(i) + P(j) - P(i \cap j).$$

the Corps of Engineers problem (Tables 9.2 and 9.3), might display the variation of factors such as population growth rate or leisure over time.

Simulation models are typically either analytic or empirical–phenomonological (recall Ayres' hierarchy) depending on their subject matter. Models of physical phenomena are more apt to be analytic than those of social systems simply because better developed theory exists for the former. Social systems models generally rely heavily on the use of mental constructs to provide data and hence are usually informal models.

Simulation models can be either deterministic or probabilisitc. Deterministic models, such as KSIM[11] (Kane 1972), assume that the system behavior for a given set of circumstances is fixed. Probabilistic models (e.g., probabilistic systems dynamics, Gordon and Stover 1976) assume that a number of system responses are possible and that the actual behavior is determined from among them probabilistically. Thus, whereas the behavior predicted by a deterministic model will be the same each time it is run, that of a probabilistic simulation will vary according to a statistical distribution.

In terms of the problem space of Figure 9.1, deterministic simulation models lie between corners 2 and 7, whereas probabilistic models approach corner 8.

The number and diversity of simulation models is large. An excellent overview of both the range and application of available models to TA is provided by J. Coates (1976a). A review for the House Committee on Merchant Marine and Fisheries (1975) compares a number of models and considers their usefulness from a policy perspective.

It is not feasible to provide a thorough overview of models here. Models are developed, changed, and applied to new problems so rapidly that to do so would be as difficult as Lincoln's task of "shoveling flies." Therefore, in the following sections one deterministic simulation model of wide applicability is considered in depth, and several other modeling approaches that may be potentially valuable in TA/EIA activities are briefly introduced.

9.4.2 KSIM: A Simple Cross-Impact Simulation Model

The KSIM model (Kane 1972) has been selected for consideration in depth because it is a relatively simple and flexible, deterministic, simulation model. Further, it is based on cross-impact matrices, which were considered in Section 9.3. This model is designed to provide the assessor with means to structure and analyze relationships in a sociotechnical system. It allows the assessor to investigate the impact of policy

[11]Kane's SIMulation model.

interventions on system behavior. The following discussion draws heavily on an especially good description of KSIM provided by Kruzic (1974).

Using KSIM in an interactive computer mode, an assessment team can easily and rapidly

- structure a problem so as to incorporate diverse team perspectives;
- incorporate "hard" data *and* subjective knowledge;
- formulate alternative policies and assess their consequences;
- examine the validity of basic assumptions as to system behavior; and
- systematically structure discussion of the problem and document assessment team activities.

The KSIM model uses a single form of ordinary differential equation to describe the dependence of the system variables on time. The properties of this differential equation and its solution describe the characteristics that the system being modeled is assumed to possess.[12] The following conditions are assumed for the model.

---

[12] The form of this equation is

$$\frac{dX_i}{dt} = \sum_{j=1}^{N} \left(\alpha_{ij} X_j + b_{ij} \frac{dX_j}{dt} \right) X_i \ln X_i ,$$

where

X_i = variable whose behavior is being described
N = the total number of variables that impact upon X_i
X_j = the individual impacting variables
α_{ij} = the long-term magnitude of the impact of event j upon i
b_{ij} = the short-term magnitude of the impact of event j upon i.

The solution is

$$X_i\,(t + \Delta t) = X_i\,(t)^{\,P_i(t)},$$

where

$X_i\,(t + \Delta t)$ = value of the variable at the end of the time period
$X_i\,(t)$ = value of the variable at the start of the time period
$P_i\,(t)$ = parameter describing the totality of impacts of the various variables upon X_i

$$P_i\,(t) = \frac{1 + \Delta t \text{ (mag. of sum of inhibiting impacts on } X_i)}{1 + \Delta t \text{ (mag. of sum of enhacing impacts on } X_i)} ;$$

explicitly

$$P_i\,(t) = \frac{1 + \tfrac{1}{2}\Delta t \displaystyle\sum_{j=1}^{N} \left[|I_{ij}\,(t)| - I_{ij}\,(t) \right] X_j\,(t)}{1 + \tfrac{1}{2}\Delta t \displaystyle\sum_{j=1}^{N} \left[|I_{ij}\,(t)| + I_{ij}\,(t) \right] X_j\,(t)} ,$$

(continued)

All variables are normalized to range from 0 to 1.

If the net effect of impacts on a variable is inhibiting (or enhancing), the level of that variable decreases (or increases) with time.

All else being equal, the larger the variable causing the impact, the greater the magnitude of that impact will be.

The impacted variable will change more slowly for a given impact as it approaches its upper or lower limits (1.0 and 0, respectively).

Variables interact in pairs and each variable affects all others.

The process for constructing a KSIM model is illustrated in Figure 9.2. The first step of this process is to discuss thoroughly the problem to be modeled. Consensus must be reached as to the scope and boundaries of the simulation, including *level of aggregation, spatial boundaries,* and *time frame.*

Next, variables must be selected to represent the principal elements of the problem. In the case shown in Table 9.2, for example, population growth, leisure time, level of urbanization, and the amount of household formation could be selected as variables to represent the demographic factors of the state of society. Measures and limits for each variable are then specified. These are then normalized on a 0–1 scale (by dividing each value by its maximum). An initial value must also be specified. For example, in computing the initial value of leisure time, one might assume 12 hours per day of discretionary time. If the work schedule were assumed to average 40 hours per week and 50 weeks per year, then an initial value for leisure time would be

$$\text{Leisure}_{\text{init.}} = \frac{4380\ (\text{hours/year}) - 2000}{4380} = 0.54$$

Cross-impact matrices are constructed to describe the manner in which changes in each variable affect all the others. Two matrices can be constructed, one to represent long-term (Table 9.4) and the other, short-term effects. Generally, however, long-term impacts are the more severe and, given the preliminary and relatively aggregated nature of

---

where

$$I_{ij}(t) = \alpha_{ij} + \frac{b_{ij}}{X_j(t)} \left(\frac{dX_j(t)}{dt} \right).$$

See Kane (1972) or Kruzic (1974) for elaboration. The KSIM model program is available from various sources including Pamela Kruzic of SRI International, Palo Alto, California and A.T. Roper of Rose-Halman Institute, Terre Haute, Indiana.

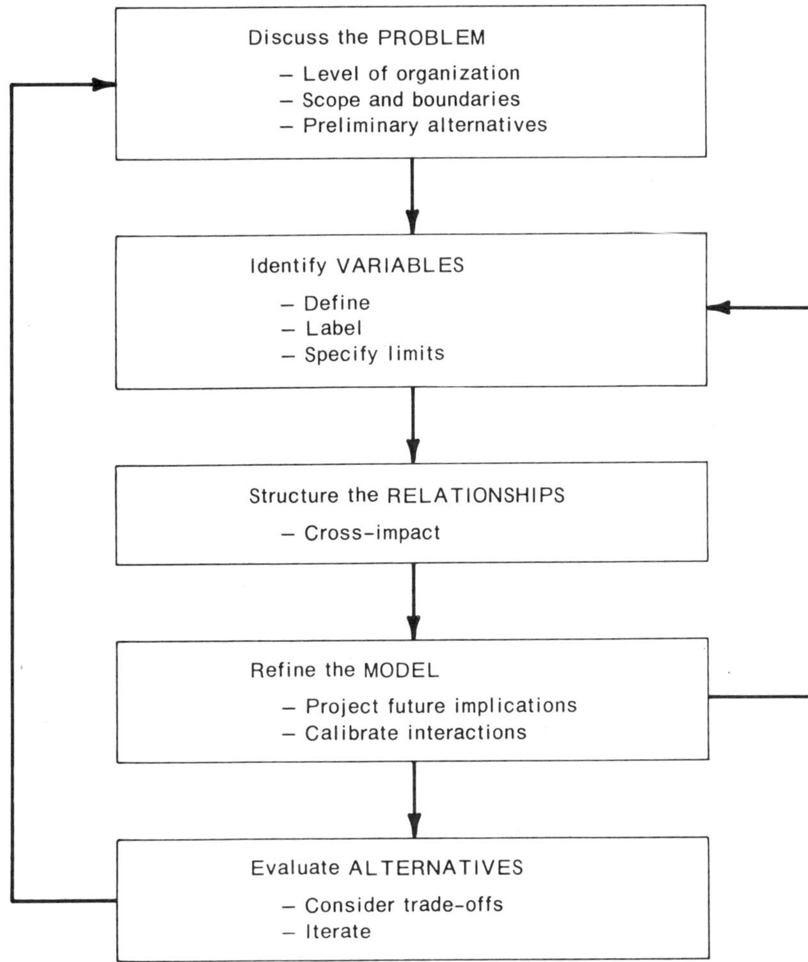

FIGURE 9.2. Flowchart depicting the construction of a KSIM model.
Source: Kruzic (1974).

most KSIM modeling efforts, short-term impacts are often neglected. The direction and magnitude of interrelationships are estimated. Experience indicates that it is generally easier to determine the direction of an impact, prior to estimating its magnitude, by asking "If variable *A* were to increase, would it inhibit, enhance, or leave unaffected the level of *B*?" Magnitudes may be scaled from -3 to $+3$, where $|3|$ is strong, $|2|$ is significant, $|1|$ is mild, and $|0|$ means no impact. Impact magnitudes must be normalized, as were initial variable values. For

TABLE 9.4. Long-Term Cross-Impact Matrix for Effects of Water-Resource Development on Urban Factors[a,b]

B	*A*							
	US(0.27)	ED(0.4)	FOC(0.5)	RD(0.3)	H_2O(0.6)	SAT(0.7)	S(0.2)	OW(1.0)
Urban services (US)	1.25	0.75	1.4	−1.9	0.0	−1.0	1.6	
Employment dispersal (ED)	1.63	2.5	0.87	−1.6	−0.12	0.0	0.0	
Flexibility of choice (FOC)	1.5	1.4	1.25	−0.25	0.0	0.0	1.0	
Residential density (RD)	−2.0	−1.0	−0.5	0.5	0.75	0.0	0.0	
Water demand (H_2O)	2.0	2.0	1.87	−1.4	−0.5	0.0	0.0	
Satisfaction (SAT)	0.75	A	1.0	−0.25	−1.12	B	−1.3	
Costs (S)	1.0	1.5	1.0	−0.75	1.6	−1.0	1.0	

[a] From Kruzic (1974).

[b] Entries represent the impact of the column variable *A* on the row variable *B* scaled from −3 (inhibiting) to +3 (enhancing). Letters indicate impacts represented by mathematical functions. The initial marginal values are shown in ( ). The letters OW = outside world, a column in which to represent the influence of an exogenous variable of interest.

instance, if an additional 8 hours per work week for leisure time were deemed a strong effect ("3"), this would amount to

$$\text{Leisure} = \frac{4380\ (\text{hours/year}) - 2000 + 400}{4380} = 0.63.$$

The impact as a change from the initial value would be (0.63 − 0.54 = 0.09). If "3" represents a strong impact, all impacts on leisure time would be scaled by 0.09/3 = 0.03 (e.g., a significant impact would be an increase of 0.06; a mild one, 0.03). The estimation process can be a valuable vehicle for group dialogue on the development and its implications.

An example of a completed long-term cross-impact matrix is presented in Table 9.4. The matrix shown (along with a short-term cross-impact matrix not shown) was developed by an eight-member group studying the effect of water resources development on urban factors (e.g., the provision of services, employment patterns, and changes in residential preferences).

Once the cross-impact matrices have been completed, the KSIM model can be run. The output of this run, referred to as the *base case*, represents the behavior of the system over time with no policy intervention or significant external events. The base case is closely examined to verify that it is realistic by comparing its behavior to historically similar situations or theoretical behavior. Initial values and/or impact magnitudes are altered until model behavior is satisfactory. The model may be further refined by expressing some impact factors as functions of time or of the level of the impacting variable. A sample base case output is presented in Figure 9.3 for the model described by Table 9.4.

When model behavior is satisfactory, the variation of each problem variable is analyzed to determine the severity and timing of impacts. The characteristics of these impacts, in turn, suggest policies that may be implemented to modify them. Policies are selected and formulated as variables that are then added as columns, but *not* rows, to the impact matrices. When policies are included, therefore, the impact matrices are no longer square. The assumption behind this arrangement is that policies impact problem variables, but problem variables do not impact policies. Exogenous or external events can be added to the model in the same fashion as policies (cf. column labeled OW in Table 9.4).

Policy impact factors are determined and scaled in the same manner as other entries in the cross-impact matrix. By examining the manner in which the problem variables respond to the policy intervention, policies can be altered and refined to produce the desired results.

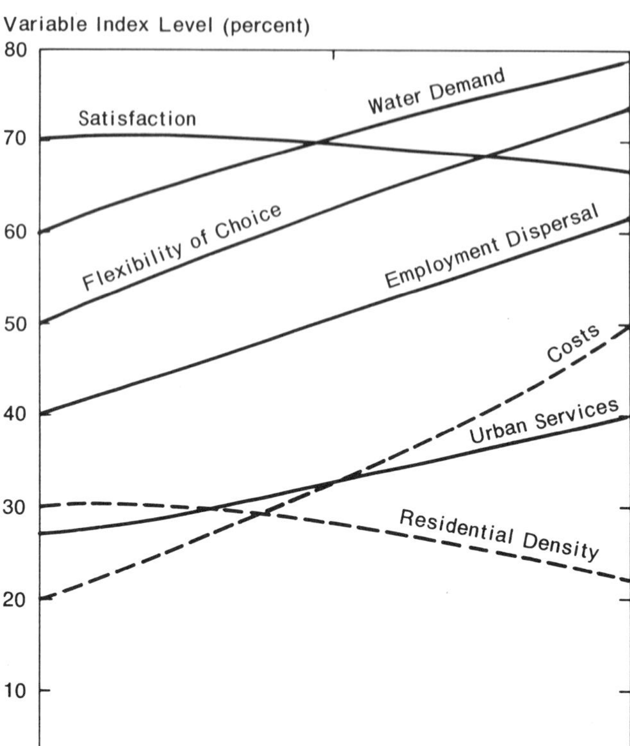

FIGURE 9.3. Sample KSIM output for base case of water-resource-development model described in Table 9.4.
Source: Kruzic (1974).

A number of modifications to the basic KSIM model are possible. For example, external events can be selected for occurrence or nonoccurrence based on their estimated probability. Alternatively, the KSIM software can be changed to allow policy selection to be altered during the model time frame. Roper and Dekker (1978) have developed an extension by coupling a problem variable × societal value matrix to the basic model. Output consists of a nonnumeric value map that represents inhibiting and enhancing impacts as areas of different shading.

A model comparable to KSIM in simplicity and cost, QSIM2, has been developed (Wakeland 1976). Advantages claimed for this approach over KSIM are that it uses input parameters more easily understood by lay users, requires nominal refining during the model-verification phase, allows most input parameters to be determined by regression analysis of historical data, and permits easy incorporation of refinements in the conceptualization of variable behavior.

There is a tendency to regard the output of system models as a prediction of future events rather than as the simulation of system behavior under restricted conditions. This tendency is dangerous since deterministic models fail to mirror the probabilistic nature of the systems they purport to describe. They are vehicles for understanding system behavior rather than predictors of future system states.

9.4.3 Physical Models

Working physical models have long been used in engineering and the physical sciences to study the behavior of systems too complex to be modeled theoretically. Using this approach, scaled models of systems ranging from airplanes to river deltas have been constructed and observed under controlled conditions. Physical models have been rarely used in TA/EIA activities. One example, however, is in the Kennedy Airport assessment (Jamaica Bay Environmental Study Group 1971).

Joseph Coates (1976a) mentions an especially interesting use of a physical model in urban planning. Optical pickups were mounted on miniature gantry cranes to allow users to view a model-city mock-up as though they were on a scale equivalent to the model (Appleyard and Craik 1976).

The most fundamental restriction on the use of physical modeling is the cost of construction, verification, instrumentation, and the gathering and analysis of data. The relatively large number of river and waterway models in existence would seem to hold promise for use, particularly in EIA studies of river canalization, flood-plain projects, and water-resource development. Wave tanks, beach erosion, harbor, and groundwater models are also promising sources of information.

9.4.4 Planning Models

Models of assorted complexities have been developed to portray the behavior of physical and social systems. Chapters 10 and 11, in particular,describe major topical models dealing with environmental and economic sectors. This section briefly illustrates the genre of planning models that may be helpful in TA/EIA. Gass and Sisson (1975) provide a useful guide to these models. This section draws heavily on their treatment.

An elaborate land-use and transportation model named EMPIRIC has been developed by Peat, Marwick, Livingston, and Company (1969). This model has been applied to a number of American cities, including Washington, D.C., Atlanta, and Dallas. The EMPIRIC model utilizes a set of simultaneous linear equations to simulate the behavior of population distribution and employment by zones within an urban area. The regional distribution and growth of these activities is related to their original distribution, overall growth, and planning policy and transportation decisions. Policy measures appear as changes in highway and transit systems, utilities, restrictive zoning, and the like.

Gaming simulation models have been widely used in urban and transportation planning. These models provide a base simulation on which stakeholders (or those playing the role of stakeholders) can play out hypothetical policy actions and observe the resultant system behavior. Many of the "games" in use today are direct descendents of the "Community Land-Use Game" (Feldt 1966) or "Metropolis" (Duke and Burkhalter 1966). "City Model," "River-Basin Model," and APEX [all described by Gass and Sisson (1975)] are examples of such gaming models. House (1971) provides an excellent review and summary of urban gaming models.

A number of planning models have been developed in conjunction with studies of sewage, water, and flood-control systems. For example, Levis (1973) presents a model that examines the criterion of least cost in the light of political and social constraints in an urban county.

9.4.5 Systems Dynamics

Forester has developed a modeling methodology known as *systems dynamics*. It has been applied to such diverse problems as solid waste generation, merchant shipbuilding, sports and recreation, urban planning, and even world futures (Forrester 1968, 1971; Meadows et al., 1972).

In this methodology, there are two different sorts of system: *open* and *closed* (or *feedback*) systems. The output of an open system is completely controlled by the input and thus no modifications in process are made by the system to correct its behavior to meet a predetermined goal. In contrast, a closed system compares output to some norm and attempts to modify its input to achieve the appropriate performance through a feedback loop. Feedback relationships are emphasized by systems dynamics models.

Systems dynamics modeling portrays behavior in terms of two fundamental types of variable: *level* and *rate*. For example, the wealth of a society is a level, whereas its income is a rate. The variables are linked by feedback loops that can be either *negative* or *positive*. Negative feedback seeks a certain goal and adjusts rate variables as necessary to attain that goal. Positive feedback conversely initiates action that progressively intensifies rate variables. Negative feedback is illustrated by a thermostat that turns on the heat only when the room temperature drops below the desired level (the goal). Positive feedback is exemplified by the ever-increasing growth of a bacteria culture.

A systems dynamics model is composed of a number of feedback loops linked together in a manner thought to represent the behavior of the system being portrayed. Equations are developed to quantify the levels and rates embodied in the system, making use of the widely available DYNAMO simulation language. Information for these equations is gathered from the study of available system behavior, expert opinion, and/or mental constructs. The behavior of the system is then simulated by solving the equation sets under various conditions.

A systems dynamics model portraying some factors in the short-run crude oil-supply problem is shown in Figure 9.4. The major characteristics of this partial system [an example taken from a more complete model described by Sage (1977: 222.)] are as follows:

> Crude oil supply derives from two sources, domestic and imported. The rate of production from each of these sources is controlled based on multiple factors. In particular, as shown in the Figure, the amount of crude oil supply desired and the actual crude oil supply available influence the domestic well production rate and the crude oil importing rate. Two important feedback loops are therefore present.

A number of criticisms have been raised concerning systems dynamics modeling efforts [e.g., Cole et al. (1973)]. Specifically (U.S. Congress, House, 1975: 47), they have

- made insufficient use of measured, supporting empirical data;
- excluded the possibility of unprecedented events (e.g., scientific breakthroughs and revolutions);
- been inadequately verified (in particular, assumptions of model structure are insufficiently justified through comparison with past history);
- overemphasized the mechanistic view of socioeconomic systems; and
- not adequately handled random events.

The extreme aggregation of systems dynamics models and the lack of sufficient data and theory to verify their structure in social systems limit their utility in policy analysis. At present systems dynamics offers

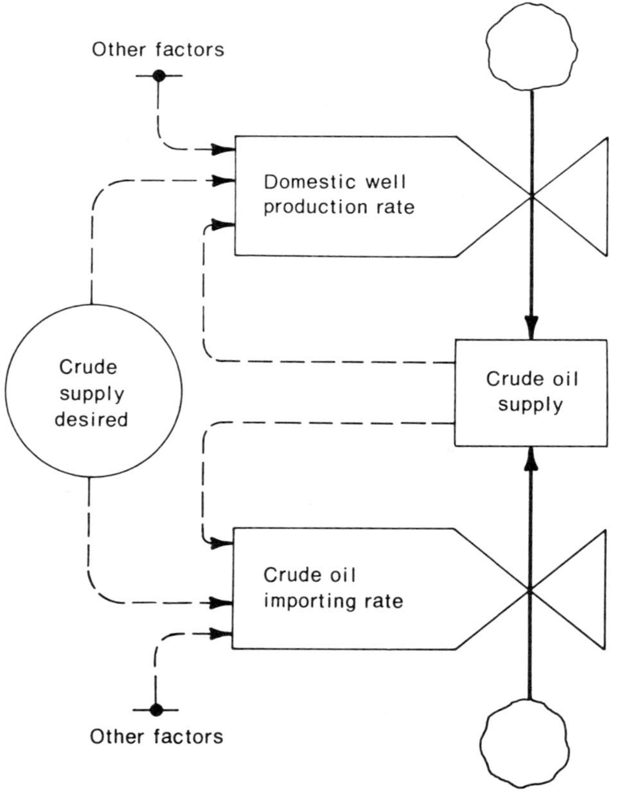

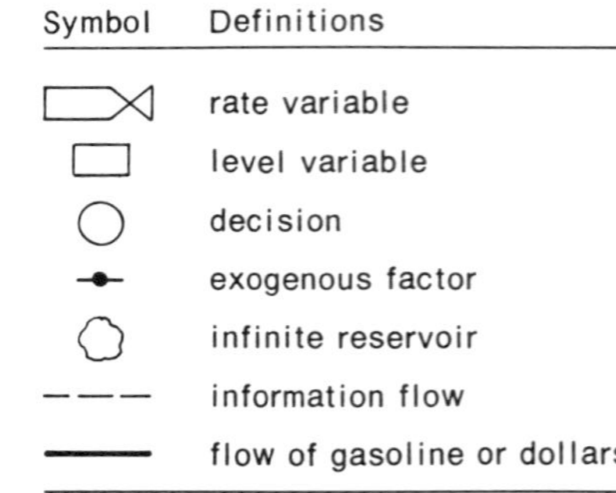

FIGURE 9.4. Feedback relationships in crude oil supply.
Source: Based on Sage (1977: 222).

a useful heuristic device to specify the important elements of a TDS and their workings, but it should be used with caution in impact analysis.

9.4.6 Stochastic Modeling Approaches

We now turn to brief descriptions of a few modeling approaches that incorporate the stochastic, or probabilistic, nature of real-world events. These models more nearly approach corner 8 of the problem space (Figure 9.1) than do deterministic approaches such as systems dynamics.

Gordon and Stover (1976) describe probabilistic systems dynamics, which allows possible impacts of future events to be incorporated in systems models through the use of the concepts of cross-impact analysis. In a national energy model, for instance, technological developments fostered by future shortfalls could be considered.

The Monte Carlo method can be used to generate probability distributions for the outcomes of a stochastic model by rerunning the model repeatedly and tabulating the results (cf. our use of it in cross-impact analysis, Section 9.3.2). It can also be used to sample a probability distribution for the occurrence of an event to determine a specific outcome. Either use can be of value in stochastic modeling approaches.

The basis of the Monte Carlo method is the approximation of the solution of a problem by sampling from a random process. The method is thus "analogous to submitting the problem under consideration to a roulette wheel," hence its name (Bruno 1974: 348). It has been used in a variety of studies, including cost–benefit analyses, the projection of future values of variables, and the approximation of physical problems where analytical solutions cannot be obtained. An excellent summary and overview of the Monte Carlo method and its applications is provided by Bruno (1974: 348).

Queuing theory provides a means for statistically treating decisions related to the provision of products or services when demand is not completely predictable. For example, it can provide a means to forecast the impact of any of a number of policies for determining which demands will receive priority treatment. Thus, queuing theory could play a useful role in analyzing possible responses to future demands and shortfalls in energy. Mitchell et al. (1975: 88) provide a useful description of queuing theory.

A *Markov chain* is a mathematical representation of a process that moves sequentially and stochastically through a number of states. This method has been widely used in hydrology. It has also been used in areas of interest to TA/EIA, for example, in the description "of social processes such as voting, resolution of conflicts, formations of friendship, and social mobility" (McNamara 1974: 301).

Bayesian statistics can be profitably applied to certain TA/EIA situations (recall Section 9.3.2). In essence, Bayesian probability formulations provide mechanisms for combining prior knowledge (including subjective judgment) with new information to revise preliminary estimates. For example, an assessor can obtain an estimate of the probability distributions appropriate to the problem and then seek further information to reassess and revise them (Gohagan 1975). The BDM Corporation's use of a minisurvey Bayesian test as an inexpensive alternative to a full-scale survey is of potential interest to assessors (Finsterbusch and Wetzel-O'Neil 1975).

Few, if any, of the methods introduced in this section are extensively used in TA/EIA studies. They nonetheless may have a potential for application in particular situations in TA/EIA.

9.5 SENSITIVITY ANALYSIS

The repeated operation of a model with successive alterations of its state (i.e., the variables, their values, or interrelationships) to observe the dependence of its behavior on state factors is called *sensitivity analysis*. There are two basic motivations for such analysis: (1) determination of the dependence of model output on assumptions and the accuracy of input data and (2) the identification of leverage points on which to base strategies for modification of system behavior.

Sensitivity is the ratio between the fractional change in a parameter that serves as a basis for decision to the fractional change in the simple parameter being tested. The procedure requires recomputing the value of the decision parameter using the changed value of the simple parameter. When *F(x)* is the decision and x the simple parameter being tested,

$$\text{Sensitivity} = \frac{\dfrac{\Delta F(x)}{F(x)}}{\dfrac{\Delta x}{x}} = \frac{\dfrac{F(x_1) - F(x_0)}{F(x_0)}}{\dfrac{X_1 - X_0}{X_0}}$$

with

x_0 = base value of the variable being tested;

x_1 = new value of x;

$F(x_0)$ = value of decision parameter when the base value of the variable is used; and

$F(x_1)$ = value of decision parameter at new value of the variable being tested.

Variables having sensitivities equal to or greater th most sensitive. Variables that are sensitive need to l mated; those that are insensitive do not require such plex models of TA/EIA it is necessary to select the uncertain variables for analysis. There is no foolpro selection. However, if the behavior of a particular to the assessment sponsor, the public, or the conc then variables directly affecting that parameter are analysis. Elements of the model particularly sen the modeler and especially "soft" or uncertair targets (see Frank 1978).

Figure 9.5 illustrates the sensitivity of $P(3)$ a and accompanying discussion) to the choice of $P(4|1)$. No is three times more sensitive than $P(3)$, although in both cases the sensitivity is less than one.

9.6 CONCLUDING OBSERVATIONS

Impact analysis links the identification of significant impacts to their evaluation and the formulation of effective policy to deal with them. This chapter focuses on generic modeling techniques for impact analysis.

> Models are systematic arrangements of elements that are intended to represent real-world systems in structure and/or behavior.

Models range in sophistication and power from analytic through empirical–phenomenological, quasi-models, and analogies, to pure conjecture. All may prove useful in impact analysis. Certain physical or economic phenomena are amenable to analytic or empirical modeling; social concerns must generally be approached through less formal models.

Any model represents a simplification of reality. The assessor must decide whether a particular model captures those characteristics of the real system that are central to the analysis. Three significant dimensions to consider are portrayed in Figure 9.1:

complexity (few to many variables),

time (static to dynamic), and

uncertainty (deterministic to probabilistic).

In general, benefits of using models include

implicit assumptions made explicit;

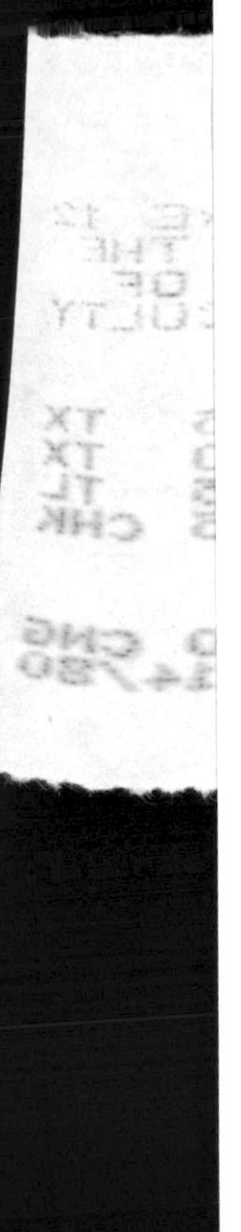

systematic analysis;

a structure that may allow human judgment to be applied at the most important points; and

replication, evaluation, and review facilitated.

On the other hand, dangers include the following:

sophisticated models consume large amounts of time and study resources;

quantification may be required even where it is not appropriate;

assumptions required may be simplistic and naive;

the model may not be credible to the sponsor and interested users of the study;

the model may substitute for critical thought.

Cross-effect matrices provide orderly cross-comparison of various important factors (Table 9.1). Either qualitative or quantitative treatment may be appropriate. We illustrate a quantitative approach with a detailed discussion of one way to compute cross-impact probabilities (Section 9.3.2).

A particular interest in TA /EIA is to anticipate effects over time. This suggests potential value of dynamic simulation. We explain the workings of a simple cross-impact simulation, KSIM, which may be useful in impact analysis.

Various other model formulations are briefly sampled. Although these have not been widely used in TA /EIA in the past, there are indications of increasing use now and in the future. We call assessors' attention to the following:

physical models;

planning models, such as for land use in urban regions;

gaming models that allow "stakeholders" to act out their inclinations and see the implications;

systems dynamics models that can show counter-intuitive workings of feedback relationships;

probabilistic systems dynamics models that can incorporate the influences of uncertain future events;

Monte Carlo methods to sample probability distributions;

queuing and Markov approaches to stochastic situations; and

Bayesian statistics to combine prior knowledge with new information.

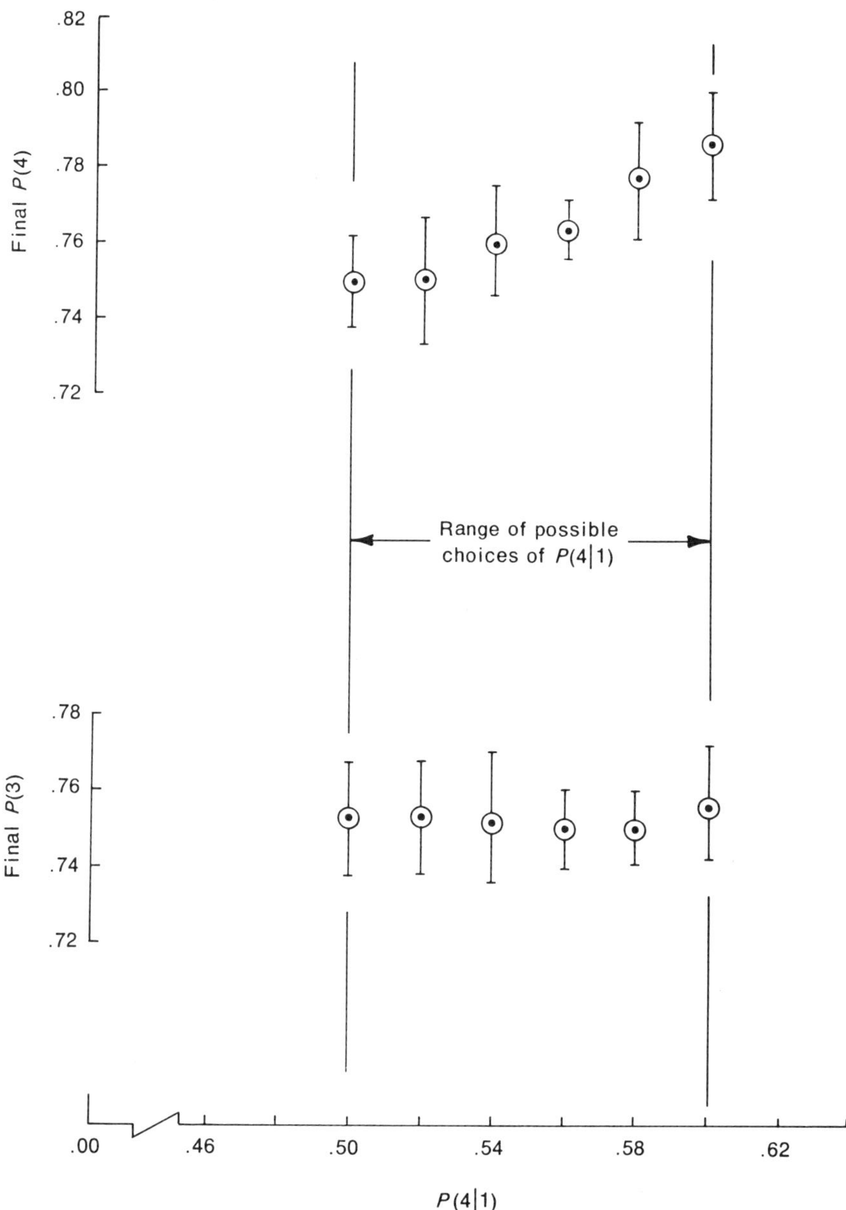

FIGURE 9.5. Sensitivity of final probabilities $P(3)$ and $P(4)$ to choice of $P(4/1)$ (Table 9.3; see also footnote 5, p. 191). Each value shown is an average computed for 10 cases of 1000 runs each ± one standard deviation.

Finally, the importance of varying key parameters to see the effects on analysis results through sensitivity analysis is recognized.

RECOMMENDED READINGS

Coates, J. F. (1976) "The Role of Formal Models in Technology Assessment," *Technological Forecasting and Social Change* 9: 139–190.

An excellent overview of the current status of models and their application to TA activities.

Frank, P. M. (1978) *Introduction to System Sensitivity Theory*. New York: Academic Press.

A treatment of sensitivity analysis.

Gass, S. I., and Sisson, R. L. (1975) *A Guide to Models in Governmental Planning and Operations*. Potomac, MD: Sauger.

A broad policy-oriented survey of formal models by topic area (air, water, solid waste, urban development, transportation, energy, etc.).

Kruzic, P. G. (1974) *Cross-Impact Simulation in Water Resources Planning*. Report 74–12 to U.S. Army Engineer Institute for Water Resources, Fort Belvoir, VA. Palo Alto, CA: Stanford Research Institute.

The clearest and most detailed explanation of KSIM modeling available. Excellent for those who do not have an extensive mathematical background.

Mitchell, A., Dodge, B. H., Kruzic, P. G., Miller, D. C., Schwartz, P., and Suta, B.E. (1975) *Handbook of Forecasting Techniques*. Report 75-7 to U.S. Army Engineer Institute for Water Resources, Fort Belvoir, VA. Palo Alto, CA: Stanford Research Institute.

An excellent summary of modeling techniques that also includes examples of their application.

Sage, A. P. (1977) *Methodology For Large Scale Systems*. New York: McGraw-Hill.

A thorough and comprehensive review of the mathematical aspects of cross-impact, KSIM, and systems dynamics modeling approaches.

U.S. Congress, House (1975) Committee on Merchant Marine and Fisheries. "Computer Simulation Methods to Aid National Growth Policy." Washington, DC: U.S. Government Printing Office.

A thorough description of macroscale models including a review of nomenclature and an excellent bibliography of articles pertaining to the subject.

EXERCISES

9-1. In Exercise 8-1 you were asked to consider the construction of a major freeway through your campus or near your place of business. Discuss which of the various modeling approaches introduced in this chapter might be appropriate for the analysis of social and economic impacts resulting from such a project. Assume that expertise in modeling and software are available and that the analysis is funded at a modest level for a 2-month period.

9-2. Discuss, compare, and contrast cross-effect, KSIM, and systems dynamics modeling techniques. Use Ayres' hierarchy and Howard's problem space as the basis for your discussion.

9-3. (Project-suitable) Assume that you are analyzing the impacts of a major coal strike on your area (consider either county, state, or region as appropriate). The strike is assumed to occur 10 years from now. As a portion of the analysis, complete the following cross-impact analysis. (You may select any topic you wish and revise the exercise accordingly.)

1. Construct a cross-impact matrix using the following three events: (a) major reduction in electrical service (−25% or more), (b) major increase in area unemployment (+2% temporary or permanent), and (c) major shift in public sentiment in favor of nuclear power-plant construction in the area (+10% or more). (You may substitute other events if you wish. If you have access to cross-impact computer software you may wish to expand the number of events.)
2. Following the procedure suggested in the text, complete occurrence and nonoccurrence matrices by estimating the initial probabilities of the three events, estimating those conditional probabilities necessary, and computing the remainder.
3. Follow the procedure outlined in the text to compute the final probabilities of the three events. Use a deck of playing cards with the face cards removed to generate the random numbers that you will need. Play the game five times and determine the final probabilities of the events.[13]

9-4. Construct a KSIM matrix for a local or regional mass-transit system with which you are reasonably familiar. Identify the principal elements of the system that relate to its viability as a means of transportation for the poor, elderly, and handicapped. Also determine the level of aggregation, spatial boundaries, and time frame you feel are appropriate.

1. Select variables to represent the elements you have chosen (service, accessability, etc.). Carefully define these variables, specify measures and determine minimum and maximum values for them. Finally set an initial value for each.
2. Arrange the variables into a matrix and estimate the direction and magnitude of the long-term impact of each variable on the others. Use an arbitrary scale [(e.g., strong (3), significant (2), mild (1), none (0)]. Finally, scale the arbitrary impacts to represent reality.

[13]Clearly, five plays are insufficient to determine accurately the final probabilities. However, if you are doing this exercise with a deck of cards, requiring more than five plays would lead to an open inquiry into the authors' parentage and/or sanity.

3. Select one exogenous event and one or two policy options, add them as columns to the matrix, and estimate their impact on the problem variables.

If KSIM software is available, you may wish to run your matrix and verify its behavior.

9-5. Select and review a completed TA/EIA study. Discuss the model(s) used in impact analysis by the study team. Do you think the model(s) used, if any, was (were) appropriate? If so, why? If not, what other techniques might have been applied? If no models were used, why was this appropriate or inappropriate?

9-6. Use the concepts of systems dynamics described in the text and the symbols present in Figure 9.4 to construct a graphical representation of a large-scale irrigation system. (Alternatively, choose a different TDS.)

10

Environmental Analysis

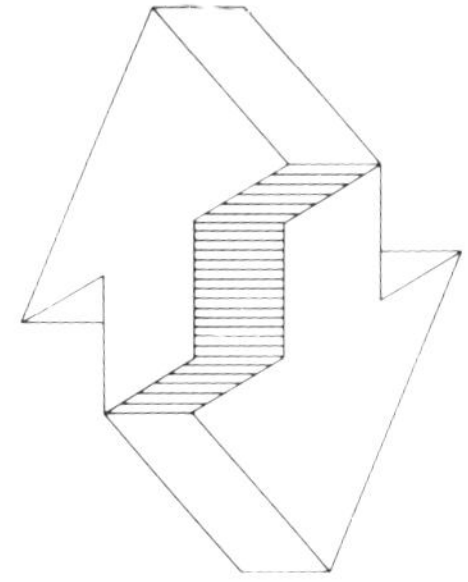

This chapter is the first of a series of chapters dealing with the analysis of specific classes of impact. It treats impacts on the natural environment and links this treatment to the processes involved in the preparation of EISs. On completing this chapter the reader should understand the general requirements for an EIS and have sources for additional information for specific types of project. This chapter also presents the most important classes of impact on the environment: ecological systems, land, water, air, noise, and radiation. General approaches to analyzing them are introduced as well as references to specific methods and examples. Other issues in environmental analysis are briefly considered.

10.1 INTRODUCTION

This chapter is entitled "Environmental Analysis"—a process distinct from environmental impact analysis (EIA), which is the total process involved in preparing an environmental impact statement (EIS) (see cameo entitled "An Environmental Impact Ditty"). Additionally, it discusses the analysis of impacts on the natural environment. As such, it is the first of a number of chapters dealing with specific classes of impact. Whereas both TA and EIA contain analyses of impacts on the environment, EIA typically emphasizes them more. Impacts on the physical environment are likely to be particularly serious for physical technologies, whether localized (e.g., a dam) or national in scope (e.g., supersonic transport development). Our treatment focuses on the relationship between environmental analysis and EIA.

We begin with a set of regulations for the preparation of environmental impact statements established by the U.S. Council of Environmental Quality (CEQ) (1978) to fulfill the requirements of the National Environmental Policy Act of 1969 (NEPA). A treatment of the substantive and methodological issues in environmental analysis follows. Finally, six major areas of environmental impact are treated: ecological systems, land, water, air, noise, and radiation.

On completing this chapter, the reader should understand the structure of environmental impact analysis. Although specific agency guidelines and detailed measurement procedures are beyond the scope of this book (entire books have been written on each of the six major areas of environmental impact), the general framework of such an analysis will become apparent. With the help of more technical guidance provided in the references, the reader should be in a position to analyze a particular situation.

For those who want to know how to perform an EIA leading to an EIS, a number of other chapters contain relevant material. These include

Chapter 4—generic study considerations for EIA

Chapter 5—particular study considerations for EIA

Chapter 8—impact identification

Chapter 9—impact analysis

Chapter 11—economic impact analysis

Chapter 12—social impact analysis

Chapter 13—institutional, legal/political, and technological impact analysis

AN ENVIRONMENTAL IMPACT DITTY

"EIS, EIR, EI, EIA
Old MacDonald had a farm,
EI, EI, Oh!"

Gordon MacDonald was the first chairman of the Council on Environmental Quality (CEQ). CEQ is mandated to receive and review all federal EISs, although its review is advisory in nature. The Environmental Protection Agency (EPA) was instituted by executive order, separately from NEPA, in 1970; it has no general authority over other agencies' environmental actions or EISs.

The most consistent usage of these terms is as follows:

EIS—environmental impact statement, required by *NEPA*, Section 102C, first as a *draft* EIS, then following review as a *final* EIS

EIR—environmental impact report, a study preliminary to an EIS or one done for other purposes, such as a state environmental protection act (a *SEPA*), although these may be referred to as EISs too

EIA—environmental impact analysis (or assessment), generic term for the study efforts

Oh!—initial reservations expressed by industry and the agencies, have now considerably subsided as procedures have been integrated into agency operations

Chapter 14—impact evaluation
Chapter 15—policy analysis

10.2 GENERAL CONSIDERATIONS

10.2.1 Procedural Issues

The initial CEQ Guidelines for Preparation of Environmental Impact Statements have played a significant role in structuring the preparation of EISs. They appeared in the Code of Federal Regulations in Title 40, Chapter V, Part 1500. They required each EIS to contain:

1. a description of the proposed action, a statement of its purposes, and a description of the environment affected;
2. the relationship of the proposed action to land-use plans, policies, and controls for the affected area;

3. the probable impact of the proposed action on the environment;
4. alternatives to the proposed action;
5. probable adverse effects that cannot be avoided;
6. the relationship between local short-term uses of humankind's environment and the maintenance and enhancement of long-term productivity;
7. any irreversible and irretrievable commitment of resources that would be involved in the proposed action should it be implemented; and
8. an indication of what other interests and considerations of federal policy are thought to offset the adverse environmental effects of the proposed action identified in (3) and (5) above.

The substantial correspondence among the eight CEQ guideline points and the 10 assessment components introduced in Chapter 4 is portrayed in Table 4.3 (p. 56).

The CEQ has revised the EIS preparation requirements, as ordered by President Carter, to decrease time and paper work in EIS preparation and improve decision making. These revisions take the form of regulations, rather than guidelines, and call for the EIS to use generally a standard format including the following substantive components (U.S. CEQ 1978: 55996): (1) purpose of and need for action, (2) alternatives, including proposed action, (3) affected environment, and (4) environmental consequences. Components 2 and 4 are viewed broadly to include policy considerations. The regulations cover essentially the same ground as the earlier guidelines.

Within the scope of these regulations, each federal agency responsible for issuing EISs may have its own particular set of guidelines that should be consulted by preparers of EISs under its jurisdiction (e.g., Federal Aviation Administration guidelines for an EIS dealing with an airport, or those of the Federal Highway Administration for a highway EIS).

10.2.2 Substantive Issues

Categorization of potential areas of environmental impact is a necessary step in structuring environmental analysis. The six major areas of environmental impact are usually considered to be ecological systems, land, water, air, noise, and radiation. We focus on these areas in this chapter. Some of these considerations overlap with other generic impact areas. For example, noise has social effects and land involves economic consequences. In addition, first-order environmental effects may contribute to higher-order effects that may be economic, social, and/or political.

The Construction Engineering Research Laboratory (U.S. Army 1975) has amplified the initial CEQ Guideline points. A modification of their work, shown as a cameo, provides an outline for an EIS. It can also serve as a checklist to assure that all relevant environmental topics are considered.

For more detailed treatment of environmental impacts, one could turn to the "granddaddy" of all environmental impact matrices (Leopold et al. 1971). This consists of cross-listing of 100 possible actions with 88 impact areas (See Tables 8.1 and 8.2, pp. 163, 164). Certain impacts listed there could be considered social or economic instead of environmental. We include a discussion of noise effects here, for example, but choose to treat energy issues in the economics chapter. More importantly, the major impact categories will vary considerably from study to study. The crucial environmental effects of a nuclear power plant will be different from those of a highway or a dam. The decision as to which impacts deserve major, minor, or essentially no attention is crucial. Higher-order impacts (effects of the first-order effects) deserve serious consideration, something usually lacking in EIAs and TAs. For instance, unspecified secondary health effects largely underlie concerns about pollution.

10.2.3 Range of Methods and Approaches

There is a wide assortment of analytical approaches from which to choose in dealing with potential environmental impacts. The range of expertise, the sorts of analysis involved, and the types of model used vary in sophistication from complex mathematics to simple description. This is indicated by the following arbitrary sampling from EIAs[1]:

- dissolved oxygen models for stream assimilative capacity, requiring solution of explicit equations based on rather precise data;
- description of levels of radioactivity and their effects on humans, animals, and plants;
- micrometeorological modeling of the air pollution from a stretch of highway based on limited traffic data and a meteorological handbook;
- ecological project study including system mapping, interviews of local naturalists, and sampling survey of species and habitat (Findley 1975);

[1]General examples—specific reports of such models may be found in Gass and Sisson (1975) and the Annual Environmental Engineering and Science Conferences held since 1971, such as Findley (1975).

SUGGESTED OUTLINE OF AN EIS [after U.S. Army (1975)]

I. Project description
  A. Purpose of the action
  B. Description of the action
  C. Environmental setting
    1. Biophysical Systems
      a. Ecology
        i. Vegetative types
        ii. Wildlife
        iii. Agricultural crops
      b. Earth science
        i. Regional topography
        ii. Scope of land
      c. Surface water
      d. Ground water
      e. Air quality
    2. Socioeconomic systems
      a. Land use
        i. General pattern of present land use
        ii. Existing trends
        iii. Existing transportation and communication network
        iv. Water supply, sewage treatment, solid waste collection, and energy resources
      b. Community health profile
        i. Population size and geographic distribution; serious public health problems or potential problems; known human stressors other than public health problems; location of schools, hospitals, convalescent homes, and other susceptible populations; known safety hazards related to area or activity; and adequacy of social-welfare facilities
        ii. Ambient noise levels

II. Land-use relationships
  A. Conformity or conflict with other land-use plans, policies, and controls
  B. Conflicts and /or inconsistent land-use plans

III. Probable environmental impacts
  A. Positive and negative effects
    1. National and international environment
    2. Environmental factors
    3. Impact of proposed actions
  B. Direct and indirect consequences
    1. Primary effects
    2. Secondary effects
IV. Alternatives to the proposed action
  A. Reasonable alternative actions, including no action, delay, and significantly different actions
  B. Analysis of alternatives
    1. Benefits
    2. Costs
    3. Risks
V. Probably adverse environmental effects that cannot be avoided
  A. Adverse unavoidable impacts
  B. How unavoidable adverse impacts will be mitigated
VI. Relationship between local, short-term use of man's environment, and maintenance and enhancement of long-term productivity
  A. Trade-off between short-term environmental gains at expense of long-term losses
  B. Trade-off between long-term environmental gains at expense of short-term losses
  C. Extent to which proposed action forecloses future options
VII. Irreversible and irretrievable commitments of resources
  A. Labor
  B. Materials
  C. Natural resources
  D. Cultural resources
VIII. Other interests and considerations of federal policy that offset adverse environmental effects of proposed action
  A. Countervailing benefits of the proposed action
  B. Countervailing benefits of alternatives

an integrated multimedia pollution model for an urban area, requiring extensive data (Paik et al. 1974);

national total environmental model, the EPA Strategic Environmental Assessment System (House and Tyndall 1975).

In a more systematic presentation, Table 10.1 provides an overview, with references concerning the state of the art in several main areas of environmental analysis. This table differentiates *sketch* techniques (relatively general, less quantitatively precise) from *detailed* ones (relatively specific, quantitatively precise). Note that some areas are well modeled, such as noise and air pollution, and others are less so, such as ecosystems, the biotic community, and groundwater. Specific areas are considered in greater detail in subsequent systems sections.

10.3 ECOLOGICAL SYSTEMS

10.3.1 General Considerations

"Ecology is concerned with all the organisms living in a habitat, the physical chemical characteristics of that habitat, and the processes which unite these two categories into a functioning system" (Warner et al. 1974: 88). The ecosystem includes the biotic (living) community and its abiotic (nonliving) environment. Essentially any intrusion into this complex system, whether an introduction of a new chemical or a change in water temperature, carries potentially widespread ramifications.

The complexity and interconnectedness of ecological systems mean that individual disturbances may produce systemic effects. Bigelow-Crain Associates (1976) specify several factors important in understanding ecosystem interactions [see also ecology textbooks, e.g., Odum (1971) and Miller (1975)]. Going beyond inventories of individual environmental effects, one should consider:

limiting factors—any factor that presents a threshold of tolerance for an organism, or group of organisms, can limit it (e.g., pond salinity);

triggering factors—any factor that upsets system balance can lead to a series of drastic changes (e.g., eliminating tidal flow from a marsh);

carrying capacity—the population that can be supported adequately by available resources;

microclimate—localized conditions affect particular organisms (e.g., humidity under rocks affected by loss of forest shade); and

food webs—transfer of food energy among ladders of plants, microorganisms, and various animals.

Having emphasized complexity and interaction of factors as vital elements of ecosystem understanding, we now return to some individual factors.

In treating these factors it is important to evaluate the relative significance of the various impacts on them. One of the prime objections to early EISs was their tendency to overwhelm the reader with a lengthy description of impacts without providing guidance as to their importance.

10.3.2 Ecosystem Parameters and Analysis

In considering impacts on ecosystems we are concerned with the number and condition of life forms and the physical and chemical properties of their habitat. The impacts consist of changes in these parameters caused by the technological intrusion in question. Table 8.2 (p. 164) gives a typology of life forms. Jain et al. (1977) offer a more compact list (see Table 10.2).

As indicated in Table 10.2, each life form requires particular considerations. The nature of the threats to each may vary drastically; for instance, large animals tend to be highly sensitive to human presence, whereas small game animals are not. Measurement of the present status may be difficult; for example, counts of predatory birds are difficult to obtain directly. Possible ways to mitigate impacts typically relate to the most sensitive features of habitat.

Technological intrusions may directly affect the number, distribution, and condition of life forms. However, effects on habitat often have the more significant long-term effects on life forms. The cameo entitled "Some Potential Ecological Impacts" illustrates some effects on habitats caused by water resource projects. Although this list is limited to one type of intrusion, and hence is not exhaustive, it is indicative of how habitats may change.

There are various phases in the typical development, such as planning, construction, implementation, secondary developments, and eventual disposal. In each of these phases the impacts will be different and their effects will accumulate. Hence adequate ecosystem analysis demands consideration of how each life form may be affected by each phase of technological development.

TABLE 10.1. A Survey of Techniques Appropriate to Various Environmental Issues[a]

Major issues	Indicators	Measurement techniques	References
Environmental health	Air pollution	Sketch and detailed analyses	
		Generation, dispersion, and impacts	Perkins (1974) CLM Systems, Inc. (1972) Holzworth (1972) Turner (1967) Sklarew et al. (1973) U.S. Federal Highway Administration (1972)
	Water pollution		
	Surface water	Sketch only	
		Source and concentration	Sartor and Boyd (1975) Sylvester and Dewalle (1972)
		Dilution	CLM Systems, Inc. (1972)
		Standards for health impacts	Field et al. (1975)
	Groundwater	No generally accepted methodology	Walton (1970)
	Noise	Sketch	
		Generation and attenuation	Bolt, Beranek and Newman, Inc. (1971) Loucks et al. (1973) Wesler (1972) U.S. Federal Highway Administration (1973) Gordon et al. (1971) Kugler and Piersol (1973)

		Detailed	
		Generation and attenuation	Bolt, Beranek and Newman, Inc. (1971) Loucks et al. (1973) Gordon et al. (1971) Kugler and Piersol (1973)
Biotic disruption	Population—plant and animal	Sketch	
		Geobotanical surveys using photogrammetry and infrared techniques	Kuchler (1967) Chikishev (1965)
		Detailed	
		Critical area delineation	Smithsonian Institution (1974)
		Species surveys using quadrants or transects	Smith (1974) Poole (1974) Watt (1968)
		Microorganism survey	Kevan (1962) Odum (1971)
	Community		
	Food webs	No numerical techniques	Odum (1971)
	Succession	No numerical techniques	Odum (1971) Platt and Griffiths (1964)
	Carrying capacity	No numerical techniques	Development Economics Group (1971)
	Pollution impacts on communities	Scattered observation: predictive techniques inadequately developed	Lagerwerff and Specht (1970)

(continued)

TABLE 10.1. A Survey of Techniques Appropriate to Various Environmental Issues *(continued)*

Major issues	Indicators	Measurement techniques	References
Abiotic disruption	Soil pollution	Detailed	
		Soil surveys and analysis	Platt and Griffiths (1964) Smith (1974) Lagerwerff and Specht (1970)
	Water pollution	Water quality monitoring and analysis	U.S. EPA (1971b) Goodwin and Niering (1974) Kaill and Frey (1973)
	Geochemical cycles	Chemical analysis	Platt and Griffiths (1964)
	Climate—macro, micro	Biometeorological methods	Geiger (1957) Munn (1970)

[a] From Bigelow-Crain Associates (1976: Dii–iii).

TABLE 10.2. Impacts on Major Life Forms[a]

Life forms	Typical threats	Measurement	Mitigation possibilities
Large animals (wild and domestic)	Reduced range Reduced food supply and cover within range	Direct counts (sample plots) Aerial photographs Local wildlife biologist's estimates	Minimize intrusion on range
Predatory birds	Reduced prey populations Destruction of habitat (brush, trees, etc.)	Local wildlife biologist's estimates	Protect nesting areas Minimize habitat reduction
Small game animals and birds	Destruction of habitat	Local wildlife biologist's estimates	Minimize disturbance of land contour and vegetation
Fish, shellfish, and waterfowl	Dissolved oxygen reduction (e.g., untreated sewage) Acidity level and toxic materials (e.g., dumping, erosion) Draining wetlands	Direct counts	Minimize pollution Alter the land–water interface as little as possible
Field crops	Change in groundwater level Excessive herbicides	Direct tally	Maintain acreage
Natural land vegetation	Land clearing and erosion Excessive herbicides	Aerial photography (acreage and character of vegetation)	Restrict the area affected
Aquatic plants	Changes in water level and quality (e.g., drainage, erosion)	Aerial photography (area) Local ecologists (water quality necessary)	Limit input of heat and erosion products Minimize wetland drainage
Threatened animal or plant species			Minimize disruption of habitat (under the guidance of ecologists)

[a] Based on Jain et al. (1977).

SOME POTENTIAL ECOLOGICAL IMPACTS ASSOCIATED WITH WATER RESOURCE PROJECTS (Warner et al. 1974)

Terrestrial biological impacts
- Inundation of habitat
  - Loses vegetation-producing lands
  - Displaces natural wildlife populations
  - Interrupts migration routes of animals

- Change in the land/water interface
  - Stimulates growth of normally suppressed organisms, possibly to detriment of others
  - Causes appearance of new dominants and population complexes
  - Forms localized marshes or swamps
  - Restricts shoreline vegetation species due to reservoir level fluctuation

- Change in downstream habitat:
  - Reduces natural fertilization from flooding
  - Changes vegetation profile

Aquatic biological impacts
- Reservoir pool formation
  - Causes instability for several years following construction
  - Increases bacterial decay
  - Reduces dissolved oxygen
  - Displaces natural river habitat and organisms
  - Alters fish distribution, interrupts migration routes
  - Creates a rich organic food supply for insect larvae
  - Changes host–parasite relationships of organisms
  - Increases aquatic growth at reservoir edges

- Alteration of downstream flow
  - Reduces environmental quality from insufficient dissolved oxygen
  - Reduces flow of food downstream from lake sedimentation
  - Changes thermal profile
  - Modifies spawning areas downstream
  - Increases hatchery success due to altered flow variability

Ecological system analysis typically involves two steps: (1) obtaining data on existing systems and (2) anticipating impacts of development on the systems.

Obtaining data on existing systems largely depends on *ecological surveys*. These may be augmented through interviews with naturalists and literature searches, depending on the needs of the TA/EIA. There are many aspects and variations to ecological surveying. One approach (Bigelow-Crain Associates, 1976: 92) suggests three basic survey steps:

1. *Reconnaissance* is an initial examination of an area, usually undertaken quickly and conveniently. It includes the most obvious and general features, such as overall topography, gross soil types, dominant vegetation, characterization of major communities present, habitat types, and some initial estimate of animal types.
2. *Primary survey* details generalizations made on initial visits. These specifics may include vegetation distribution, major topographical features, and animal species makeup.
3. *Intensive survey* usually covers smaller areas than the first two steps. More detailed maps are developed, and many quantitative data are gathered. Diagrams, profiles, lists, and charts are produced.

These steps can be expanded by planned experimental manipulations, monitoring of important indicators, and management actions to ameliorate negative project impacts. Maps of the of topography, vegetation population distributions, and climatic characteristics can be invaluable. These may be generated from aerial and terrestrial photographs, remote sensing technologies, site pacing and observation, and standard survey methods. Alternative development maps may be very useful in conveying policy-relevant ecological information (Findley 1975). Surveys naturally must vary to address the more significant impact concerns. For instance, water-resource projects may entail fish, microorganism, and aquatic plant surveys. Wetlands may require special study in some instances, and so on.

For all ecological systems, it is important to acquire descriptive baseline data on the system in question. This generally implies some form of survey, as quantitative as possible. But this is not enough. One must develop reasonable projections of the likely effects of alternative developments. Expert opinion and reasoning based on analogous cases appear to be the best methods at hand. Modeling of ecological systems and the effects of specific intrusions is not sufficiently developed for practical TA/EIA application in most cases. However, this is not a reason for allowing ecological system analysis to avoid specifics. Instead, careful incorporation of field data with logical analysis is appropriate.

10.4 LAND

Depending on the nature of the technological development, one may be concerned with: *land use*, *solid-waste* accumulation, *erosion*, and *pollutants* (e.g., pesticides). The methods of analysis depend greatly on the situation and the relative importance of particular impacts.

Land use with respect to a project, its induced developments (e.g., land used in construction of a highway and related commercial developments), and their compatibility with the environment is commonly considered in TA/EIA. The extent of discussion depends on the importance of the use—public parkland disturbances exemplify a sensitive issue (Johanning and Talvitie 1976). One should discuss alternative sites, likely impacts, and steps taken to minimize harm to the land.

Land-use changes are one of the few environmental impacts sometimes claimed to be beneficial. For instance, reclamation of swampland for agricultural usage or commercial/residential development on previously agricultural land may be considered advantageous. Naturally, both presumedly positive and adverse effects should be carefully documented. For example, the review of the Garrison Diversion project (Pearson et al. 1975; see also Chapter 5 Appendix) noted that it would reduce cropland by 8148 acres, grassland by 39,172 acres, and trees by 6276 acres. Coördination with land-use planning efforts is important, particularly where special regulations apply, as, for instance, in coastal zone areas.

The federal government is involved with solid waste through the Solid Waste Disposal Act of 1965. The Resource Conservation and Recovery Act of 1976 deals with the sanitary disposal of solid waste. National standards are presently under development by the EPA administrator. However, enforcement is on the state level with the EPA dealing with states to ensure that their standards are sufficiently stringent. For projects creating solid waste, U.S. EPA Region X (1973a: 33–34) offers the following useful consideration:

> Projects that will result in creation of solid waste, either during construction or as a result of operation of the completed facility, should address the following information.
>
> 1. The quantities and composition of solid waste which will be generated both in the construction process and as a result of operation of the facility.

2. Will any hazardous wastes be produced as a result of the proposed action?
3. Discuss the forecast for long term future waste loads resulting from the project. That is, what additional waste loads from population influx can be anticipated? Increased solid waste loads may overload existing facilities for handling residential, commercial and industrial wastes. Have local waste authorities been made fully aware of the new waste loads that will result from a rapid increase in population?
4. What plan has been developed for the storage, collection, and disposal of all the different types of waste that will be generated?
   a. Where and how will wastes be stored?
   b. When will collections be made?
   c. What is disposal method?
5. Has the potential for recycling or reuse of wastes generated by project been fully investigated?

In considering the impacts of solid-waste disposal, the Resource Conservation and Recovery Act considers a sanitary landfill. Brunner and Keller (1972) discuss sanitary landfill design and operation in detail. The main potential negative impacts are on ground and surface-water quality. Leachate (water that has percolated through solid waste carrying with it soluble and suspended substances) from solid-waste disposal sites can cause serious and expensive problems (Garland and Mosher 1975). Unfortunately, knowledge of leaching mechanisms is presently weak (Cook 1976). Harmful effects by landfills on water purity can be avoided by proper site location, design, and operation (Brunner and Keller 1972). Incineration is an alternative to landfill. Energy can be recovered from solid waste through such available technologies as waterwall incineration (U.S. EPA 1977). To reduce the amount of solid-waste landfill, materials recovery systems have been designed. An EPA pilot system at Franklin, Ohio, in which only about 10% of the incoming waste needs to be landfilled, has been described by the U.S. EPA (1977: 102–104).

Erosion, in which soil particles are dislodged and carried off, results from interactions of water and air with soil. Land clearing sets the stage for erosion. Ground cover, slope, and soil characteristics interact with the physical development to produce erosion. Models of soil loss have been developed (U.S. Department of Agriculture 1975) that depend on these factors. Erosion can be minimized by maintaining and replacing adequate ground cover. Maps are a useful means for displaying land use and erosion impacts.

Soil, as well as air and water, is susceptible to pollution from chemical, biological, and radiological agents. Pesticide use, either as a major project feature (e.g., area-wide eradication of fire ants) or an incidental one (e.g., clearing of brush along a highway) is of particular concern. Pollution may lead to a loss of groundcover and hence indirectly cause erosion. Finally, soil quality affects land-use possibilities.

10.5 WATER

The modeling of aquatic impacts is often quite complex. The scope of water-quality analyses precludes detailed coverage here. Instead, we will identify standards and the range of water-quality considerations, briefly discuss some influences on water quality, and mention some modeling efforts.

First, it is important to be aware of the existence of *legal* standards. The 1972 amendments to the Federal Water Pollution Control Act (PL 92–500) take precedence over NEPA in establishing effluent limitations [see U.S. Congress, Congressional Research Service (1973a)]. These require the states to revise their water-quality standards in consultation with EPA. Standards are determined by types of use of streams and bodies of water, such as fish and wildlife habitat, agriculture, and public water supply. Each body of water is then classified appropriately.

Figure 10.1 shows some of the various water quality parameters that may be of concern and possible water bodies to which that concern may be directed. Particular issues will naturally vary according to the assessment involved.

Hydrologic characteristics include such information as high and low streamflows, occurrence of floods, flood-plain characteristics, groundwater flows, tributaries, natural drainage channels, and alterations to natural hydrologic conditions that will result from a project's construction and operation (U.S. EPA 1973a: 25).

The most studied of the physical properties is temperature (Warner et al. 1974). Temperature affects many of the other environmental properties. The level of biological activity, for example (as measured by gross productivity), doubles for every 10°C increase in water temperature, creating a corresponding increase in oxygen demand. In reservoirs or other substantial water bodies, thermal stratification occurs, and surface waters receive a greater amount of solar heating. These effects and their consequences merit close attention (Table 10.3).

The quantity of dissolved oxygen (DO) is a primary indicator of biological activity. The solubility of oxygen in water decreases with increases in temperature. The form of aquatic life depends on DO levels—absence may produce anaerobic conditions (and associated odors). The 3–6-mg/liter range contains the critical level of DO for

	Water bodies[a]							
Water-quality parameters	Oceans	Estuaries	Lakes (reservoirs)	Streams	Wetlands	Irrigation	Urban runoff	Groundwater
Hydrologic characteristics								
– Water quantity (flows)								
– Drainage patterns, etc.								
Physical properties								
– Temperature (stratification, diffusion)								
– Other (density, turbulence, light, solubilities)								
Chemical constituents								
– Dissolved oxygen (stratification, various BODs, anaerobic conditions)								
– Dissolved solids (salinity, leaching of minerals, sediment)								
– Nutrients (nitrogen, phosphorus)								
– Toxic chemicals (pesticides)								
– Other (trace elements, CO_2, pH)								
Biological constituents								
– Algae								
– Bacteria (coliform)								
– Vegetation (weeds, woody debris)								
– Fish								

[a] Usage is an important consideration also; water-quality standards differ for drinking-water supplies, swimming, fishing, irrigation, industrial use, and so on.

FIGURE 10.1. Water-quality parameters and potential points of concern.

TABLE 10.3. Selected Reservoir Water Temperature Models[a]

Model type	Major model assumptions	Parameters measured	Reservoir modeled: correlation of predicted with measured	Simulation method	References
TIDEP III (turbulent diffusion)	1. Horizontal temperature homogeneity 2. Inflow and outflow are advective heat sources and sinks 3. Coefficient of vertical turbulent diffusion constant with depth 4. Change of area with depth is an exponential function 5. Coefficient of thermal exchange is a cosine function	1. Wind speed 2. Air temperature 3. Air humidity 4. Net incoming radiation 5. Surface water temperature 6. Equilibrium temperature 7. Heat exchange coefficient 8. Vertical turbulent diffusion coefficient 9. Change of area with depth 10. Inflow and outflow temperature	Priest reservoir—fair to good correlation	Short computer simulation	Burdick and Parker (1971)
Cornell model (turbulent diffusion)	(Same assumptions as 1, 4, and 5) 2. Inflow and outflow are advective heat sources and sinks 3. Coefficient of vertical turbulent diffusion varies with depth and friction velocity 6. Equilibrium temperature variation a sine function	1–7 Vertical turbulent diffusion coefficient measurement Inflow and outflow advective temperature	Cayuga Lake—good correlation	Numerical analysis by computer	Sundaram et al. (1971)

Internal absorption	1. Nonhomogeneity of horizontal temperature profile 2. Net shortwave energy distribution beneath the water surface a decaying exponential function 3. Coefficient of turbulent diffusion not constant 4. Change of area with depth significant	Hydrologic, meteorologic, climatologic conditions Cloudiness, DB-temperature DB-temperature, barometric pressure, wind velocity	Fontana Reservoir, Lake Roosevelt, Hungry Horse Reservoir—good correlation	Explicit numerical analysis by computer	Orlob and Selna (1968); Water Resources Engineers, Inc. (1969); U.S. Army Corps of Engineers (1970); Bacca et al. (1973)
MIT model	1. Thermal gradient exists in vertical direction—horizontal homogeneity 2. Sides and bottom of the reservoir insulated 3. Density, specific heat, coefficient of molecular diffusivity constant 4. Solar radiation is absorbed both at the water surface as well as in the water body 5. No "intrance" complications due to density difference 6. Withdrawal can take place from several levels, singly or concurrently	Hydrological and meteorological parameter 1. Solar radiation 2. Atmospheric radiation 3. Air temperature 4. Relative humidity 5. Wind speeds 6. Streamflow rate 7. Streamflow temperature	Fontana Reservoir—good correlation	Explicit numerical analysis by computer	Ryan and Harleman (1971); Huber and Harleman (1968); Huber et al. (1972)

[a] From Warner et al. (1974: 41).

nearly all fish. Dissolved oxygen varies with time and depth; it stratifies with temperature. Biochemical oxygen demands (BODs) of several sorts can be analyzed. High BODs threaten the DO level, with attendant consequences. Biochemical oxygen demand relates to the amount of biologically degradable organic material in the water. Proper treatment of organic wastes lessens BOD. Most DO problems occur in summer when biological activity, creating the DO demand, is high, while the solubility of oxygen is reduced due to the increased temperature. BOD and DO problems are among the main threats to water quality.

Dissolved solids are significant in a variety of contexts. Sedimentation may affect streams and impoundments. Salinity increases may occur downstream from impoundments in association with reduced flows and decreased oxygen solubility. Irrigation may deposit salts in soils and leach out minerals that, in turn, affect receptacle water bodies. For potable water, 500 mg/liter of total dissolved solids is a recommended maximum (Jain et al. 1977: 221). Nutrient levels are related to biological activity (e.g., algae growth) and, hence, changes in BOD. Acidity or alkalinity (with extreme pH values destroying aquatic life), carbon dioxide in solution, trace elements, and pesticides deserve consideration also.

Biological constituents are clearly part of the aquatic ecosystem. Some important aspects are treated here in the discussion of water quality. Coliform bacteria from sewage may cause disease and are a routine public-health concern. These can be reduced by treatment of sewage. Vegetation such as aquatic weeds (floating, submersed, or emersed) can prevent use of waterways. Hundreds of square kilometers may be affected (Holm et al. 1969). Biological species interact in interesting ways—for example, carp digest sediment and excrete large amounts of ammonia and phosphate, contributing to algae growth (Pearson et al. 1975: 36).

Warner et al. (1974: 29) discuss certain factors that greatly influence water quality in reservoirs. Although these factors do not apply exactly to other bodies of water, the list suggests points worth attention in environmental analysis:

1. the quality of the inflowing water;
2. climatic controls, including precipitation, temperature, insolation, wind, and evaporation;
3. inorganic chemical reactions, including chemical precipitation, complex formation, and oxidation–reduction reactions;
4. morphometric controls, such as the size, shape, and depth of the impoundment;

5. biological controls with regard to effects of living organisms on organic and inorganic materials;
6. management controls, including factors manipulated by man to regulate conditions within the reservoir or downstream from it, such as retention time (Love and Slack 1963).

Beneficial, as well as undesirable, effects on water quality can be caused by confinement. Although the following lists cannot be readily generalized, they offer ideas for impact analysis (Love 1961).

Important beneficial effects of impoundment on water quality include:

1. reduction of turbidity, silica, color (in certain reservoirs), and coliform bacteria;
2. evening out of sharp variations in hardness of dissolved minerals and pH (a measure of acidity and alkalinity);
3. reductions in temperature that sometimes benefit fish life;
4. entrapment of sediment; and
5. storage of water for release in dry periods for the dilution of polluted waters.

In addition, impoundment also has certain undesirable effects, including:

1. increased growth of algae, which may give rise to tastes and odors;
2. reduction in dissolved oxygen in the deeper parts of the reservoir;
3. increase in carbon dioxide and frequently iron, manganese, and alkalinity, especially near the bottom;
4. increases in dissolved solids and hardness as a result of evaporation and dissolution of rock materials; and
5. reductions in temperature, which although sometimes beneficial, may also be detrimental to fish life.

Quantitative models of water quality have been developed for many of the parameters in Figure 10.1. These include temperature, DO, BOD, chemical oxygen demand, nitrogen, phosphorus, toxic compounds, salts, and turbidity. Some water bodies have been extensively studied (e.g., reservoirs), whereas others are still poorly understood [e.g., groundwater—see McCaull and Crossland (1974)].

Warner et al. (1974) illustrate the nature of the modeling efforts applicable to reservoirs or streams for temperature, DO and chemical constituents. Temperature models basically involve heat exchange, a topic well studied by chemical engineers in various contexts. The differential equation formulations vary greatly in their complexity (see

Table 10.3). Dissolved-oxygen models for reservoirs are complicated by stratification and the complex hydrodynamics. Many models apply to the top layer only (the reservoir euphotic zone, where 99% of the incident light is observed) or to flowing streams. One differential equation equates the rate change in DO to DO import and removal rates. A different approach involves an empirically derived regression equation to describe downstream DO effects from a point source of waste input. Another considers DO in the context of a set of mass transport and chemical reactions, yielding distributions of DO, BOD, algae populations, and nutrient materials. Various oxygen demand models consider algae and plant growth rates, and carbonaceous and nitrogenous components. A nitrogen model is built, for instance, upon algae effects and ammonia–nitrate dynamics. Bacca et al. (1973) discuss construction of such water-quality models.

Baseline data and information on water quality impacts can often be obtained from appropriate state and local agencies. These can inform mitigation efforts. Most of the efforts to improve water quality focus on discharges from point sources. Nonpoint source problems are usually addressed indirectly through reduction of point discharges.

10.6 AIR

Clean air, like clean water, is necessary for the survival of human beings and other living things. Discussions of air quality impacts are strikingly parallel to those relating to water quality impacts. Both have received recent national attention leading to strong standards; both have been quantitatively modeled with significant successes. As with water, the analyses of air quality extend beyond our capability to summarize here. We discuss the key parameters and sketch some analytical considerations.

Environmental analysis must consider whether the proposed action will lead directly or indirectly to air-pollutant concentrations that exceed the national ambient air-quality standards. These standards are set by EPA, pursuant to Section 109 of the 1970 amendments to the Clean Air Act. They are constantly altered as new knowledge and technology becomes available (cf. the 1977 edition of the *Federal Register*, 40 CFR, part 50). In addition, projects must be in conformance with federal performance standards for stationary sources, for example, for steam generators, cement plants, or municipal incinerators. There are also standards for allowable degradation by new air-pollution sources set specifically by Congress. These appear in the June 19, 1978 *Federal Register*. State and local standards and pollution control emission regulations for pollutants for which federal standards have not been established must also be addressed. Any proposed action should also be consistent with state-adopted plans for achieving air-quality standards (U.S. EPA 1973a, 1974a). A listing of 28 major categories of air-pollution sources appears in the August 7, 1977 amendments to the Clean Air Act.

The five major primary air pollutants (Bigelow-Crain Associates, 1976: D2–3; Jain et al. 1977: Appendix B) are as follows:

Hydrocarbons (HC) are substances whose molecules contain only hydrogen and carbon atoms. They are emitted mainly as a result of the partial combustion of fossil fuels.

Carbon monoxide (CO) is a colorless, odorless, tasteless gas resulting from the incomplete combustion of hydrocarbons. Its main source is the automobile.

Nitrogen oxides (NO_x), mainly nitric oxide (NO) and nitrogen dioxide (NO_2), are formed when nitrogen and oxygen from the air are combined under high temperature. Thus they are characteristic of any high-temperature combustion process such as occurs in an automobile engine or a fossil-fueled electric power plant.

Sulfur oxides (SO_x), mostly sulfur dioxide (SO_2) with some sulfur trioxide (SO_3), are emitted when fossil fuels containing sulfur impurities are burned.

Particulates is a broad category that includes a wide range of solid or liquid particles that are typically emitted during combustion, from the grinding of materials, during the course of moving dirt, and so on. They may range in diameter from 100 μ (microns) to less than 0.01 μ. Some of the deleterious properties of particulates result from their chemical composition, whereas others are merely a result of their size.

These five factors are commonly referred to as "primary pollutants" to distinguish them from the so-called "secondary pollutants" associated with photochemical smog. In particular, nitric oxides and hydrocarbons, in the presence of sunlight, undergo a partially understood, complex series of reactions. This results in the generation of various harmful *secondary* pollutants, including nitrogen dioxide (NO_2), ozone (O_3), and peroxyacetyl nitrate ("PAN," $CH_3CO_3NO_2$). Ozone and PAN are usually referred to as *photochemical oxidants* (Masters 1974: 177).

In addition to the substances mentioned above, a variety of other hazardous toxicants may be present in the air. These include arsenic, asbestos, beryllium, boron, copper, lead, zinc, and mercury. The federal government has established standards for asbestos, beryllium, and mercury. Airborne radioactive wastes are also of concern.

Airborne diffusion of pollutants is an area of concern. The diffusion pattern is determined by such factors as the vertical temperature gradient of the air, wind, topography, humidity, and pressure. Mathematical models of atmospheric diffusion have been developed that can be used in TA/EIAs (Ingram and Fauth 1974; Gifford and Hanna 1970; Martin and Tikvart 1968). Grad et al. (1975) used quantitative techniques to model air quality and dispersal in a TA. Cook (1976: 91) describes the state of the art:

> There are numerous validated models available for calculating air pollutant concentrations resulting from various types of projects in both the point source and non-point source categories. . . . Only SO_2, particulates, and CO, which are relatively inert, have been modeled with consistent accuracy. The other pollutants are fairly reactive and require the combination of complex chemical reaction formulas with diffusion models, which simulate atmospheric processes.

Until validated models are available for the reactive pollutants, empirical techniques must be used.

The effects of the air pollutants are not fully understood. However, serious consequences arise from excessive concentrations. Combinations of pollutants are often associated with undesirable effects not related to a single pollutant. [For a more detailed discussion, see Jain et al. (1977: Appendix B).] Carbon monoxide combines with the hemoglobin in the blood, displacing oxygen and lessening the oxygen-carrying ability of the circulatory system. Excessive concentrations (as with a running car in a closed garage) lead rapidly to death. Excessive concentrations of nitrogen oxides cause increased incidence of respiratory diseases such as bronchitis. Sulfur oxides are associated with respiratory diseases and are involved in corrosion and plant damage. The combination of sulfur oxides and particulates can increase mortalities among sufferers from certain respiratory diseases. Particulates are also responsible for corrosion to metal structures, soiling structures and clothing, and visual problems. Photochemical oxidants cause serious injury to vegetation and attack materials such as rubber and polymers. Asbestos may cause cancer. A comprehensive reference on the relationship between air quality and health is published by the League of Women Voters (1970) and distributed by EPA.

Measurement techniques for atmospheric pollutants vary from substance to substance. A summary of these techniques appears in the relevant sections of Appendix B of *Environmental Impact Analysis* by Jain et al. (1977). The basic unit of measurement is the quantity of pollutant present, averaged over some time interval, typically 24 hours or less.

The seven steps laid out by Bigelow-Crain Associates (1976: D13–15) for transportation developments give an example of a format for analysis of impacts on air quality:

1. Collect data to establish background levels of air-quality indicators likely to be influenced by transportation activities [e.g., carbon monoxide (CO), hydrocarbons (HC), oxides of nitrogen (NO_x), oxidants, and particulates].[2]
2. Determine all existing land uses likely to be influenced by air-quality changes associated with the alternative transportation proposals under consideration. Identify "sensitive receptors," (i.e., areas and activities especially sensitive to air-quality degradation).
3. Identify criteria and standards for determining acceptable air-quality levels for each land use and activity determined in step 2. This can be done with the aid of EPA's "air quality criteria" documents (U.S. EPA 1971a), and the federal and relevant state air-quality standards. [Federal standards are summarized by Masters (1974: 194–195).]
4. Estimate the air-pollution emissions associated with alternative transportation proposals for air-quality indicators identified in step 1.
5. Employ available methods and techniques to estimate the changes in concentration of various air-quality indicators caused by the emissions identified in step 4; this is done for selected meteorological conditions and future time periods.[3]
6. Identify all areas where anticipated air quality changes will exceed acceptable levels.
7. When forecasted concentrations of various air-quality indicators are unacceptable, modify facility designs (e.g., consider alternative highway routes) or proposed operations (e.g., alter flight patterns of aircraft) to bring air-quality changes to acceptable levels.

Mitigation strategies vary from pollutant to pollutant and from situa-

[2]Sources of data regarding background levels include county and state air-pollution control agencies, regional and state "air-quality implementation plans," and EPA's system for "Storage and Retrieval of Aerometric Data" (SAROAD); for information on the latter, see Chamblee and Nehls (1973).

[3]A compendium of meteorological conditions in the United States that is useful for air-quality studies is given by Holzworth (1972).

tion to situation. Emissions may be reduced by the use of technological devices for pollution control at the point of emission, the change to less-polluting processes (e.g., by changing type of fuel used), or by limiting the use of polluting processes (e.g., by limiting hours of operation). The number of entities affected by atmospheric pollution may be reduced by relocating them to areas of less pollution. Finally, climate control in limited areas may lessen exposure to pollution.

10.7 NOISE

Noise can be defined as unwanted sound. The principal sources of noise are surface and air transportation, human and animal activity, construction, and industry. Noise is a concern because of its effects on people, with community annoyance a more sensitive factor than physiological or structural induced damages (e.g., sonic boom). As with air pollution, noise levels must conform to standards, are quantitatively modeled, and are particularly significant in transportation. The effects of noise on people are diverse. They include [Jain et al. 1977: 55; U.S. Federal Aviation Administration (FAA) 1977]:

- interference with speech communication (e.g., residences, schools);
- interference with sleep (e.g., motels, hospitals);
- lessened enjoyment of recreational areas (e.g., campgrounds, wild areas);
- annoyance (e.g., adjacency to aircraft operations, truck routes);
- hearing impairment (e.g., from working around aircraft); and
- other physiological effects.

The Noise Control Act of 1972 required the EPA to establish criteria to protect the public health from noise and to set standards for various classes of equipment, such as motor vehicles and construction equipment (U.S. EPA 1973c). Current "in-use" standards are set for interstate motor and rail carriers. New product standards exist for portable air compressors and medium- and heavy-duty trucks. Each federal agency is responsible for noise standards in its areas of concern. For example, the Department of Housing and Urban Development (HUD) has its *Noise Assessment Guidelines* (1971). In addition, applicable state and local standards should be considered in any analysis of the impact of noise.

Important properties of noise are loudness, duration, and frequency content. Equipment for analyzing the frequency spectrum of noise is complex. In general, high-frequency noises such as whines are more annoying and damaging than most other noises. Noise is measured using a sound-level meter at the point where the noise will be heard. The meter has an "A"-weighted network that simulates the response of the human ear. The unit of measure is decibels on the A-weighted scale (dBA). Peterson and Gross (1972) and the American National Standards Institute (1971) offer detailed treatment of noise measurement.

Noise must be considered both in absolute and relative terms. The sound excellence level is denoted by L_X, where X is the percentage of time it is exceeded. For instance, recommended criteria that enable people to converse at 10 feet are L_{50} and 55 dBA outside. To protect 50% of sleepers from awakening, an outside L_{10} 60-dBA level is recommended (EPA-NTID 300.7). But just as importantly, an increase over preexisting levels of over 10 dBA is likely to generate a substantial number of complaints, whereas few complaints are likely if the increase is under 5 dBA (U.S. EPA 1973a).

The dBA$-L_X$ noise-description scheme is not the sole choice. For instance, the FAA favors an effective perceived noise level (EPNL) scale that takes into account the subjective response to aircraft noise, the presence of pure tones, and the duration of exposure (Anthrop 1973; Sperry 1968). However, the EPA objects that the resultant Aircraft Sound Description System forecasts cannot be used to evaluate compatible land-use or noise impact areas (Cook 1976).

The following are procedural recommendations for analyzing noise impacts from one important noise source, transportation projects (Hawker 1973: 503; Bigelow-Crain Associates 1976: D63–64).

1. Collect data to establish existing background noise levels (EPA has a research arm that may assist other governmental units in collecting data on noise levels).
2. Determine all existing and proposed land uses likely to be influenced by noise from the alternative transportation facilities under consideration; identify areas or activities especially sensitive to noise.
3. Identify criteria for acceptable noise levels for each type of land use or activity specified in step 2 [for information on criteria, see, e.g., U.S. EPA (1973c), U.S. Federal Highway Administration (1973)].
4. Employ available methods (e.g., computer models) for determining noise production from the operation of the alternative transportation facilities.
5. Identify all areas where anticipated noise levels exceed acceptable levels.

6. When noise levels exceed acceptable levels, modify facility designs (e.g., use of noise barriers) or proposed operations (e.g., rerouting of airport flight patterns) to reduce noise to acceptable levels.

Transportation EISs often use quantitative noise models (Johanning and Talvitie 1976), following the procedures suggested by the National Cooperative Highway Research Program.[4]

Highway-noise calculations based on vehicle volume, average speed, and truck-to-auto ratio projections give L_{10} and L_{50} at reference locations under various simplifying assumptions. Monographs have been developed for simple manual computations (Wesler 1972). For more complex situations, widely available computer programs exist (Transportation Research Board 1976).

For airports, "noise footprints" (contours of the effective perceived noise decibels) for particular types of aircraft are entered with various factors into computer models to project airport noise (Bartel et al. 1974; U.S. FAA 1977). Most planners compute both the aircraft sound description system and the noise exposure forecasts, preferred respectively by FAA and EPA.

A specific analysis of noise impacts could use an approach such as the following. The project may be divided into two phases of activity—construction and operation. Each phase may be broken down into activities performed. Noise sources should be listed and described. The time of day and duration of each should be determined and noise contours constructed. Applicable state standards should be investigated to assure compliance. Additional noise arising from transportation in the area should also be taken into consideration. Particularly sensitive areas, where noise problems might arise, should be determined, and strategies for mitigating the impacts developed.

The most basic means to reduce noise impacts is to reduce the noise level at its source. This is accomplished by phasing out or retrofitting old equipment. Replacement by new equipment with lower noise levels will solve the problem in time. Alternatively, or additionally, individuals can be removed from proximity to the sources of noise (e.g., by relocating airport neighbors). Finally, sound insulation can be used to protect individuals from noise.

[4]See especially Gordon et al. (1971), Kugler and Piersol (1973), and Transportation Research Board (1976). Other pertinent sources of guidance in assessing noise impacts include Bragdon (1972), Peterson and Gross (1972), U.S. Department of Housing and Urban Development (1971), and the sources noted in Table 10.1.

10.8 RADIATION

Radiation effects are generated by cosmic radiation, terrestrial nuclear radiation, x-rays, and microwaves. In the United States the regulation of radiation is divided among a number of agencies. Maximum available radiation levels are set nationally by the Nuclear Regulatory Commission (NRC). The NRC has primary responsibility for assessing both the radiological and nonradiological impacts of nuclear facilities and processes. Recently EPA has been working with the NRC to develop joint regulations and licensing procedures for nuclear reactors. These regulations are becoming increasingly more stringent. The Bureau of Radiological Health of the Food and Drug Administration has responsibility, in collaboration with the states, for regulation in the areas of x-rays and microwave devices. Additionally, they are responsible for federal product standards affecting microwave emissions by such consumer goods as television sets and microwave ovens. Finally, EPA has a generic interest in microwaves and performs research on their effects.

Radiation effects are of the utmost concern in connection with the development of nuclear power. The NRC has established EIA guidelines (Part 51 of NRC regulations). They require prospective licensees to submit a comprehensive environmental report from which NRC staff develop a draft, then final, EIS. The NRC Regulatory Guide 4.2, Rev. 1, "Preparation of Environmental Reports for Nuclear Power Stations," provides guidance for such license applications. The GESMO statement (see Chapter 5 Appendix) encompasses radiation effects associated with recycling of fuels in normal modes of operation, in accidents, and under threat of sabotage or theft.

Introduction of a new technology can sometimes indirectly lead to an alteration of radiation levels. Two recent controversial technologies, the supersonic transport (SST) and spray-can propellants, threatened a potential disruption of the atmospheric ozone layer with accompanying increases in cancer-producing radiation.

The radiation dose received by a target is a reflection of the energy imparted by the incident radiation per unit mass of the target (Eichholz 1976). One *rad* is equal to 100 ergs per gram. The *rem* is a rad multiplied by Q, a quality factor expressing the effect on biological tissue of the absorbed dose in terms of a scale common to all types of radiation. Natural radiation, caused by cosmic rays and natural terrestrial sources of radioactivity, is about 50 mrem/year. This is normally higher than the amount of artificial radiation.

The principal impact of high doses of radiation is the destruction of the cells in the vicinity of the irradiated tissue. There are two generic types of impact. Somatic impacts lead to radiation-induced diseases, including leukemia, and cancer of the lung, breast, and thyroid. Genetic impacts consist of gene mutations and chromosome aberrations. Gene mutations are the more serious since they persist through generations. Estimates of mutations range from 30–1500 per 10^6 people per rad (Eichholz 1976: 123). Standards have been set for radiation exposures (acute and cumulative) aimed at those in occupational exposure and for the general public. Monitoring of exposures of workers is maintained.

The issue of whether the current standards are sufficiently stringent is a topic of debate. In particular, one may either assume that small dosages over extended periods must be considered additively or that there exist thresholds below which no significant dangers occur. These opposing assumptions lead to an order of magnitude difference in projected health effects. Gofman and Tamplin (1970) raised this issue luridly in terms of thousands of predicted additional cancer deaths annually from low-level exposure of the general population. They also voiced concerns that cancer mortalities may lag many years after exposure. They charged that the Atomic Energy Commission (now succeeded by the NRC and the Department of Energy) was an advocate of nuclear development at the expense of determined assessment and regulation.

Another forecasting problem involves the proper way to calculate the accident risk. In the case of nuclear power plants, one is dealing with complex systems with redundant subsystems, each having small individual probabilities of mechanical failure, but also the less predictable human factor in system operation (see Section 11.5.1 on risk–benefit analyses). Predictive models also have difficulty in taking account of intentional threats to the security of nuclear facilities (note GESMO, Chapter 5 Appendix).

Finally, there is the impact on the environment due to the radioactive wastes from nuclear fission reactors, weapons systems, and other sources. Some radioactivity will always escape. It is dispersed into the food and water supplies and through the air (Eichholz 1976). Disposing of radioactive wastes is one of the major problems associated with nuclear power. Present regulations call for the wastes to be solidified and stabilized within 5 years, then shipped to a federal repository within 10 years. There are various types of storage, including retrievable storage in tunnels and permanent underground disposal in deep wells. All current approaches concern environmentalists because of the long-lived radioactivity of some of the waste products and the uncertainties associated with their storage.

Damage to human beings from microwave and x-rays may also be significant. Standards for maximum exposure exist, but Brodeur (1977) has written a popular indictment of government and industrial practice in determining allowable exposure to microwaves. He cites such proven and possible dangers as cataracts, genetic damage (including mongoloidism), and behavioral changes. Because of the strong differences within the U.S. medical community and throughout the world, it would seem appropriate to learn more about the problem through research and to take precautions against exposure to x-rays and microwaves in the home, in industry, and in the military. Appropriate precautions include improved design and use of shielding, and periodic inspection of radiation-emitting devices.

10.9 OTHER ISSUES

Two areas that deserve mention but are not covered in detail here are historic preservation and light intrusion. *Historic preservation* is the responsibility of the Department of Interior's Heritage Conservation and Recreation Service. Various laws from the American Antiquities Act of 1906 to the present deal with this area. The National Historic Preservation Act of 1966 requires that any federal agency having direct or indirect jurisdiction must take into account the effect of the undertaking on anything that is included or eligible for inclusion in the National Register of Historic Places. At present the Advisory Council on Historic Preservation is given a reasonable opportunity to comment on all such undertakings. Executive Order 11593 of May 1971 ordered the development of criteria and procedures to be applied by federal agencies in determining the effects of their actions on historic preservation.

Historic preservation involves not only artifacts of relatively recent times, such as houses and cemetaries, but also archeological remains of past cultures. Each site has its own past. Efforts by historians, anthropologists, and archeologists are required to analyze the significance of sites.

Light intrusion is a localized phenomenon, analogous to sound except that it is less pervasive. It impacts buildings and their surroundings by putting vegetation into shadow and altering heating and cooling requirements. Were solar energy to come into widespread use, the blockage of sunlight would become a factor of even greater concern.

At this point, we need to stress the importance of *higher-order en-*

vironmental impacts. For example, it is clear from the previous discussion that water quality impacts affect ecosystems. Higher-order impacts induced by nonenvironmental impacts are also important. For example, greater economic resources in a region may lead to increased recreational facilities that may spawn additional environmental impacts. In the case of the sample highway EIS (summarized in Chapter 5 Appendix), the more consequential environmental effects are attributable to economic development spurred by the highway, not to the highway itself.

The assessor must also be alert to prevent slippage between the categorical cracks. Besides missing environmental aspects of peculiar importance to a given project, there are dangers of too-narrow conception. For instance, the Institute of Ecology's review of the Garrison Diversion project EIS points out apparent violations of U.S.–Canadian International Law regarding migratory birds and boundary waters. A meaningful point of comparison from an *energy analysis* (see Section 11.5.2) of this project is that 2.4 million acres could be fertilized annually if the energy inputs were redirected. It is unlikely that a preestablished set of impact categories will fully satisfy any environmental analysis.

Iteration of analysis is therefore a good idea. Preliminary results of the environmental, economic, social, and institutional analyses can augment each other. For instance, understanding changes in the ecosystem resulting from alterations in water quality may suggest further water quality changes to be taken into account.

Widely varying *levels of analysis* in terms of depth and breadth are possible. Without belaboring the issue, decision as to the appropriate level of analysis is important and may be subject to criticism. As much as anything, the Institute of Ecology's critique of the Garrison Diversion EIS (Pearson et al. 1975) pointed out inadequacies in the depth and detail of analyses.

Evaluation of impacts (see Chapter 14) deserves special attention in the environmental area because of the existence of standards for air quality, radiation, and so on. However, impacts must be assessed even if standards will not be violated. The valuation issues, which can be highly polarized, need to be faced. Environmental quality indices (Kimball 1973) offer some sense of general baseline conditions and trends.

Finally, environmental impact assessment can be augmented by other sources of information such as *monitoring* (see Section 6.4.2 and Bigelow-Crain Associates 1976: D-V). Given the current questions as to the validity of predictive techniques, identification and monitoring of critical environmental parameters attendant to a technological development may offer early warning of potentially significant impacts.

10.10 CONCLUDING OBSERVATIONS

This chapter is concerned with the analysis of impacts on the natural environment. Because these impacts are so very important in EIAs, the preparation of environmental impact statements is discussed as well. The 1978 CEQ regulations apply to all federal EISs. In addition, each federal agency may prepare specific guidelines for particular areas under its jurisdiction. The Construction Engineering Research Laboratory EIS outline is summarized as a good guide to topical coverage.

Environmental quality standards for various impact categories are mandated by federal and state law and must be taken into account. They are particularly important considerations in the assessment of air and water quality, radiation, and noise impacts.

The six major categories of environmental impacts are ecological systems, land, water, air, noise, and radiation. These impacts should not be treated in isolation because they are interactive, producing higher-order impacts. For example, air and water quality may affect ecological systems.

Various analytical techniques are available to treat environmental impacts. These range from sophisticated, quantitative models to qualitative, sketch techniques (Table 10.1). Water quality and quantity, air quality, and radiation are among the most modeled areas.

Some key considerations for each of the types of environmental impact include the following:

Impacts on ecological systems are studied by determining the population and character of the various life forms present. Habitat effects are a critical ecosystem concern.

Impacts on land are dealt with under the categories of land use, erosion, solid-waste disposal, and pollutants.

Water-quality parameters of importance include temperature, DO, BOD, sedimentation, pH, algae concentration, and coliform bacteria count. Figure 10.1 lists water-quality parameters and the variety of settings to which they may pertain.

Major air pollutants are carbon monoxide, nitrogen oxides, sulfur oxides, hydrocarbons, particulates, and photochemical oxidants.

Noise impacts include physiological, psychological, and performance effects as well as problems relating to communication and social behavior. Human response to noise can be measured on the dBA scale, and L ratings provide a measure of sound excellence characteristics.

Radiation effects arise from artificial and natural nuclear radiation as well as x-rays and microwaves. They include damage to tissues and cells and, more importantly, cumulative genetic damage.

Historic preservation should be specifically addressed in an EIA.

Light intrusion may be significant; it is somewhat analogous to noise.

Several general considerations deserve special attention. Iteration of environmental analyses in conjunction with other impact analyses (i.e., EPISTLE) can uncover higher-order, interactive effects. Evaluation of impacts is necessary to distinguish the important from the trivial. Finally, monitoring of environmental conditions over the course of a technological development can usefully augment TA/EIA efforts.

RECOMMENDED READINGS

Bigelow-Crain Associates (1976) *State and Regional Transportation Impact Identification and Measurement, Phase I Report.* Prepared for National Cooperative Highway Research Program, Transportation Research Board, National Research Council.

This report emphasizes the understanding of impacts, primarily environmental, due to transportation projects. It nicely summarizes a variety of techniques and gives numerous references.

Heer, J., Jr. and Hagerty, D. J. (1977) *Environmental Assessments and Statements.* New York: Van Nostrand Reinhold.

This book gives an overview of environmental impact assessment with an emphasis on the legal basis of assessment and procedural requirements.

Jain, R. K., Urban, L. V., and Stacey, G. S. (1977) *Environmental Impact Analysis: A New Dimension in Decision Making.* New York: Van Nostrand Reinhold.

This book offers a coherent treatment of many environmental "attributes" whose understanding is essential to EIA.

Warner, M. L., Moore, J. L., Chatterjee, S., Cooper, D. C., Ifeadi, C., Lawhon, W. T., and Reimers, R. S. (1974) *An Assessment Methodology for the Environmental Impact of Water Resource Projects.* Prepared for the Office of Research and Development, U.S. Environmental Protection Agency.

This report covers a wide variety of impact considerations for water resource projects.

Warner, M. L., and Preston, E. H. (1974) *A Review of Environmental Impact Assessment Methodologies.* Prepared for the Office of Research and Development, U.S. Environmental Protection Agency.

This work summarizes 17 different general impact analysis methods and compares them on a number of criteria.

EXERCISES

10-1. (Project-suitable) Select a topic for environmental analysis. It may be local or national in scope, an ongoing TA/EIA project or not, a real or hypothetical case.

1. Step lightly through the Construction Engineering Research Laboratory outline (U.S. Army 1975), identifying the more significant environmental impacts which you might anticipate in the selected case.
2. Drawing upon the text discussions and your impressions as to the more significant impacts, describe your analytic strategy. Specifically, indicate promising data sources, what sorts of models you might use, and the reasons for your choices.
3. If a group has dealt with the same topic, discuss and compare strategies.

10-2. Referring to Exercise 10-1, identify several instances where environmental impacts may interact to cause higher-order effects. For instance, you may have suggested that construction of a dam will lead to agricultural land irrigation and also to reduced stream flow. It is possible that runoff from the irrigated land will greatly increase toxic element concentrations in the low flow stream.
Also identify several instances where environmental impacts will interact significantly with economic, social, legal, and other factors.

10-3. Pick one of the major environmental impact areas (ecological systems, land, water, air, noise, or radiation). Choose one important parameter and a given type of setting (e.g., dissolved oxygen concentration for water quality in a flowing stream). Research in some detail the available techniques to *measure* this parameter (e.g., water sampling).

10-4. Research in some detail the available techniques to *analyze impacts* on the parameter chosen in Exercise 10-3 (e.g., in the DO instance, detail available models). If the analytical techniques for the selected parameter are elaborate (e.g., sophisticated models), focus on a single one described in the literature. (The text references a number of analytical techniques.)

10-5. Discuss the important provisions of laws (including federal and state) that affect one of the generic classes of impacts in your state. Mention any applicable standards and procedures for their enforcement.

10-6. Interview a person or persons affected by some major environmental action. What effects did they think were most important? What was their overall evaluation of the action? If an EIS was prepared, were they aware of it? Check the EIS (if one exists) to see to what degree it treated the expressed concerns. Were comments on those concerns received and answered?

11

Economic Impact Analysis

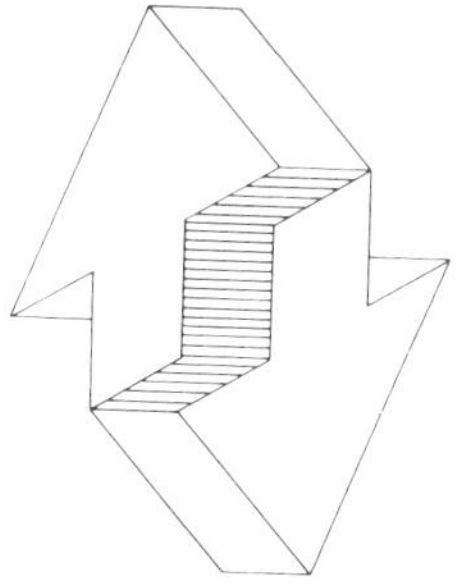

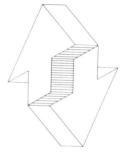

Any technological development hinges on its economic feasibility. Economic analysis in TA/EIA concerns both the potential profitability of a development and its implications for the broader economic interests of the affected publics. This chapter presents cost–benefit analysis as the main framework for economic analysis in TA/EIA. The reader should gain a sound understanding of the components of cost–benefit analysis, become sensitized to the conceptual issues, and learn how to perform the basic analysis. Selection of economic evaluation criteria (i.e., how to discount future costs and benefits) receives special attention as an important issue. Particular attention is also paid to several techniques that can contribute to the understanding of indirect economic impacts. The chapter provides a general accounting of the principles of economic base models, input–output models, and regression analyses. Risk–benefit analysis and net energy analysis, two techniques that complement the cost–benefit approach, are also introduced.

11.1 INTRODUCTION

A common sequence for the assessment of a proposed technological development is to consider its (1) technological feasibility, (2) economic consequences, and (3) other impacts. In other words, the economics of a situation speak loudly—an uneconomic option rarely receives further consideration. In this chapter we address the analysis of the economics of a technological development.

In the private sector, economic analysis focuses on the profitability of a development. However, a thorough analysis requires attention to a wide range of factors that could affect profitability beyond simple market considerations. For instance, governmental incentives directly alter the balance of costs and benefits to be weighed. For public-sector developments, even the focus of economic analysis must be broad. There, one is attempting to gauge the social advantages of developing a particular technology or policy. Cost–benefit analysis (CBA), the central topic of this chapter, is a response to the need to assess such developments in the public sector.

The Institute of Ecology (Andrews et al. 1977) has addressed the difficulty in understanding economics from an ecological perspective. Economics values entities such as air, water, land, and energy by their utility to man. Thus an acre of urban wooded land is valued economically more highly than an acre in a national forest. The guiding concept in the development of economics is the law of scarcity. The goal of economics is to determine how to provide for the demands of the human population from a stock of scarce resources.

The economist responds to scarcity by allocating resources using three criteria: efficiency, effectiveness, and equity. *Efficiency* attempts to maximize output of an activity while minimizing input. Efficiency answers the questions asked initially about a new endeavor such as how much it will cost and whether it will be profitable. *Effectiveness* is concerned with whether an activity will achieve its goals within the time allowed. Effectiveness is primarily an evaluative criterion to be used during the development of the project. Together, effectiveness and efficiency speak to the economic attractiveness (profitability) of a development. *Equity* focuses on the distribution of costs and benefits among different sectors of society. Winter cloud seeding to increase water flow in the Colorado River Basin may lead to effects in Colorado resorts, for example, raising equity-related issues. Economic analysis in TA/EIA involves all three of these criteria.

A number of methods have been developed to evaluate considerations of efficiency, effectiveness, and equity. Discounting analyses, for example, concern economic attractiveness (efficiency and effectiveness); consideration of externalities concerns equity. Private decision typically rests on economic attractiveness; public decision requires knowledge of both the attractiveness and equity of an option. The cost–benefit framework serves to integrate these concerns.

Indirect impacts of a development are a major concern of TA/EIA. More than in other disciplines, economists have developed rigorous methods to treat indirect effects of technological changes. Such indirect effects certainly accrue as costs or benefits and should, therefore, be incorporated into CBA. The methods involved are rather intricate and will be treated in detail in Section 11.6. Economic base (or export base) models, input–output models, and the broad family of other regression-based techniques are introduced there.

Two concerns not routinely treated in CBA have raised considerable recent attention—analysis of risks and energy usage. Risk–benefit analysis deals with probable risk situations in which costs and benefits depend on particular outcomes of a development (e.g., an accident at a nuclear power plant). Net energy analysis considers a development in terms of energy use rather than monetary implications. Both these techniques can be seen as variations of CBA with potential value in TA/EIA (Section 11.5).

Economic analysis can provide a great deal of information about the potential for a technological development. In most cases, economic impact analyses tend to be the most quantitative of the impact analysis categories. This should not, however, be taken as the most important or accurate part of TA/EIAs. Quality assessment requires the artful blending of the quantitative and the qualitative, of the objective and the subjective, and of science and politics.

The next section sets out the principles of CBA; the subsequent sections raise analytical issues concerning measurement of costs and benefits and the use of discounting in CBA.

11.2 COST–BENEFIT ANALYSIS

11.2.1 The Cost–Benefit Framework

The basic objective of an economic analysis is to determine whether the benefits derived from a project outweigh the costs incurred. The analysis of economic impacts for a specific project or technology has its origins in the investment decisions of private firms. There, the investment commitment is weighed against the expected return before a decision is made on whether or not to invest. Such a narrow consideration of costs and benefits can be readily expressed in dollar values. TA/EIA considers costs and benefits far beyond those to a particular firm. They include costs to the environment, to society at large, and to those institutions that plan, finance, regulate, and otherwise interact with the development.

Cost–benefit analysis incorporates a broad array of loosely structured economic analysis techniques in a common framework. The following series of steps comprise the structure of a CBA.

1. Define the subject of the analysis, including the parties at interest.
2. Identify and describe the anticipated costs and benefits.
3. Measure the costs and benefits, including:
   a. direct costs and benefits;
   b. externalities;
   c. public goods; and
   d. secondary and higher-order impacts.
4. Evaluate the economic impacts:
   a. select decision criteria;
   b. select discount rate;
   c. perform the economic analysis; and
   d. do sensitivity analysis.

These steps serve to structure the remainder of Sections 11.2–11.4. Mishan (1976) and Sassone and Schaffer (1978), among many others, provide detailed descriptions of cost–benefit analysis.

11.2.2 Definition of the Subject of the Analysis

The first step in CBA is to define the subject of the analysis. This is most important as it provides scope and direction for the entire analysis. If the project is not clearly defined, the results of the analysis may be meaningless. Subject definition consists of the following:

1. A description of the subject of the assessment at the level of technical detail required to perform the economic analysis.
2. The time period over which the analysis will be made, including important milestone events.
3. The location, if any, of the subject of the assessment.
4. Background assumptions relating to the technical and social context in which the development will take place.
5. A "without project" description, that is, a description of how things would be if the development being assessed did not take place. This provides a baseline against which the project is com-

pared. This description should be presented in a level of technical detail consistent with (1).

6. The parties who will pay the costs and reap the benefits, and their interest in the subject of the assessment.

The TA of cloud seeding in the Rocky Mountans to increase snowpack (Weisbecker 1974) provides an illustration of a well-documented CBA discussion. The cameo entitled "Colorado Cloud Seeding CBA" in Section 11.2.3 gives a sense of the scope of subject for that CBA.

11.2.3 Identification of Costs and Benefits

Identification of economic costs and benefits may seem a straightforward task. It is not. The breadth of assessment considerations and the concern with higher-order and unintended consequences often make this identification quite complex. Cost–benefit analysis must not only address direct monetary factors (e.g., costs and profits), but also the economic aspects of other impacts (e.g., a dollar value of recreational use of a lake created by a new dam). By implication, sound economic analysis in a TA/EIA must consider the impacts identified in the other topical analyses (EPISTLE).

Two useful distinctions can be made in considering economic effects: (1) *private vs public* goods and (2) *internal costs vs externalities.* These will prove important in distinguishing among measurement strategies in the next section.

Private (individual) goods are those that each individual can choose to enjoy in varying amounts (e.g., ice cream). Public (collective) goods are those affecting many individuals at the same time, and for which each individual has little freedom in the matter of his or her own consumption (e.g., the air we breathe). The measurement of costs and benefits for public goods presents special problems because there is no market for them.

Internal costs and benefits are those that are routinely chargeable to a given development. In contrast, externalities are "spillover" effects. They are not readily charged, or credited, against the development that causes them. For example, the use of solar energy for home heating in a given area could result in a partial deforestation (to give solar collectors access to the sun's radiation). This could lead to externalities such as more rapid paint deterioration on neighboring houses and reduced natural facilities for children's play. These may have negative implications (e.g., children lose treehouses and tire swings) and positive ones (e.g., children suffer fewer accidents). Clear specification of the effects to be included in the economic analysis is important. The cameo entitled "Colorado Cloud Seeding CBA" illustrates the sort of economic impacts one might include.

COLORADO CLOUD SEEDING CBA

The TA of Winter Orographic Snowpack Augmentation (WOSA) (Weisbecker 1972) addresses the effects of cloud seeding over the Rocky Mountains to increase water supply in the arid West. The study uses, but does not overplay, cost–benefit analysis in a nicely balanced perspective.

A wide array of benefits and costs was identified in the TA. Benefits result primarily from the increased water supply. Major costs are due to the actual WOSA operations, impacts of increased snowfall, and impacts of increased waterflow in the Colorado River Basin. "External" costs to citizens east of the Colorado Basin region from decreased precipitation are also noted. The following five types of impacts are identified:

1. changes in the cash flows of individuals, whether local residents or residents of other areas;
2. changes in costs, levels of output, employment, or profits of locally operating firms and industries;
3. impact on visitors and tourists;
4. impact on local government finances;
5. impact on intangibles.

For example, the expense of removing the additional snow affects government finances. Also, changes in economic activity (e.g., increased skiing) may change the demand for local services. Changes in local property values would change property-tax revenues. Changes in the population may result in a reallocation of state and federal aid money. Impacts on intangibles include effects of WOSA on scenery, personal inconvenience, and ecological concerns (e.g., changes in the structure of local ecosystems).

11.3 MEASUREMENT OF COSTS AND BENEFITS

Costs and benefits may or may not be quantifiable. This chapter emphasizes the process of quantification of economic impacts. *This should not be taken as an indication of greater importance for the quantifiable effects.* Indeed, the Council on Environmental Quality (CEQ) NEPA regulations specify that an EIS that uses CBA shall not

Table 11.1 summarizes selected costs and benefits of WOSA. In many cases the magnitudes of the impacts are indicated only by pluses or minuses. These impacts are either relatively insignificant, not quantifiable in monetary terms, or not quantified due to the uncertainty of their occurrence.

The WOSA CBA distinguishes impacts by region. For example, if a company moves away from Colorado due to increased snowfall, a detrimental impact will result for the regional economy. (If that company relocates in the United States, the net impact on the national economy would be minimal.)

The decision criterion used in the WOSA study is the benefit:cost ratio, traditional in evaluating water projects. In evaluating WOSA, the benefit:cost ratio from a national welfare perspective is 4.43:1. However, the bulk of the benefits hinge on construction of five projects (e.g., dams) to hold back and transfer water primarily for irrigation. The individual project benefit:cost ratios range from 0.89:1 to 1.17:1, despite assuming low discount rates of approximately 3%. Were these projects not constructed, the national benefit:cost ratio would reduce to a marginal 1.34:1. The local and regional benefit:cost ratios are not computed. Rather than attempting to arrive at an overall measure of the total net benefit or cost of WOSA, the researchers chose to describe the economic consequences of the proposed technology for different localities and regions.

Joseph Coates (1976a: 169) said that the WOSA study

> highlighted the fact that cost benefit evaluation is not a clear cut technique. In this particular case (the WOSA TA) costs were inextricably embedded in major assumptions of continuing federal subsidies in relation to water and also contingent upon a variety of assumptions about sunk and non-sunk costs. The exercise illustrates the subtleties and value of applying cost benefit analysis to public policy decision making.

weigh the merits of alternatives by monetary values alone, when there are important qualitative considerations involved (U.S. CEQ 1978: 55997).

It is useful to quantify what can be appropriately quantified. Qualitative treatment of other effects must accompany the quantitative analysis. We suggest the framework of Sassone and Schaffer (1978) to accomplish this delicate balancing (Figure 11.1).

TABLE 11.1. Selected WOSA Costs and Benefits[a]

WOSA Costs		
Cost item	Basis	Cost (millions of dollars per year)
WOSA systems costs	Annual operating costs plus 10% of investment	5.424
Avalanche control	Sum of active control program and $50,000 per year for construction of passive protective structures	0.84
Flood forecasting	The sum of annual installation and OM&R costs for the period 1966–1980; as the system is established, the costs will decrease	0.25
Environmental monitoring	Costs were not estimated for a specific program; it is assumed that within the allowance, an adequate program could be developed based on needs and priorities that will change with time	1.000
In-basin economic detriments	The loss of one day's income over the entire area	2.000
Out-of-basin economic detriments	One-half of the in-basin estimate	1.000
Total		9.533

WOSA Benefits

Location	Condition	Water use	Basis	Benefit (millions of dollars per year)
In basin	Only existing facilities	Water quality	Cost of alternative-salinity management program	2.5
			Sediment trapping	–
		Water-supply reliability	Short-term, local—rescue benefits	+
			Cost of alternative-storage facility	+
		Hydropower generation	Fuel savings of marginal thermal plants	5.3
		Agriculture		+
		Municipal and industrial		+
		Intermountain diversion		+
	Construction of additional facilities—regardless of WOSA	Mexican Water Treaty	Cost of transfer from agriculture	30
	Additional facilities built only if WOSA is implemented	Prospect uses	Combine WOSA costs with project benefits and costs	0
Out of basin	Only existing facilities	Hydropower generation		0.9
		Agriculture		2.0
		Municipal and industrial		2.1
Total				42.8

[a] From Weisbecker (1972).

FIGURE 11.1. Cost–benefit analysis accounting work sheet.
Line 3 = Line 1 + Line 2; Line 4 = Line 1 + Line 2.
Source: Sassone and Schaffer (1977).

	BENEFITS (NPV)									COSTS (NPV)							
		NATIONAL	REGIONAL			INCOME CLASS					NATIONAL	REGIONAL			INCOME CLASS		
	PROJECT EFFECT	I + II + III - L + M + U	I	II	III	L	M	U		PROJECT EFFECT	I + II + III - L + M + V	I	II	III	L	M	U
MONETIZED EFFECTS	REAL FINAL GOODS AND SERVICES																
1	SUM REAL BENEFIT								1	SUM REAL COSTS							
2	TRANSFERS								2	TRANSFERS							
3	REGIONAL ACCOUNT								3	REGIONAL ACCOUNT							
4	INCOME DISTRIBUTION								4	INCOME DISTRIBUTION							

NON-MONETIZED EFFECTS	ENVIRONMENTAL	AESTHETICS								AESTHETICS							
		IMPORTANT SITES								IMPORTANT SITES							
		WATER AND AIR QUALITY								WATER AND AIR QUALITY							
		IRREVERSIBLE CONSEQUENCES								IRREVERSIBLE CONSEQUENCES							
		OTHER								OTHER							
	SOCIAL	LIFE								LIFE							
		HEALTH								HEALTH							
		SAFETY								SAFETY							
		EDUCATION								EDUCATION							
		CULTURE								CULTURE							
		RECREATION								RECREATION							
		EMERGENCY PREP.								EMERGENCY PREP.							
		OTHER								OTHER							

Quantification of the costs and benefits is a difficult task. Many variables defy quantification, particularly in monetary terms. For example, the costs of an oil spill that kills birds and temporarily devastates beaches are impossible to quantify in a manner acceptable to all. As indicated in Figure 11.1, these should be treated qualitatively. Other variables are amenable to assignment of the dollar values. This section discusses a variety of applicable economic quantification techniques.

An important consideration in the measurement task is the unit in which costs and benefits are quantified. The measurement basis used in virtually all CBA studies is monetary (e.g., U.S. dollars). Other measurement bases have been proposed (e.g., labor value or energy). Energy units are sometimes used to supplement the use of dollars (see Section 11.5.2).

The three subsections that follow discuss the measurement of direct costs and benefits, externalities, and public goods. Direct costs and benefits are private and internal. Externalities and public goods are addressed separately because of the peculiar measurement issues they raise.

11.3.1 Direct Costs and Benefits

Direct costs and benefits are those clearly accountable to the development being assessed. For example, a direct cost of a new airport is the contractor's fee for laying a runway. Direct benefits are the revenues received from airlines who pay a fee to use the new facility. Direct costs and benefits are the primary consideration in private-investor decision making.

It is important to recognize the typical flows of costs and benefits caused by a development. In most cases, a development incurs a high capital cost required to pay for the initial construction or implementation. Once implementation is completed, annual operating costs are incurred for labor, raw materials, and maintenance. At the same time, benefits, usually in the form of revenues, are received. The benefits may increase over time as the development increases in visibility and popularity. The flow of costs and benefits may be represented as a time stream, as shown in Figure 11.2. Techniques used to determine whether a particular flow of costs and benefits is economically attractive are discussed in Section 11.4.

Pricing costs and benefits is not an automatic process. The price of an item is determined in the marketplace through the workings of supply and demand. For instance, a project that leads to a substantial increase in the supply of a product is likely to lower the price. The

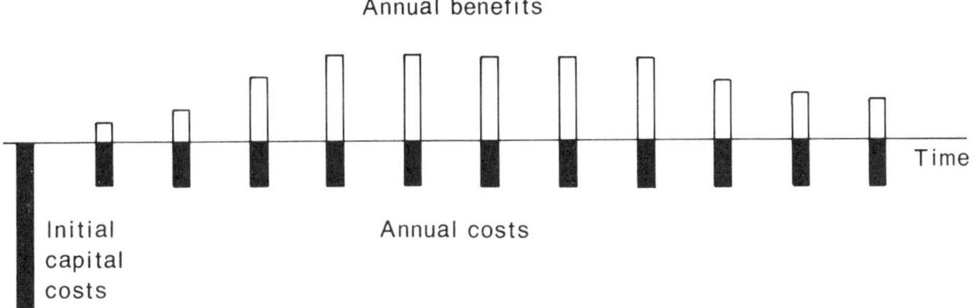

FIGURE 11.2. Typical flow of costs and benefits.

cameo entitled "Supply and Demand Curves" discusses the computation of expected price changes due to changes in supply and demand.

11.3.2 Externalities

Direct costs and benefits often fail to include many variables important in a TA/EIA. External effects result from existence of the project but are not included in the private investor analysis. For example, health impairment due to air pollution is an external cost. An increase in property values in existing homes due to renovation of nearby blighted areas is an external benefit.

Externalities can result either from requirements of the development (inputs) or from effects caused by the development (outputs). An example of an input is electricity used for power in a project. Of course, the generation of electricity causes a wide variety of external effects (e.g., air and water pollution). Whether these externalities should be considered in a project that simply consumes electricity depends on the effect the added demand has on the total amount of electricity generated. Three cases exist: (1) if the power company supplies the new demand by reducing supply to other sources and does not increase generation, the project is not responsible for these externalities of the electrical output; (2) if the power company increases its generation output to match the requirements of the project, the externalities resulting from the increased generation are attributable to the project; or (3) if the power company increases its generation to a level below that in (2) but greater than in (1), the project is responsible only for the increases in externalities due to power generated for its own use. Case 1 reflects completely inelastic supply; case 2 displays a completely elastic supply; and case 3 is somewhere in between. Refer to the cameo on "Supply and Demand Curves" for a discussion of elasticity.

SUPPLY AND DEMAND CURVES

Figure 11.3 shows a typical set of supply and demand curves. The demand curve slopes downward, reflecting a tendency for consumers to purchase less as the price of the product increases. The supply curve slopes upward, showing the sellers' willingness to put more products on the market as the price increases. The intersection of the two lines is the equilibrium point, where supply and demand match. In an ideal, perfectly competitive market, the equilibrium price would be the market price.

The notion of *consumer surplus* can be derived from Figure 11.3. Some consumers would be willing to pay a higher price for the product than the equilibrium price, P_0. For example, Q_1 products could be sold at price P_1. The price at which buyers are willing to purchase a product reflects its value to them. Thus, by buying the product at P_0, a consumer willing to pay price P_1 is saving $P_1 - P_0$. This is the consumer surplus. The area of the shaded triangle shows the total consumer surplus.

Another important concept reflected in the supply and demand curves is *elasticity*. Supply elasticity is the relationship between the quantity supplied and the price of a product. Similarly, demand elasticity is that between the demand for a product and its price. Elasticity can be computed as follows:

$$\text{Elasticity} = \frac{\text{Change in quantity}}{\text{Change in price}} \times \frac{\text{Average price}}{\text{Average quantity}}$$

Sassone and Schaffer (1978) suggest four ways of assigning external costs of outputs (none of which they find completely satisfactory). We paraphrase these.

1. Conduct a survey of *willingness to pay* among the affected individuals and hope that true preferences are revealed or that the exaggerations cancel out the understatements.
2. As an estimate of willingness to pay, compute the *costs of avoiding* the externality. The avoidance cost has no special significance to a CBA beyond its intuitive appeal as a reasonable number to look at, and perhaps as an indicator of the order of magnitude of the true cost.

A commodity with a high demand elasticity (>1) shows a proportionally larger drop in demand for an increase in price. Products with readily competitive substitutes, or luxury items, tend to have high demand elasticities (e.g., caviar and European vacations). Commodities with low demand elasticities (<1) show proportionally smaller drops in demand in response to an increase in price. Necessities or nonsubstitutable items fall in this category (e.g., salt, gasoline, and matches).

Another important use of supply and demand curves is to determine the impact of an increase in the price of a product on the demand for that product. Imagine that P_1 in Figure 11.3 represents a case in which the price of a good increases due to imposition of some form of regulation (e.g., water-quality standards) requiring investment (e.g., pollution-control equipment). Demand responds by decreasing the quantity sold to Q_1. The revenues derived from sales before the price increase are equal to the product of price and quantity sold (P_0Q_0), which is the area of the rectangle $0P_0EQ_0$. If P_1Q_1 is greater than P_0Q_0 the revenues will actually increase. Any additional revenues will be offset by the additional costs required, thus affecting net income. Also, decreased production resulting from lower demand will probably cause unemployment.

3. All the social benefits, and all *other* social costs, can be computed to yield a *net social value* of the project before the inclusion of the value of the externality. Thus if the value is already negative, or lower than some alternative project, the project is definitely not worth undertaking, and an exact computation of the loss due to the externality need not be attempted. If the value turns out positive, a *critical value* can be computed for the externality, and the judgment left to the decision maker as to whether the actual social cost of the externality exceeds the critical value.
4. As a last resort, a qualitative description of the impact of the externality is presented to the decision maker. In many cases it is better to describe the effect than to give it an arbitrary value.

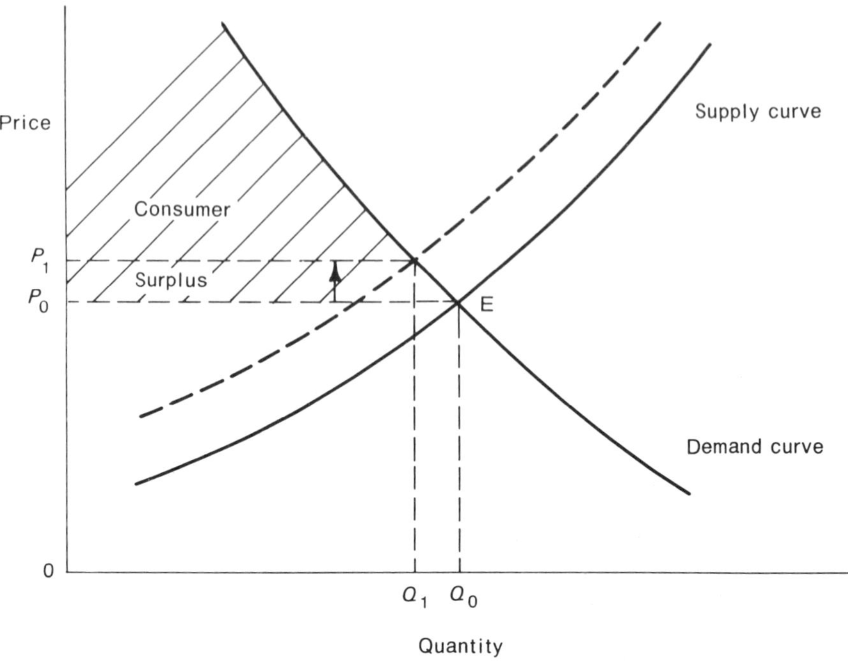

FIGURE 11.3. Supply and demand response to a price increase.

11.3.3 Public Goods

Public goods have two important characteristics. First, one person's consumption does not significantly diminish the amount available to any other person. Second, it is not feasible to exclude any individual from consuming the good once the good is provided. Since one does not have to pay explicitly to consume the good, there is insufficient incentive for the good to be provided by the private sector. Thus market conditions for the good do not exist, and there is no price. Examples of public goods include clean air, highways, and parks. Public goods are often difficult to quantify in monetary terms.

One approach to the problem of establishing a measurement base is to "*shadow price*" the costs and benefits to which an immediate dollar value cannot be assigned. A shadow price may be defined as "a value associated with a unit of some good which indicates by how much some specified index of performance can be increased or decreased by the use or loss of the marginal unit of that commodity" (Sassone and Schaffer 1978: Chapter 4). There are no comprehensive and foolproof procedures for shadow pricing. Subjective judgment weighs heavily.

Sassone and Schaffer (1978: Chapter 5) suggest a number of possible approaches to the shadow pricing of public goods, but, as with externalities, they find none fully satisfactory. We paraphrase:

1. *Surveys of willingness to pay.* The questions are asked in a manner disguising their purpose. The problem remains that responses to survey questions may not honestly reflect one's values.
2. *Analogy to private goods.* Where the public good is related to some marketed good, the price of the latter may be a guide to the value of the former. The concept of consumer surplus (see Figure 11.3), however, reveals that market prices do not reflect the total value of a marketed good.
3. *Experiments.* Individuals might be asked to participate in "realistic" games designed to reveal their true preferences. Such information is costly and usually of questionable reliability.
4. *Public referenda.* These provide a number of output-cost levels to vote on, where the means of financing the project can reasonably be claimed to be currently unknown. Since no one knows how the costs will be shared, it is hoped the votes do not reflect strategic behavior.

Lovins (1977) eloquently critiques such quantification of costs and benefits. He cites the case of a study on the relocation of the London Gatwick airport in which homeowners were asked how much they would accept as compensation for loss of their house. In two willingness-to-sell surveys, a significant percentage (8% in one case and 38% in another) of respondents said that no amount of money could replace their home. The analysts of one survey decided that three times the market price of the house was a fair assessment of its value to the owner. Lovins lambasts this arbitrary assignment of values.

Much of Lovins' argument is directed against the misuse of CBA. Many of the detractions can be circumvented, if CBA is properly employed. Perhaps the major point to be made is that quantitative CBA results should not be the sole criteria on which a project is judged.

A special concern is the assignment of a "social discount rate" to public sector costs and benefits. Treatment of this topic accompanies consideration of the basic discounting issues of the next section.

11.4 EVALUATING ECONOMIC IMPACTS

After measuring the quantifiable costs and benefits, the assessor must decide how to evaluate them. Variables must be quantified on three scales—magnitude, time of occurrence, and place of occurrence.

Magnitudes are treated by using a common metric (i.e., dollars or energy units). Techniques for evaluating time streams of costs and benefits are numerous. Approaches for reconciling divergent regional impacts, however, are not well developed. This section emphasizes analytical techniques for numerically evaluating time streams of costs and benefits.

11.4.1 Discounting

The discount rate is a crucial concept in evaluating the economics of projects. Discounting is the means by which one compares current dollar investments with future dollar returns. Basically, the discount rate measures how much more resources used now are valued over resources used in the future, that is, the time preferences for consumption. For instance, if an individual felt that $106 a year from now was as acceptable as $100 today, the discount rate would be 6%. (Even if that person did not want to spend that $100 now, it could be invested to earn interest.) At that $6 discount rate, the individual would expect $112.36 for a $100 investment in two years ($100 × 1.06 × 1.06). Costs and benefits are discounted at an increasing factor equal to $(1 + \text{discount rate})^t$, where t is the number of years in the future.

Discounting exerts a tremendous impact on the value of costs and benefits. For instance, at the same 6% discount rate, $100 in 20 years would be worth $31 today. At a 20% discount rate, the $100 in 20 years would be worth only $3 today. Obviously, at a 0% discount, $100 in 20 years is equivalent to $100 today.

Inflation is related to the choice of a discount rate. The value of all currencies, in terms of purchasing power, has deteriorated over time in recent years. Given this state of affairs, in order to normalize monetary values, it is sometimes useful to convert current dollars to constant dollars. *Current dollars* refer to the prices of products at any point in time. *Constant dollars* represent the corresponding amount that would have been paid for the same products in a base year. For example, if the inflation rate were 6% per year between 1970 and 1980, a device purchased for $2000 in 1980 would be valued at $2000 in both current- and constant-dollar 1980 dollars (which are the same). However, the device would cost $1117 (or $\$2000/1.06^{10}$) in constant 1970 dollars.

For costs that increase at a rate different from the standard inflation rate, *escalation rates* are used. These are the differences in the projected rate of increase for cost variables and the standard inflation rate. For example, if fuel prices are projected to increase at an annual rate of 9.5% while the standard rate of inflation is 6%, the escalation rate for fuel prices is 3.5%.

A CBA must specify whether prices are in current or constant dollars. If the analysis is in constant dollars, the appropriate discount factor is

$$\frac{1 + \text{current dollar discount rate (e.g., loan rate)}}{1 + \text{inflation rate}} - 1.$$

Where appropriate, inflation rates should be replaced by escalation rates.

In practice, the usual procedure is to perform the analysis using *several discount rates*. For example, the evaluation of a public sector project might use discount rates of 2%, 6%, and 10%. Lovins (1977), in his critique of cost–benefit analysis, accuses analysts of varying the discount rate to justify those programs they favor. By selecting the guidelines for the analysis before any results are obtained, the analyst is less likely to interject bias in the outcome. By showing the results for different discount rates, the user of the assessment can see the sensitivity of the CBA to the discount rate.

Public decision making uses the social discount rate, which measures the willingness of society as a whole to give up present consumption for future consumption. Most people agree (Sassone and Schaffer 1978) that the social discount rate is lower than the private discount rate. One basis for this conclusion is that the discount rate reflects the willingness of investors to accept risks. If the investment requires only a small part of the organization's resources, then the cost of risk-bearing to the organization as a whole is negligible. Another rationale for a low discount rate is the value of conserving nonrenewable resources. For instance, society may value a barrel of scarce oil more highly for future use than for present consumption. (In the extreme, this would imply a negative discount rate.) The difference in public and private discount rates also reflects the government's willingness to enter into potentially nonprofitable projects that provide nonmonetary benefits to citizens. However, when the public sector enters into projects that private industry can perform, it must be careful to avoid low rates of return, since taxation, which reduces the private rate of return, has paid for the public investment. Of course, one could argue that tax money requires no principal or interest payments and, hence, zero discount rate is justified. Still, one would want to compare alternative projects, considering the time course of the return on investment as possible opportunities forgone.

Calculation of a discount rate for a public-sector CBA is an uncertain task. Based on the preceding arguments, the range of discount rates to consider extends from the current private sector levels [reflecting return on investment and risk considerations, e.g., perhaps 10–13% in the late 1970s, to conservation levels (0%)]. A government agency may specify the discount rate to be used in a study. We recommend that analyses ascertain the implications of a wide range of discount rates for the CBA.

11.4.2 Alternative Decision Criteria

Many techniques have been devised to evaluate the economics of developments. The techniques reflect alternative decision criteria by which a development can be evaluated. The decision criterion selected depends on the characteristics of the project or policy itself, on the parties it will affect (intentionally and otherwise), and on the viewpoint adopted (public or private; local, regional, or national). The CBA decision criteria reduce the time stream of costs and benefits to a single number. The numerical results of the use of a particular decision criterion on several projects provides a ranking of their respective desirability from an economic efficiency standpoint. The following decision criteria are notable (however, many other possibilities exist): (1) net present value, (2) annual equivalent, (3) internal rate of return, (4) payback period, (5) benefit:cost ratio, and (6) composite economic and factor profiles.

We discuss some general factors to consider in selecting economic decision criteria. Each of the six approaches is introduced, but the reader is referred to textbooks on engineering economy (e.g., Thuesen et al. 1977; J. A. White et al. 1977) for detailed development. The reader is also referred to Chapter 14 for treatment of overall impact evaluation in TA/EIA.

Economic impact analyses may be categorized in several ways that have a bearing on the decision criteria to be used. First, they may pertain to either a private or a public perspective. We focus on the latter. Such techniques as *discounted cash flow* analysis are used in the former. This approach takes specific account of direct costs (outlays) and benefits (receipts), depreciation, taxes, interest, and salvage values—discounting entries to yield an after taxes annual cash flow (Thuesen et al. 1977).

A second element for economic analyses pertains to the breadth of focus—interest in all relevant economic effects or focus on a specific mission. We are concerned with the former, but note two potentially useful focused approaches. *Cost-effectiveness analysis* postulates specific developmental objectives and then considers the relative cost and effectiveness of alternatives in meeting those objectives. The primary distinction from CBA is that the specific focus allows for more clear-cut operationalization of developmental benefits—in terms of effectiveness in meeting the specified objectives. *Life-cycle costing* can be viewed as a special case of cost effectiveness in which the effectiveness of alternative developments is considered equal. Then alternatives (e.g., solar vs oil home heating) are compared in terms of their costs over the life of the systems. Discounting is a major consideration in cost effectiveness and life-cycle costing, as it is in the approaches we emphasize.

Another way to look at economic analyses is in terms of available comparisons. Sometimes a development can be compared with an existing alternative (e.g., in the life-cycle cost example of solar vs oil home heating). In others, the development performs novel functions, so its costs and benefits must be compared with a "do-nothing" alternative.

All of the six criteria share the notion of comparing time streams of costs and benefits. The first five are comparable ways to treat the monetary effects and are discussed together. The sixth augments the consideration of quantitative effects with nonquantitative ones and is considered later.

The five quantitative decision criteria differ in the way they present the results of discounting the time stream of costs and benefits. Generally, the higher the discount rate used, the less favorable the indication because project costs tend to be high early in the development. All of the five approaches are inherently comparable, but they differ in their advantages and weaknesses. The cameo entitled "Calculation of the Five Economic Criteria" briefly describes the calculation of each.

Net present value (present worth) provides a concise way to depict the time stream of project economics. It suffers from a difficulty in relating the calculated value to other parameters of interest, such as the initial investment.

As the computation (see cameo) suggests, annual equivalent is similar in nature to net present value. It can prove quite useful in relating specific units. For instance, suppose the oil and solar home heating systems postulated previously are being compared. An annual equivalent cost is intuitively useful; furthermore, it can readily be converted to a percentage comparison. For example, at a 6% discount rate the solar heating system is 16% higher in annualized cost.

Internal rate of return is informative in providing a break-even comparison, that is, with the discount rate that equalizes benefits and costs. It is somewhat more appropriate for private decision making.

The payback period also provides an intuitively attractive measure.

CALCULATION OF THE FIVE ECONOMIC CRITERIA

1. *Net present value* computes the present worth of a time stream of costs and benefits as follows:

$$\text{Net present value} = B_0 - C_0 + \frac{B_1 - C_1}{(1 + d)^n} + \frac{B_2 - C_2}{(1 + d)^2} + \cdots + \frac{B_n - C_n}{(1 + d)^n} + \frac{B_{n+1} - C_{n+1}}{(1 + d)^{n+1}},$$

where

B_0 = initial \$ benefits of the project;

C_0 = initial \$ costs of the project;

B_t = annual \$ benefits of the project in year t;

C_t = annual \$ cost of the project in year t;

B_{n+1} = \$ salvage value of the project;

C_{n+1} = \$ cost of dismantling the project;

d = discount rate;

n = life of the project in years.

When the annual benefits and costs are the same for all years, the summation can be simplified to

$$\text{Net present value} = -C_0 + (B - C)(F),$$

where

$B = B_t$

$C = C_t$

$F = [((1+d)^n - 1)/d(1+d)^n]$ and

B_0, B_{n+1}, and C_{n+1} are assumed to be insignificant.

2. *Annual equivalent* calculation involves first computing present worth and then converting it to an equivalent stream of annual amounts:

$$\text{Annual equivalent} = \frac{\text{Net present value}}{F},$$

where terms are as in (1).

3. *Internal rate of return* is the discount rate at which the present worth of a series of net benefits is equal to the present worth of a series of net costs. This is analogous to solving for d in

$$\sum_{t=0}^{n} B_t/(1+d)^t = \sum_{t=0}^{n} C_t/(1+d)^t.$$

When benefits and costs vary from year to year, more than one solution for d may exist.

4. *Payback period* is the time required for a project to recover its costs. When annual net benefits $(B - C)$ are equal in all years, the payback period is computed as follows:

$$\text{Payback period (years)} = \frac{\text{Initial costs}}{\text{Annual net benefits}} = \frac{C_0}{B - C}.$$

The payback does not use a discount rate because the payback period, in itself, is an indicator of the rate of time preference.

5. *Benefit:cost ratio* is computed as

$$B{:}C \text{ ratio} = \sum_{t=0}^{n} \frac{B_t}{(1+d)^t} \Bigg/ \sum_{t=0}^{n} \frac{C_t}{(1+d)^t}$$

$$= \frac{BF}{C_0 + CF} \quad \text{when annual benefits and costs are the same for all years.}$$

In $B{:}C$ ratios, careful delineation of costs and benefits is needed. If a benefit is identified mistakenly as a cost saving and subtracted from the denominator instead of added to the numerator, a different $B{:}C$ ratio results.

TABLE 11.2. Direct and Community Effects of Freeway Location[a]

Factor	Example	Units
Direct effects		
Cost of highway	Capital and annual costs of planning, construction, maintaining, and operating the freeway	Dollars
Costs (benefits) to highway user	Net effect (increase or decrease) to user in vehicle operation costs, commercial travel time, and accident rate	Dollars
Nonmarket values	Noncommercial travel time	Minutes or hours
Nonquantifiable, nonmarket value costs (benefits) to highway user	Pain, suffering, and deprivation costs of accidents; pleasure of scenic views from the road	?
Community effects		
Local transportation effects	Percentage reduction of through traffic on city streets; distance of freeway access from major traffic generators	Miles
Land use and environment	Creation of access to potential development land	Acres
	Increase in noise weighted by facilities exposed	Decibels (weighted)
Neighborhood and social structure	Number of housing units displaced	Number
	Cohesive neighborhoods served by freeway	Number of people
	Increase (decrease) in providing school services due to changes in bus routing	Dollars
Community economic and fiscal structure	Net change in assessed value of property on tax rolls	Dollars
	Net change in the costs of providing fire and police protection	Dollars
	Net change over normal trend in gross wholesale and retail sales	Dollars
	Net change in job opportunities in community area	Number

[a] Excerpted from Oglesby et al. (1970).

Its major flaw is in failing to consider costs and benefits beyond the payback period. For instance, a project with diminishing long-term benefits may appear more attractive than one with large benefits in future years that might have a higher net present value.

The benefit:cost ratio is used frequently in CBA (see cameo entitled "Colorado Cloud Seeding CBA"). It provides a neat metric on which to compare alternative projects and something of an absolute criterion in comparing the ratio to 1. Clearly, a project with a benefit:cost ratio less than 1 should not usually be accepted. As noted in the computation description, distinguishing benefits and costs is more critical than in the other four approaches.

The sixth approach to evaluating costs and benefits attempts to synthesize nonquantifiable effects with the quantifiable (recall Figure 11.1). We present the method of Oglesby et al. (1970) based on study of a highway project. They emphasize two principles of decision making: (1) decisions are to be based on the differences between alternatives, and (2) monetary aspects must be separated from aspects that are not quantifiable in monetary terms. The decision maker must weigh the nonmonetary terms against the monetary aspects. Project consequences are divided into direct effects (those emanating from construction and use) and indirect effects (those associated with the nonuser and community). Examples of the types of effect are shown in Table 11.2.

The methodology of Oglesby et al. (1970) utilizes a two-part analysis. The first part is an economic analysis, including all items quantifiable in monetary terms. The second is an analysis of pertinent items that are not quantifiable in monetary terms. The approach proposed for the second part is called a *community factor profile*.

The community factor profile indicates, for each alternative, the percentage of the maximum value of the measure determined for each factor. Table 11.2 presents a simplified version of the community factor profile. The factors used in the profile should include the major elements associated with the community impact. Factors should be aggregated, whenever possible. Inconsequential factors, and factors having the same value for all alternatives, should be eliminated.

As different interest groups will view the alternatives differently, it is important to determine their opinions of the importance of the factors associated with a particular project. Oglesby et al. (1970) suggest surveying as a means of obtaining this information. For instance in the highway study, air pollution was a minor consideration by engineers and planners but was quite important to citizens of the community.

Oglesby et al. propose the following steps to accomplish the economic and community impact analysis.

1. *Engineering economic analysis.* Rank the alternatives in order of preference as determined by the economic analysis (i.e., using one of the five economic decision criteria just discussed).
2. *Factor profiles.* Prepare factor profiles that show the freeway's impact on each relevant factor from the viewpoint of each interest group, and from an aggregate viewpoint.
3. *Economic and factor profile comparison.* Eliminate from consideration any alternative inferior to another based on both an economic and factor profile comparison.
4. *Paired comparison of alternatives.* Compare the more economically favorable alternatives to those having more attractive community factor profiles. The results of the survey help determine importance of factors having substantially different profiles for different alternatives. The basic question is whether the gains in community factors are worth the additional incremental costs of an alternative.
5. *Continued paired comparison procedure.* Repeat step 4 until all feasible alternatives have been included in comparisons. The end result is a preferred alternative, as well as a preference ranking among all alternatives.

In sum, we have presented five ways to gauge the financial attractiveness of alternative developments, and one way to combine these with nonquantifiable economic considerations. These results must be integrated with the impact analyses of other component parts of the TA/EIA. In addition, it may be useful to perform *sensitivity analyses* (Section 9.5).

11.5 VARIATIONS ON THE COST–BENEFIT FRAMEWORK

Two techniques provide potentially useful variations on cost–benefit analysis. Risk–benefit analysis and net energy analysis have received both interest and negative criticism. We introduce both techniques, noting potential utility for TA/EIA, as well as some criticisms.

11.5.1 Risk–Benefit Analysis

Risk–benefit analysis (also known as *risk assessment*) incorporates notions of probability and uncertainty into cost–benefit analysis. Risks may entail direct threats to human safety, to the environment, or simply the chance of economic damage. They may derive from the implementation of a technology per se, from the impacts of the technology, or even from social responses to perceived threat. The nuclear-energy field illustrates risk–benefit possibilities well (Otway and Pahner 1976: 123):

> operational risks may be reduced by expenditure on control equipment; the possibilty of disastrous, but infrequent, accidents; accident probabilities which can only be theoretically estimated and which are thus highly uncertain; the non-random distribution of risks and benefits to different groups of people; concerns about possible future (genetic) risks weighed against present benefits.

Otway and Pahner (1976) provide a suitable framework for risk–benefit analysis (Figure 11.4). There are two elements in the framework—*risk estimation* and *risk evaluation*. Risk estimation involves identification of the direct and indirect effects of a development, and calculation of the probabilities and magnitudes of the associated consequences. The notion of "risk" incorporates the probability and uncertainty of an occurrence (e.g., the chances of rain are 30%, but this is a highly uncertain estimate because of a new weather front developing), and the probability and uncertainty of a consequence if the event occurs (e.g., there is little uncertainty in the estimate that the probability of canceling the tennis match in case of rain is 90%).

Risk evaluation is the process of determining the value of the risks to those affected. Figure 11.4 offers two basic methods to scale risks. Surveys can be used to measure attitudes toward risks. These have the advantage of being current and directly focused on the issues of concern; however, they measure attitudes, not behavior itself. Analyses of statistical data, on the other hand, reflect past behavioral preferences toward existing risks; however, they assume that past preferences will hold in the future and that data relevant to the issue at hand are available.

Chauncey Starr's (1969) classic study of risk–benefit drew upon U.S. accident statistics to suggest several interesting points.

The public is roughly 1000 times more willing to accept "voluntary" risks than "involuntary risks" (e.g., private vs commercial aviation).

The statistical risk of death from disease offers a psychological yardstick for establishing the level of acceptability of other risks to life (e.g., the percentage of the public using automobiles increased proportionately as its risk declined to that from disease).

The acceptability of risk appears to be crudely proportional to the third power of the perceived benefits (e.g., the wages of miners in more hazardous jobs go up relatively faster than the risk increases).

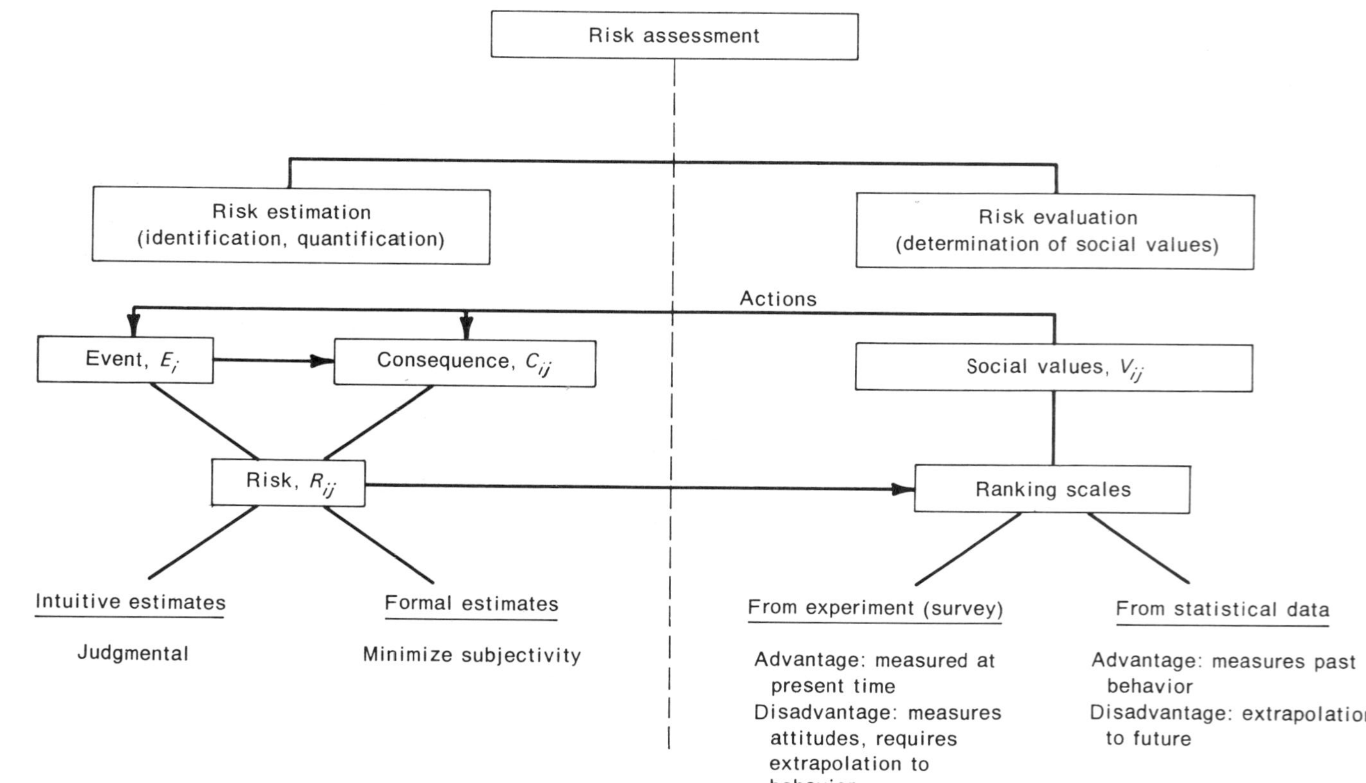

FIGURE 11.4. General structure of risk assessment.
Source: Otway and Pahner (1976).

Nuclear energy has presented the most appealing subject for risk–benefit analysis. A major study sponsored by the U.S. Nuclear Regulatory Commission (1975) attempted to estimate the risks of a nuclear plant accident and their consequences. It concluded that the relative risk was acceptably low. However, it failed to incorporate the risk evaluation of diverse societal groups (who rejected the conclusions on various grounds). Otway et al. (1978) found that public attitudes toward nuclear power and its use are highly complex. For example, their results suggest that people perceive risks and benefits independently, rather than as commensurable, as in CBA. Further, people differentiate among types of risk—some might believe that nuclear power entails certain risks (e.g., sociopolitical) without necessarily believing that it leads to others (e.g., environmental). Some interesting anomalies related to risk perception arise in the nuclear case. Sometimes expenditures on nuclear safety appear to be several orders of magnitude higher than justified by biological risks. Yet, even then, many facilities cannot gain public approval (Otway and Pahner 1976). This suggests the practical importance of understanding societal responses to risk (Lind 1972).

Critiques of risk–benefit analysis have focused on both risk estimation and evaluation. In the nuclear instance, one could question whether the laws of probability pertain for very low-probability events and very serious consequences. Lovins (1977) suggests that all of the risks may not be identifiable, especially before the development is implemented. Green (1975) recognizes the "Catch 22" nature of this situation. If it were decided that for an action to be implemented, it must be proven safe, and if we were not to know whether it is safe until it is implemented, technological development would be at a standstill.

Lovins (1977) raises some risk-related issues that are unanswerable by science.

Are solutions to safety problems of some technologies to be found by engineering answers, or are the solutions limited by human fallibility?

Who accepts the burden of examining the level of risk of a technology—do the proponents have to prove it acceptable, or the opponents have to prove it unacceptable?

Is an action presumed safe until proven harmful, or potentially harmful until proven safe?

11.5.2 Net Energy Analysis

Since our future energy supply is uncertain, an increasing amount of importance has been placed on the energy value of materials, in both manufacturing and use. A justification for this analysis is that market prices do not reflect the tremendous and nonrenewable investment by nature in providing energy resources. Gilliland (1975) describes energy as the ultimate limiting factor since (1) energy is the only commodity for which a substitute cannot be found; (2) energy is required to run every type of system; and (3) energy cannot be recycled without violating the second law of thermodynamics.

Various techniques of energy accounting have been developed to determine the energy content of materials.[1] These techniques estimate the amount of energy required to construct a product and to run it during its lifetime. For example, in performing a net energy analysis of an automobile, one must include energy consumed by the plant where steel is milled into automobile parts, and by the final assembly of the finished product. The analysis then calls for estimation of automobile usage, which, when considered with gas mileage, is used to compute fuel usage. Maintenance requirements also represent an energy investment and should be included in the analysis. Another factor of major importance is the lifetime of the automobile. The car will be replaced once it becomes obsolete, and the entire process of energy-intensive fabrication must be redone. If recycling of parts occurs, subsequent investments in fabrication will differ from the initial investment.

One major point of net energy analysis is that the discount rate is zero. The reason for this is that energy is a capital resource and should be treated as an investment, rather than an operating cost. As such, it can never be replaced and yields no greater return in the present than in the future.

Huettner (1976) criticizes net energy analysis as an exercise based on erroneous assumptions and compares it to the labor theory of value developed in the 19th century. He argues against Gilliland's contention that energy is the ultimate limiting factor. Other resources (e.g., water) may prove to be more crucial. Langham and McPherson (1976), as well as Heuttner, criticize its use as a strict policy determinant, but allow its importance as a technique that can inform policy making. Its use in TA/EIA has been restricted primarily to assessments of alternative energy technologies. As a complement to CBA, it has potential relevance to many TA/EIAs.

[1]Different measures can be used, depending on the purpose of the analysis. For instance, in considering energy systems one can deal with marginal or cumulative gross energy requirements, energy balance, various net energy requirements, or the energy output:input ratio. Energy quality and energy opportunity costs may require consideration. Gilliland (1975) provides a starting point. Slesser (1974) provides examples for a TA context per se; Bechtel Corporation (1975) illustrates energy calculation in a TA.

11.6 SECONDARY AND HIGHER-ORDER IMPACTS

Economic impacts show a ricocheting, chain-reaction effect throughout the economy as, in the words of novelist Thomas Wolfe, "the pebble whose concentric circles widen across the seas." New businesses mean new employment, which creates the need for other businesses, and so on. The net effect of these higher-order impacts is complex. A project may cause decline in some industries and growth in others. Some regions may be penalized, while others benefit.

Comprehensive cost–benefit analysis requires that one consider the indirect effects of a development. The techniques discussed in this section can aid in that task. In addressing effects beyond the specific context of a development, we enter into macroeconomics.

11.6.1 Macroeconomic Concerns

Study of the secondary economic effects in TA/EIA is macroeconomic, considering the impact of that development on the whole society. Macroeconomics is closely linked to the work of John Maynard Keynes. In a Keynesian framework, one classifies expenditures as either consumption (for immediate gratification) or investment (toward later gratification). A distinction is also made between the public and private sectors. As an example of this approach, consider Victor's (1972) discussion of the issues that may arise in a Keynesian analysis of the implications of pollution control. Several of these follow.

Employment is likely to be boosted by expenditures on waste-treatment equipment.

Unemployment may occur as particular industries are hard hit and become relatively uncompetitive; government policy to assist the mobility of displaced labor may be helpful.

Average *price levels* are likely to increase as industry tries to recoup its costs, leading to subsequent increases in prices and wages as various sectors of the economy adjust to the new circumstances.

Unless other governments take similar pollution control policy actions, the *balance of payments* will shift toward deficit.

There are other aggregate economic issues not addressed by Keynesian macroeconomics (Victor 1972). Marxists emphasize the class inequities of the pollution situation. For instance, the laborers and the poor are likely to be most subject to the effects of pollution (e.g., living in more polluted neighborhoods) and to the impacts of pollution control (e.g., suffering from any unemployment and general rises in prices). Growth economists, such as Kenneth Boulding, worry about the inherent waste discharges associated with economic activity. All processed materials must eventually be discharged as wastes (a process slowed by durability and recycling) and are potential sources of pollution. Thus, policies should be geared to the maintenance of economic and ecological stability, rather than economic growth.

Whereas our analytical tools derive from a Keynesian framework and address such concerns as employment levels, prices, and regional economic advantages, it is quite appropriate to keep the other questions (viz., "Who pays?" and "What are our objectives in terms of economic growth?") vividly in mind.

Several forms of macroeconomic analysis potentially useful in TA/EIA are introduced in the remainder of this section. These techniques are (1) economic-base models, (2) input–output models, and (3) regression analyses. Economic base models and input–output models are approaches included in *regional economics*. The basic concept of regional economics is that the economic system of a given region (which can be defined as narrowly as, say, the Midtown Neighborhood Planning Unit in Atlanta, Georgia, or as broadly as the European Common Market) depends on interaction between industries and consumers in that region, and on imports from, and exports to, other regions. Economic-base analysis techniques vary in complexity but, at all levels, address the impact of economic activity on a region. Input–output analysis uses highly complex, computerized models in evaluating how a specific economic activity influences other economic activities. It incorporates first-order impacts and successively higher-order influences as well. Quantitative economic analyses (econometrics) apply regression analyses, in a number of ways, at various levels of sophistication. We emphasize relatively simple applications pertinent to TA/EIA.

11.6.2 Economic-Base Models

The primary objective of economic-base modeling is to estimate changes in two principal sources of socioeconomic impacts—regional income and regional employment. These are of interest in themselves, and they also provide a basis for estimating regional population, another central concern.

The keystone of the analysis is determining *economic-base multipliers*. The concept of multipliers appears in two different forms: (1) the "ricocheting" effect of money spent for purchasing goods and (2) the secondary employment and population that is supported by a basic

INCOME MULTIPLIERS

The multiplier effect can be defined as the amount by which a change in investment acts to produce a change in income and output—or employment and population levels (see the text and cameo on employment multipliers). The effect can be described through the sequence of events that occur after an investment. Suppose that an increase in investment (e.g., a new development) of $100 occurs on day 1. This amount is paid to the wage earners of the industry involved. If the wage earners spend $80 and put $20 in savings, their marginal propensity to consume (m) is 0.8. If it is assumed that m is 0.8 for day 2, the wage earners paid by the $80 spent on day 1 spend $0.8 \times (0.8 \times \$100)$ on consumption goods. On day 3 these expenditures become income to others, who spend $0.8 \times (0.8 \times 0.8 \times \$100)$, and the cycle continues. Thus

$$\text{Multiplier} = 1 + m + m^2 + m^3 + \cdots .$$

The multiplier computed depends on the time period of concern. For an equilibrium situation, in which time is essentially infinite,

$$\text{Multiplier} = 1/(1 - m).$$

Thus if $m = 0.8$, the increase in income resulting from a $100 increase in investment is $\$100/(1 - 0.8) = \$500$; the multiplier is 5. The multiplier is specific for a defined region.

Interregional income models trace economic activity in a multiregion economy. Interregional multipliers are quite different from simple economic-base multipliers. The general equation for interregional multipliers is as follows:

$$\text{Interregional multiplier} = [1 - m_i (1 - T_i) + \sum_{j=1}^{K} m_{ji} (1 - T_i)]^{-1},$$

where

m_i = marginal propensity to consume in region i;

T_i = tax rates in region i; and

m_{ji} = marginal propensity for region i to import from region j (for K regions).

For further discussion of economic base income multipliers, see Richardson (1969), Isard (1960), or Tiebout (1962).

EMPLOYMENT MULTIPLIERS

There are various ways to compute an economic-base multiplier. The following procedural example is abstracted from Schaffer and Davidson (1972); see also Thompson (1959). The example describes one way to calculate the economic-base employment multiplier for Atlanta over the period 1961–1970.

1. Employment in 1970 in each industry for (a) Atlanta, (b) Georgia, less Atlanta, and (c) the continental U.S., less Atlanta, is obtained from the U.S. Bureau of Labor Statistics (1971a,b).
2. "Location quotients" for each of the industries are computed as follows:

$$\text{Location quotient} = \frac{\text{Industry employment as percentage of total employment in Atlanta}}{\text{Industry employment as percentage of total employment in the prime candidate market regions.}}$$

   The two prime candidate market regions for Atlanta are (a) Georgia, less Atlanta, and (b) the United States, less Atlanta.
3. Location quotients are compared. If the location quotient for either market region is greater than one, the industry is considered to have some export employment. The market region yielding the largest location quotient is designated as the benchmark.

industry. The former notion is used to compute *income multipliers*, which are discussed in the cameo entitled "Income Multipliers." The latter concept is used to compute *employment multipliers*, treated in the cameo of that name. Employment multipliers differ from income multipliers in that they consider only impacts on jobs, rather than on income changes in a region. Since jobs differ in salary and wage scales for employees, employment and income multipliers may differ substantially.

The starting point for consideration of regional multipliers is to distinguish between *basic* (or export) *employment* and *secondary employment*. Basic employment is that involved in producing goods and services, for sale to other regions (e.g., for Japan, that portion of the automobile manufacturing for sale in other countries, such as the United States). Secondary employment is that supported by the income

4. The specialization ratio for each export industry is then computed, using the location quotient for whichever region is the benchmark, as

$$\text{Specialization ratio} = 1 - (\text{location quotient})^{-1}$$

This ratio indicates the proportion of employment in that industry, in Atlanta, producing for export.
5. Employment in each export industry in Atlanta is multiplied by its specialization ratio and summed. The resulting figure is export employment in Atlanta.
6. Employment data are then obtained over the relevant time period—in this Atlanta example, 1961–1970. The relationship between total employment and export employment is then calculated using a simple regression equation:

$$\text{Total employment} = B_0 + B_1 \times (\text{export employment}).$$

In the Atlanta case,

$$\text{Total employment} = 16{,}920 + 3.3\ (\text{export employment}).$$

The employment multiplier in this example is 3.3. This is interpreted to mean that each additional person employed in a basic (export) industry is associated with 3.3 additional persons employed in total.

from the basic industry (e.g., local retailers sell to the automobile workers).

One can carry the notion of multipliers a step further to estimate population as a function of employment (basic and secondary). This requires determination of the *labor-force participation rate*, that is, the proportion of the population that is employed. For instance, a typical family of four might have only one employed.

An example of the application of economic base modeling in a TA is provided by Ryan [paraphrased from Dickson et al. (1976b: 412–415)]. In this instance the task was to anticipate the socioeconomic effects likely to result from the advent of coal mining and processing in Cambell County, Wyoming. Several alternative development scenarios were first postulated. In particular, one included a constrained rate of growth at 5% annually. First, the base population of the county was estimated at 17,000 in 1975. A 5% annual growth rate curve was used to project the population to the year 2000. The population levels were divided by a population/basic employment multiplier (6.5) to determine the basic employment allowable each year to stay within the 5% growth rate. Then an appropriate level of coal mining and processing facilities was devised that would utilize the basic employment allotments by year.

The population/basic employment multiplier of 6.5 is the product of (1) a total/basic employment multiplier (see cameo entitled "Employment Multipliers") and (2) a population/total employment multiplier (the inverse of the labor-force participation rate). A total/basic employment multiplier (the ratio of total/basic employment) of 2.6 and a population/total employment ratio of 2.5 are multiplied to obtain the composite multiplier of 6.5. Higher multipliers (in the 6.7–7.3 range) are used for future growth because the anticipated population influx will be able to support a wider range of service activities than currently available.

The economic-base analysis in this TA thus serves to relate development of coal mining in the area studied to (1) *regional employment* (and economic) gains, (2) *population growth* (in turn, a major factor in determining demand for services, environmental stresses, etc.), and (3) *planning constraints*.

Several qualifications to the economic-base-multiplier approach need to be mentioned. First, as Davidson and Schaffer (1973) point out, input–output models offer more sophisticated economic impact analyses. (However, the advantage of the multiplier is that it is feasible to compute from limited data, even if it has not been previously estimated by others.) Second, economic base models tend to grossly aggregate effects and may not accurately represent a situation. Refinement may be needed to consider differences in multiplier relationships during construction and operation phases, by type of basic industry (e.g., mining and manufacturing), by urban character of the region and distance from trade centers, and by likely age and sex characteristics of immigrant labor. Third, the approach does not readily accommodate the effects of temporal and spatial distortions. For instance, increased government revenues necessary to provide services for the increased population may not accrue to the governmental units serving that population (e.g., industry locates in one locale to which it pays taxes, but employees reside in another). Provision of services such as schools and sewers may lag the demand (e.g., boom towns may need services immediately, but will not retain the population to support them over the long term). Fourth, this approach neglects the effects of *displacement*. Within a region, and outside the region, new developments may displace alternative resource uses. As The Institute of Ecology's review of the Garrison Diversion EIS (see Chapter 5 Appendix) points out, there are major opportunity costs associated with the use of water for irrigation, thus displacing hydroelectric and industrial uses.

11.6.3 Input–Output Models

Input–output (I–O) analysis provides a more complete tool for evaluating the effects of alternative forms of economic activity on a region. It uses relationships between sectors of the economy (Table 11.3 exemplifies sectors considered in one TA) to identify their interdependencies and enables the user to trace the effect of change in a given industry's activity through the entire economy. Input–output models can be used to measure the long-run effect of a change in one sector, such as coal mining, on all other sectors of the economy, through both direct and indirect buyer–seller relationships.

Lacking time and resources to construct I–O models, it is likely that assessors will make use of available national, state, or substate I–O models to determine likely sectoral economic impacts, and the regional income, employment, and population effects of the proposed developments. Input–output models usually are computerized data banks with

TABLE 11.3. Economic Sectors Providing Inputs to the Coal-Mining Sector, Ranked by Size of 1967 Total Requirement Coefficient[a]

Rank	Industry title	Coefficient	Input–output sector code
1	Coal mining	1.148	7.00
2	Real estate	0.075	71.02
3	Blast furnaces and basic steel products	0.037	37.01
4	Wholesale trade	0.034	69.01
5	Miscellaneous business services	0.034	73.01
6	Electric utilities	0.031	68.01
7	Mining machinery	0.026	45.02
8	Petroleum refining	0.020	31.01
9	Screw machine products and bolts, nuts, rivets, washers	0.017	41.01
10	Miscellaneous chemical products	0.017	27.04

[a] From Dickson et al. (1976b).

accompanying subroutines for manipulating data. Small models involve approximately 35 industry classifications with 1225 production coefficients. Large models can involve more than 600 industry classifications. The static nature of I–O models (they are constructed for a specific year, usually at least 5 years earlier due to lags in data accessibility) limits their usefulness in assessment. Tiller (1979) presents an approach to use the existing I–O models in a TA/EIA context at modest cost, taking into account projected changes in the interindustry structure.

To illustrate the nature of I–O models, we exhibit an I–O transactions table, in a generalized format, in Figure 11.5, which describes the flow of commodities through the economy. Quadrant I depicts con-

FIGURE 11.5. The transactions table as a picture of the economy.
Source: Schaffer et al. (1972: 21)

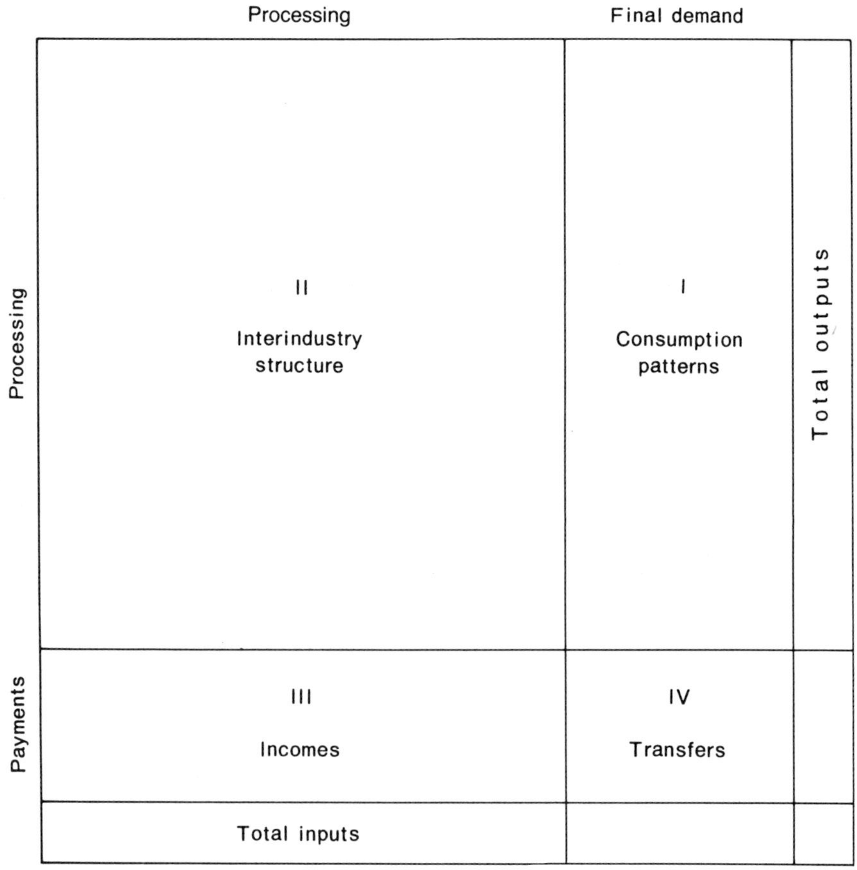

	Processing	Final demand	
Processing	II Interindustry structure	I Consumption patterns	Total outputs
Payments	III Incomes	IV Transfers	
	Total inputs		

sumer economic behavior by describing purchases of final goods by households, investors, and governments. Quadrant II describes interindustry transactions and shows how raw material and intermediate products are consumed to produce goods for sale. A sample cell entry in Quadrant II would represent the sales of industry *i* to industry *j*. Quadrant III shows the incomes of the economic sectors, including households, industries (depreciation and retained earnings), and taxes (paid to different levels of government). Quadrant IV includes nonmarket transfer payments between sectors of the economy. Examples include gifts, savings and taxes of households, surpluses and deficits of governments, and their payments to households and other governmental bodies.

The I–O model is essentially a complex, cross-effects matrix (see Chapter 9). In operation, it can show the effects of changes in one sector (e.g., opening a major coal mine in Wyoming) on other elements of the region's economy (e.g., increased sales by mine suppliers and shifts in energy purchases from other sectors to coal).

The "Synfuels" TA (Dickson et al. 1976b: 383–402) exemplifies a simple use of an I–O model. National I–O information was used to identify the likely economic implications of production of synthetic fuels from coal or oil shale. The I–O analysis identified the needed supporting and supplying industries for synthetic fuels production, and determined which regions of the country would be impacted. Table 11.3 lists the 10 largest coal-supplying sectors with the largest total requirements coefficients. The coefficients specify the direct and indirect output of other industries needed to produce a dollar's worth of coal delivered to final demand in 1967. For example, to produce $1000 worth of coal in 1967, $26 of mining machinery was needed, as well as an additional $148 of coal production, used primarily as an energy source for mining operations. Table 11.3 can be used to project resource requirements for expanded coal production. However, interindustry transactions do vary over time, due to technological change and other factors. The coefficients in Table 11.3 are static, and projections that rely on them are approximations.

11.6.4 Regression Analysis

Regression analysis (Johnston 1972; Theil 1978) is a basic statistical technique that is used in many disciplines. It has a number of potential applications in TA/EIA, but it is discussed here because econometricians can be credited with leadership in applying it in policy contexts.

A regression equation describes the relationship between a dependent variable and one or more independent variables. The complexity of the analysis depends on the number of independent variables in the analysis and the form of the functions used as variables (e.g., linear, x; higher-order, x^m; or transcendental, ln x or sin x).

A multiple linear regression equation that describes how one variable (Y) depends on selected other variables (Xs) is written as follows:

$$Y = B_0 + B_1X_1 + B_2X_2 \cdot \cdot \cdot + B_KX_K + \epsilon,$$

where

Y	=	a dependent variable;
X_i	=	independent variables;
B_0	=	value Y takes when all Xs are 0;
$B_1, B_2, \ldots, B_K$	=	regression coefficients; and
ϵ	=	error.

The value B_0 is interpreted as the mean effect on Y of all the independent variables that are left out of the equation. In econometric analyses, it usually cannot be interpreted as an intercept (Rao and Miller 1971: 4–6). Values $B_1, \ldots, B_K$ can be interpreted as the partial derivatives of Y with respect to $X_1, \ldots, X_K$, respectively. Thus B_1 indicates how much Y will change in response to a unit change in X_1 while the other Xs are held constant. The value ϵ is the residual, or error, term.

Notice that because of the error term, there is no exact solution to the equation. Instead, one estimates the values of the coefficients (the Bs). Various methods are used, but least-squares estimation is the most common (for a simplified discussion, see Section 6.4.3.1). Ordinary least-squares estimation is based on minimizing the sums of squares of residuals (i.e., the error term). It requires a number of critical simplifying assumptions (e.g., error terms have zero mean value and are uncorrelated with the independent variables). Furthermore, there are a number of crucial considerations about the assumed form (e.g., linear), the specification of the correct independent variables (neither too few nor too many), proper measurement, underlying statistical distributions (normal), and relationships among the independent variables (Rao and Miller 1971). Some statistical or econometric expertise is needed to carry out a complex regression analysis.

In TA/EIA, one is likely to use regression formulations in three distinguishable ways: (1) treating individual firm and macroeconomic effects (supply and demand, etc.), (2) developing and utilizing relatively simple regression formulations, or (3) using a highly developed, complex econometric model to see the effects of a possible policy action. We now briefly consider each sort of application.

Evaluating the relationships among supply, demand, price, and other variables is relevant to assessment studies. For instance, in considering pollution-control alternatives for a power plant, understanding the financial structure for that enterprise could indicate the likely increase in product price from each alternative. This would shift supply and demand (see cameo entitled "Supply and Demand Curves" in Section 11.3.1), implying changes in production levels. Assuming production reductions, this could mean energy conservation, loss of jobs, and decreases in tax bases for affected communities—issues of importance in impact assessment.

To turn to TA/EIA experience, in The Institute of Ecology's review of two EISs (Pearson et al. 1975; Fletcher and Baldwin 1973), specific criticism was leveled at the absence of supply and demand price-elasticity considerations. In the first case (Garrison Diversion), they noted that the increased irrigated acreage in staple crops such as potatoes could lead to reduced income to the producers because of the inelastic demand. In the second case (oil shale), they estimated that the rate of exploitation (supply) will be sensitive to changes in the prices of oil and alternative fuels.

To reemphasize, regression analysis is a quite general tool. Accordingly, one may wish to develop simple regression equations for assessment purposes. For example, one may wish to anticipate the effect of a given development through observation of similar prior developments; or one may wish to derive relationships among policy alternatives and various dependent measures of interest. Even nonlinear relationships can sometimes be simply accommodated. For instance, if population were taken as increasing exponentially, a semilogarithmic transform will yield a linear regression relationship:

$$\text{Population} = B_0 e^{B_1 t}$$
$$\text{In Population} = \text{In}(B_0 e^{B_1 t}) = \text{In } B_0 + B_1 t,$$

where e is a mathematical constant, t is time, and B_0 and B_1 are coefficients.

The regression equation would have the form

$$P' = B'_0 + B_1 t + \epsilon,$$

where P' is equal to ln(population) and B'_0 is equal to ln B_0.

In contrast to the ad hoc development of a simple regression equation, one may make use of an elaborate, existing regression model. For instance, one might insert specific policy alternatives into a regional or national econometric model to predict their impacts on the economy.

There are numerous dynamic econometric models that attempt to describe the workings of a region or a nation. They vary greatly in their complexity, ease of use, access, and accuracy. Such models are built by

1. selecting the presumed important variables;
2. establishing some as exogenous (not influenced by other variables within the equation system);
3. specifying relationships among the variables (usually recursive; i.e., some variables within the system are assumed not to be influenced by others);
4. iteratively deriving solutions by testing the model on existing observations over various time periods; and
5. testing the model on observations that follow those from which solutions were determined.

An example is the Brookings Model of the U.S. economy (Fromm and Taubman 1968). This contains over 300 equations and draws upon data for more than 2000 variables. A few sample variables are (1) personal consumption expenditures on services, (2) exports of durable goods, (3) business construction, (4) government receipts, and (5) corporate profits before taxes. This simultaneous equation model could be used to study impacts of technological development, policy approval, and so on, on such factors as gross national product (GNP), consumption, employment levels, and the balance of trade.

11.7 CONCLUDING OBSERVATIONS

This chapter treats factors involved in carrying out economic analysis in TA/EIA. It raises a number of difficult issues, such as the impossibility of quantifying all effects, the difficulties in measuring costs and benefits over time, the need to weigh intangibles, and the methodological intricacies of many economic techniques. Nonetheless, it suggests ways to deal with the difficulties of producing sound analyses of economic effects.

Three concerns underlie economic analysis—efficiency, effectiveness, and equity. Cost–benefit analysis provides a basic framework to address these concerns. Figure 11.1 offers a general CBA framework within which to treat both quantitative and qualitative economic impacts. These impacts interlock strongly with the other impacts treated in TA/EIA (EPISTLE).

Four steps in performing CBA can be distinguished: (1) careful definition of the subject, (2) identification of costs and benefits, (3) their measurement, and (4) their evaluation.

Measurement of costs and benefits is aided by separate consideration of

1. direct costs (private, internal),
2. externalities (spillover effects), and
3. public goods (collective effects, not chargeable to individual consumers).

Supply–demand relationships are important in ascribing direct costs. Surrogates for market pricing (e.g., shadow pricing) are needed to measure externalities and public goods.

When there exists a time preference for consumption, discounting must be used. This is an absolutely critical factor in the outcome of CBA. Social discount rates are generally lower than private ones but are difficult to determine unequivocally. Therefore, we suggest that one perform CBA calculations using several different social discount rates.

Five approaches to discount future costs and benefits are described:

1. net present value,
2. annual equivalent,
3. internal rate of return,
4. payback period, and
5. benefit:cost ratio.

In addition, we suggest a composite approach that explicitly attempts to combine quantitative and qualitative effects through factor profiles as an aid in policy making.

Two variants of CBA are risk–benefit and net-energy analysis. Risk–benefit concerns the perceived costs associated with risks and the probabilistic treatment of risk. Net-energy analysis offers an alternative metric—energy units in place of dollars (at a zero discount rate).

Keynesian macroeconomics deals with such factors as employment, prices, and income. We describe several techniques within this framework useful in determining secondary economic impacts. We also recognize the value of asking broader questions of "Who pays?" and "What are our objectives in terms of economic growth?"

Economic implications for a region can be elucidated by economic base models. Separate cameos detail the calculation of economic-base income and employment multipliers.

Input–output analysis is a more complex approach that treats specific industrial sectors and their interrelationships for a region.

Regression analysis is a broad family of analytical techniques for determining the relationships among a set of variables. Applications of several versions, from simple to complex, are discussed for TA/EIA.

Economic analysis is a vital and highly developed part of TA/EIA. However, it should not be mistakenly taken to be the whole assessment.

RECOMMENDED READINGS

Lovins, A. B. (1977) "Cost–Risk Benefit Assessments in Energy Policy," *George Washington Law Review* 45 (August): 911–943.

A critique of cost–benefit and risk–benefit analyses as currently performed, particularly in energy studies.

Richardson, H. W. (1969) *Elements of Regional Economics*. New York: Praeger.

A solid treatment of economic-base and input–output models.

Sassone, P. G., and W. A. Schaffer (1978) *Cost–Benefit Analysis: A Handbook*. New York: Academic Press.

An excellent text on how to do cost–benefit analysis.

Slesser, M. (1974) "Energy Analysis in Technology Assessment," *Technology Assessment* 2 (3): 201–208.

Net-energy analysis examples for impact assessment.

EXERCISES

11-1. In evaluating the benefits of a public good, such as a dam, what is wrong with simply asking all of those who would benefit what the actual value of their benefit is, and then summing their estimates to get the total benefit?

11-2. If demand for a product were inelastic and the product's price doubled, what would happen to the demand?

11-3. Suppose a public power plant required 50 tons per day of coal to generate electricity. The coal supplier increased production from 1025 to 1065 tons per day in response to this demand. Externalities due to coal mining are estimated to be $2 per ton. What is the daily external cost of the coal required by the power plant?

11-4. If an analysis of an airport-expansion project showed a benefit:cost ratio of 2.0 for alternative A and 1.8 for alternative B but ranked B better than A on a community factor profile, how might a decision between them be made?

11-5. Assume the net annual benefits of a water-supply project are $1 million now. The net benefits of the water, due to projected shortages, will increase 10% annually. The projected rate of inflation is 5%. The discount rate (real dollar) is 7%. What would be the net present value of the benefits of this project if it operated 20 years? Five years?

11-6. Discuss an appropriate social discount rate for use in valuing consumption of U.S. oil reserves.

11-7. The EPA is contemplating implementation of pollution-control standards on air emissions in the pulp and paper industry. The costs to meet these standards are as follows in a plant producing 2000 tons of paper per day:

Capital cost of equipment (year 0) $= \$1.5$ million

Annual operating and maintenance costs (years 1–20) $= \$180{,}000$

Net present value (NPV) $$= -C_0 + \sum_{t=1}^{20} \frac{B_t - C_t}{(1+d)^t}$$

The expected decrease in emissions is as tabulated:

	Expected decrease	Estimated damage to human health, buildings, etc.
Particulate matter (PM)	50 tons/day	\$25/ton
Sulfur dioxide (SO_2)	10 tons/day	\$7.30/ton

1. a. Using a 5% discount rate, perform a CBA of the proposed standard on the 2000-ton/day plant. The plant operates 300 days a year. Use a 20-year analysis period. Find the NPV of the new equipment at this discount rate.
   b. At what discount rate would the costs and benefits break even? Plot the NPV against the discount rate to show the breakeven point.
2. A typical pulp and paper mill charges \$85 per ton of product. One way to estimate the percentage price increase to pay for the control equipment is to use the following formula. The discount rate to be used in present-value calculations for this computation is the after-tax cost of capital, which is 8.7%.

$$\text{Price increase (percentage)} = \frac{\text{PVP}(100)}{(1-T)\,(\text{PVR})},$$

where

PVP = present value of pollution-control equipment $C_0 + \sum_{t=1}^{n} \frac{C_t}{(1+d)^t}$;

T = tax rate (48%); and

PVR = present value of revenues $\sum_{t=1}^{m} \frac{(\text{Revenues})_t}{(1+d)^t}$.

Estimate the price increase that will result from implementation of the pollution-control equipment.

3. The supply and demand curves for pulp and paper are shown on the following graph. What is the reduction in demand due to the price increase? What is the impact of this decrease on the GNP (the total output of the national economy)?

11-8. (Project-suitable) Select a topic for impact assessment. Sketch out how you would go about each of the four steps in a CBA (Section 11.2.1). Specifically address *each* of the subpoints raised (e.g., public goods, selecting decision criteria).

FIGURE FOR EXERCISE 11-7. Supply and demand curves for pulp and paper products.

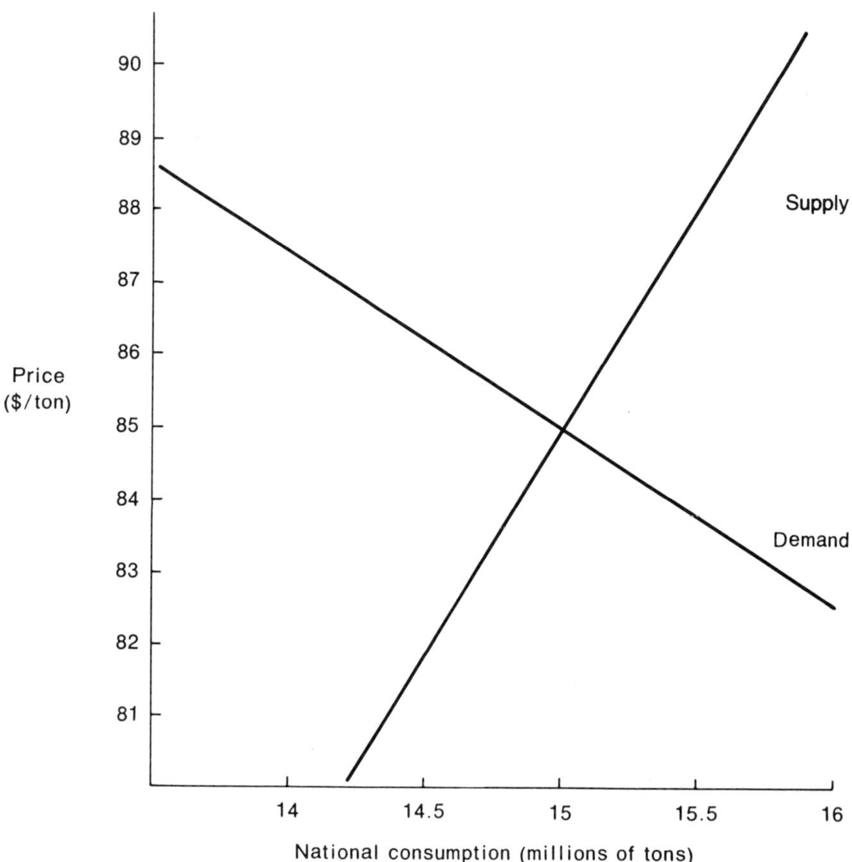

11-9. (Project-suitable) Perform a literature search and/or informal survey of experts to find an economic-base model appropriate to a topic of interest (e.g., the state model for a statewide impact assessment). Use the model to probe the secondary economic implications of the major features of the development in question.

11-10. 1. Formulate and administer a survey to at least five people concerning how safe they would feel when traveling cross-country by (a) private automobile, (b) private airplane (a four-seater), (c) commercial bus, (d) commercial airplane.

2. Obtain actual safety measures (e.g., fatalities, accidents) on appropriate bases (e.g., passenger-miles traveled) for each of these modes.

3. Compare perceived risk with actual risk. What are the implications for risk–benefit analysis?

4. Discuss the effects of measurement, particularly (a) differences due to phrasing of the survey questions, (b) differences due to scaling of the survey responses, (c) differences due to measures used for the actual statistics, and (d) sampling issues—choice of actual data and their suitability for the "population" being surveyed.

11-11. An environmental group has filed suit to prevent liquefied natural-gas tankers from entering a port. They cite the risk of explosion and the disastrous consequences that could result. The following estimates are acknowledged:

The chances of an accident are one in one million (10^{-6}).
The chances of sabotage are three in one million (3×10^{-6}).
The consequences of either an accident or sabotage are valued at $3 billion.
The benefits to the nation of the incoming fuel are $500 million per year, yet the benefit to the local community is only $2 million.
A tanker would enter the port once every month.

1. Compute the annual risk and compare this with the national and local benefits.

2. Suppose that an alternative energy source is available that yields net annual benefits to this community of $1.9 million. Discuss how you might perform risk evaluation, considering differences between local and national interests, environmental and industrial perspectives, and so on.

12

Analysis of Social and Psychological Impacts

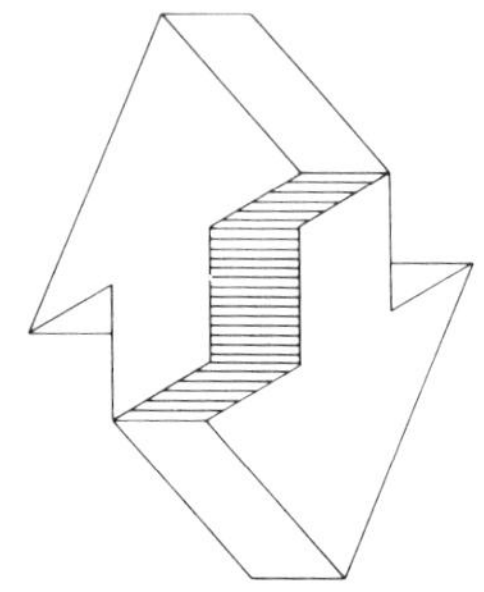

This chapter deals with impacts of technologies and projects on human beings. It treats the sometimes overlapping categories of impacts on society, on values, and on the individual. The study of these impacts is one of the major novel premises of TA/EIA. As such, it is one of the most important topics in this book. However, when compared to economic and technical analyses, social-impact assessment (SIA), considered broadly, is just beginning to develop substantively and methodologically. This chapter develops an approach to SIA based on the Corps of Engineers Guidelines. Ways to gather and use both primary and secondary data are emphasized. Means to analyze impacts on values are also discussed. Finally, impacts on individual human beings are considered. In these last two areas, the chapter suggests points to consider rather than offering well-developed analytic methods.

12.1 INTRODUCTION

This chapter deals with the analysis of impacts of technology on people. Effects across broad groupings of people are referred to as *social;* impacts on individuals are termed *psychological.* Neither has received the extensive methodological development of environmental and economic impact analysis. Along with the treatment of higher-order impacts, concern for impacts on people has been one of the key new contributions of TA and, by law, EIA. Psychological impacts focus on the quality of life and the human needs of individuals. Analytical development here and in the study of impacts on values has been less apparent than in certain other aspects of social-impact analysis (SIA).

The importance of SIA has emerged during the last decade in conjunction with that of environmental protection. Initially, SIA was either ignored (see the Garrison Diversion EIS in the Chapter 5 Appendix) or done by "seat-of-the-pants" speculation [Rossini et al. (1978) noted a TA performer who referred to it as "dreaming"]. It is beginning to progress, particularly in the direction of developing data about parties at interest. In 1974 Friesma and Culhane could criticize the absence of primary research in SIA in EIS preparation. They found a bias toward project justification and a lack of proposals to ameliorate negative impacts of a project. By 1977 primary data gathering and use was becoming more common, for example, in the TA on hail suppression (Changnon et al. 1977). The use of social indicators (see Section 7.3.2) was beginning to be seriously considered.

The boundaries of SIA are somewhat uncertain. The boundary between social and economic impacts in such areas as equity and nonquantifiable costs and benefits is unclear. An analogous problem arises with reference to impacts on "the human environment." Local, political, and institutional impacts (treated in Chapter 13) are not always readily distinguished from social and psychological ones. One basis for distinguishing among these classes of impacts lies in the training of the professionals who analyze them. Sociologists, psychologists, and anthropologists would typically be involved in dealing with the impacts considered in this chapter. Lawyers, public administrators, and political scientists would be concerned with those of Chapter 13.

It is important to note that social impacts often appear as higher-order impacts (e.g., the increase of personal mobility due to the automobile). In addition, they interact with other impacts. For example, demographic change may interact with an altered economic base in a region to change the power of the regional political institutions.

We will treat SIA by following the U.S. Army Corps of Engineers (1973) Section 122 Guidelines (which have contributed much to SIA's identity). The Corps' efforts are typical of the best current practice. It has devoted considerable attention to SIA, in part due to the legislative mandate of Section 122 of the River and Harbor Flood Control Act of 1970 (P.L. 91–611). The Corps' Guidelines emphasize context description and forecasting; and impact identification, analysis, and evaluation. However, because of the structure of this book, we shall emphasize impact analysis, looking backward and forward as necessary to the "social" aspects of other TA/EIA component tasks. The discussion of social impacts is followed by a treatment of impacts on values, which leads to a section on psychological impacts. This chapter draws heavily on Chapter 7—social description and forecasting. It relates to Chapters 8–13—impact identification and analysis—and anticipates Chapters 14 and 15—impact evaluation and policy analysis.

12.2 SIA STRATEGIES

What are the sorts of effect that come under the generic heading of social impacts? Table 12.1 lists a variety. Note that they range from the highly quantifiable, such as population density, to essentially unquantifiable, such as aesthetic effects. Some of the factors listed have been considered elsewhere (e.g., noise as an environmental impact—Chapter 10).

Social impacts need to be considered within a framework. Our choice is the U.S. Army Corps of Engineers (1973) Section 122 Guidelines because of their general relevance—they apply to other classes of impacts as well—and the leading role the Corps has played to date in SIA. The first seven of the 11 steps are listed here, followed by loose parenthetical translations into the TA/EIA components used in this book:

1. profile (social and technological description);
2. projection without project (forecasting);
3. projection with project (forecasting);
4. identification of significant impacts (impact identification);
5. description and display of impacts (impact analysis);
6. evaluation of effects with publics (impact evaluation); and
7. mitigation of adverse effects (policy analysis).

The other four steps not included refer to public hearings and official reporting sequences, ending with the preparation of the EIS.

Before beginning our discussion of these steps, we shall consider

TABLE 12.1. Types of Social Effect[a]

1. Population-density patterns
2. Population mobility
   - Growth and migration trends
   - Age composition
   - Ethnicity
   - Relationship to regional distribution patterns
3. Housing
   - Total units, single–multioccupancy
   - Owner–renter
   - Vacancy rates
   - Median rent, or value
   - Housing condition and age
   - Construction trends, relocation programs
   - Crowding—persons per room
4. Displacement of people[b]
   - Educational attainment
   - Political jurisdiction
   - Municipal finance
   - Public-assistance recipients
   - Employment characteristics
   - Income characteristics
5. Transportation
   - Highway access
   - Commuter patterns
   - Rail, airport, and pipeline facilities
6. Desirable Community Growth[b,c]
7. Institutional dynamics
8. Health
9. Community cohesion[b]
   - Religious activities and civic groups
   - Population stability
   - Media—press, TV, etc.
   - Community social-service programs
10. Noise[b]
11. Leisure, cultural, and recreational activities
    - Parks
    - Recreation facilities—outdoor, indoor
12. Aesthetic effects
13. Other

[a] From U.S. Army Corps of Engineers (1973).

[b] Mentioned specifically in Section 122.

[c] To be reviewed by the state agency designated by the Office of Management and Budget Circular A:95.

two other general strategic approaches to SIA. The first of these highlights policy considerations, and the other emphasizes SIA modeling.

Finsterbusch (1975; see also, Finsterbusch and Wolf 1977) offers an approach that deemphasizes impact analysis to stress the development of adjustments through policy modification (Figure 12.1). Figure 12.1 is best understood by beginning with social impact description (top center of the diagram). This block identifies four substantial social impact categories—households (including individuals), communities, organizations, and societal institutions. Input to the social-impact description begins with societal data (top right). Societal trend analysis is incorporated to provide better future projections for long-term and wide-scope assessments. The social-impact description, in turn, provides input to the determination of the responses of impacted parties (middle of the diagram). Identification and elaboration of these responses is vital to understand the social forces affecting the program in

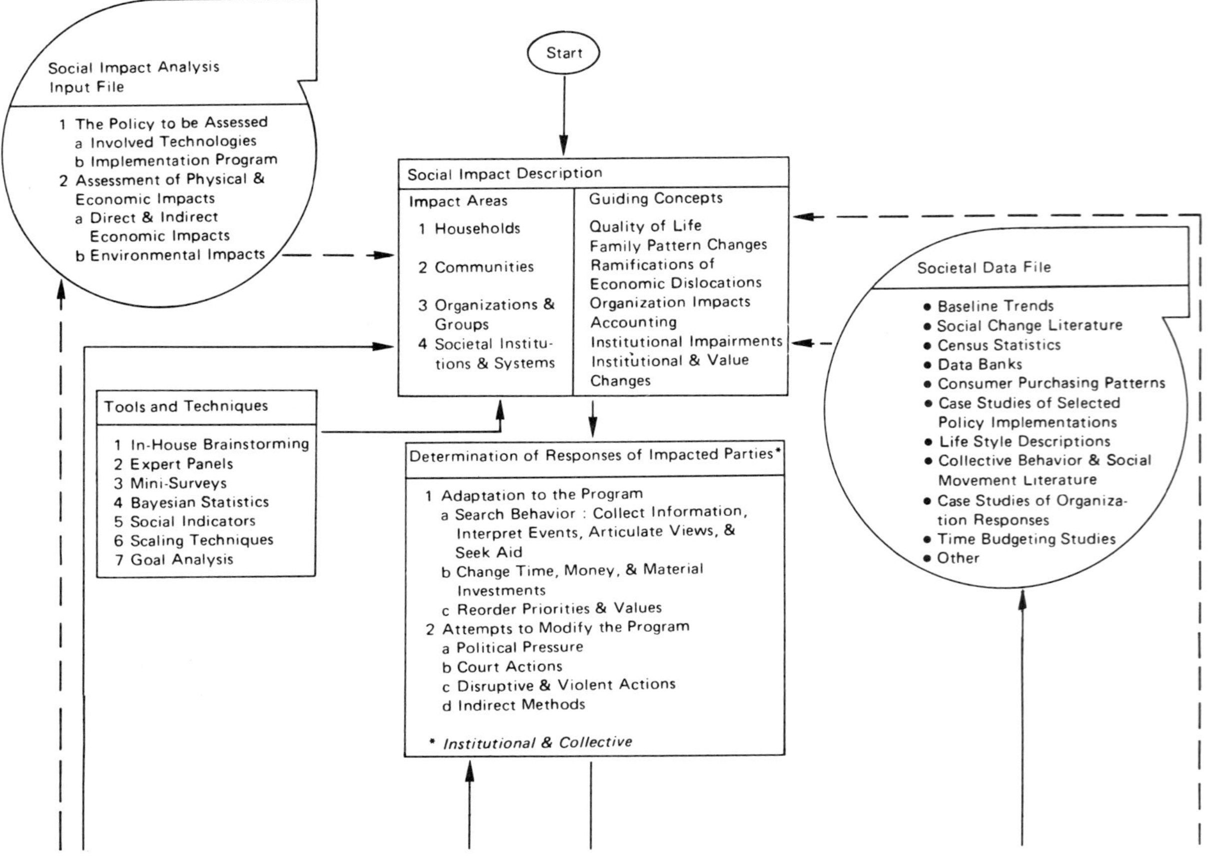

Start
Social Impact Analysis Input File
1 The Policy to be Assessed
a Involved Technologies
b Implementation Program
2 Assessment of Physical & Economic Impacts
a Direct & Indirect Economic Impacts
b Environmental Impacts
Social Impact Description
Impact Areas
1 Households
2 Communities
3 Organizations & Groups
4 Societal Institutions & Systems
Guiding Concepts
Quality of Life
Family Pattern Changes
Ramifications of Economic Dislocations
Organization Impacts Accounting
Institutional Impairments
Institutional & Value Changes
Societal Data File
• Baseline Trends
• Social Change Literature
• Census Statistics
• Data Banks
• Consumer Purchasing Patterns
• Case Studies of Selected Policy Implementations
• Life Style Descriptions
• Collective Behavior & Social Movement Literature
• Case Studies of Organization Responses
• Time Budgeting Studies
• Other
Tools and Techniques
1 In-House Brainstorming
2 Expert Panels
3 Mini-Surveys
4 Bayesian Statistics
5 Social Indicators
6 Scaling Techniques
7 Goal Analysis
Determination of Responses of Impacted Parties*
1 Adaptation to the Program
a Search Behavior : Collect Information, Interpret Events, Articulate Views, & Seek Aid
b Change Time, Money, & Material Investments
c Reorder Priorities & Values
2 Attempts to Modify the Program
a Political Pressure
b Court Actions
c Disruptive & Violent Actions
d Indirect Methods
* Institutional & Collective

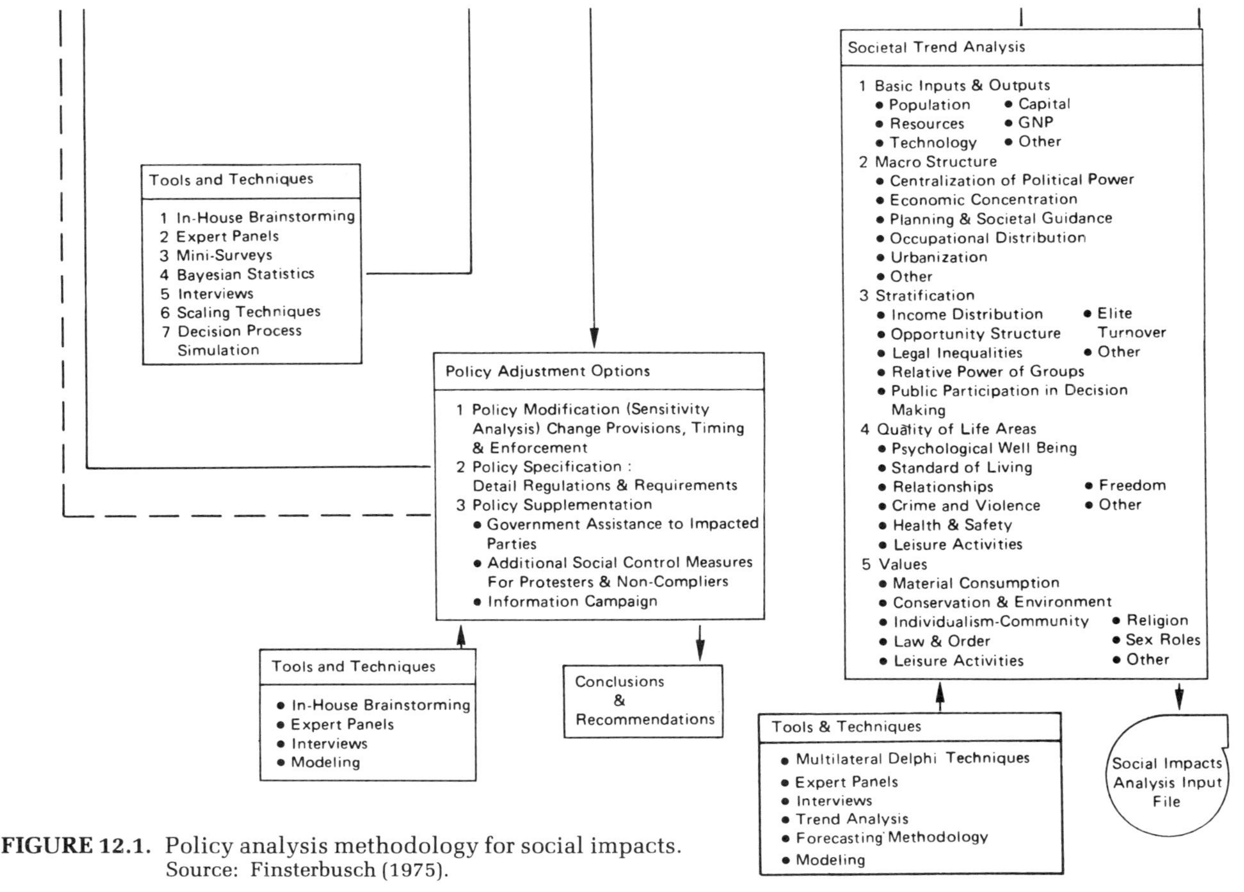

FIGURE 12.1. Policy analysis methodology for social impacts.
Source: Finsterbusch (1975).

question. In turn, these forces lead to consideration of policy options in light of the responses of the parties at interest (bottom center). Insights garnered through this step may alter the policy under assessment, thus modifying the array of physical and economic impacts that are inputs to the SIA (the social-impact analysis input file, top left). The "tools-and-techniques" blocks suggest appropriate analytical techniques, most of which are covered in this book.

The second strategic approach emphasizes *modeling*. There are many difficulties in modeling complex sociotechnical systems (see Section 7.2). For one, modeling is dependent on theory, so the limited applicability of social theory hampers social modeling. However, there is reason to hope that the SIAs' demand for such models will facilitate their development.

A review of SIA modeling activities (Battelle–Seattle 1974) makes three useful points about any social modeling effort:

1. It must attempt to determine the objective content of data in the presence of inherent biases due to collection and selection.
2. It must recognize the dynamic nature of social data, in terms of both the fleeting and perishable nature of attitudes and opinions and the constantly changing nature of human value structures, and hence social priorities.
3. It must seek solutions that are feasible and robust, rather than optimal. The best today cannot be the best tomorrow; what is optimal for one location or environment will not be optimal for another. Rather than strive to optimize, we need to learn how to select those solutions that are widely applicable in both space and time, and hence generally and lastingly acceptable.

An ambitious attempt to develop an operational SIA model is underway by Battelle (Olsen et al. 1978). They offer a general conceptual model (Figure 12.2) grounded in the structural–functional perspective of human ecology (social life as shaped by the forces acting in the human environment). They add to this the influence on social behavior of the values held by community members, and then view a community as a problem-solving system that actively responds to changed conditions. There are five resulting impact categories (demographic, economic, community structure, public service, and social well-being), and responses may either be subjective or based on the social impacts, as shown in Figure 12.2. Olsen et al. develop specific factors for each impact category and multiple indicators for each of those factors. They further offer a basis for defining a preferred value for each indicator and a method for standardizing indicator values on this basis. The resulting

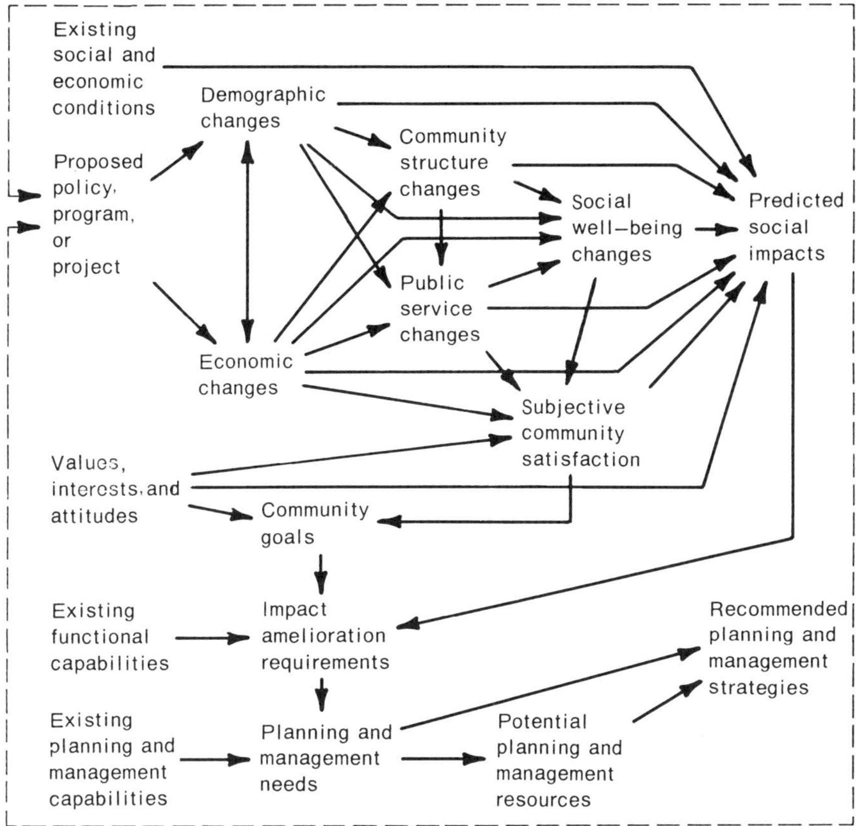

FIGURE 12.2. General social impact assessment and management model. Source: Olsen et al. (1977).

indexes can be cumulated to reflect social impacts, for example, by computing the difference in status before and after an innovation. However, the preferred values are not well established, and the relative weights to be given to each indicator in establishing factor scores, and each factor in establishing category scores, are not set. These could be chosen by obtaining subjective judgments from various stakeholders in a particular SIA context.

As the variations among these approaches indicate, there are many possible frameworks in which SIA can be done. In following the aforementioned U.S. Army Corps of Engineers (1973) Section 122 Guidelines, we emphasize the acquisition and use of social data. This strategy is suitable both for geographically localized and geographically diffuse assessments. It provides a framework in which simple analyses or sophisticated social science methods[1] can be employed to good purpose.

12.3 PROFILING AND PROJECTION

12.3.1 Profiling

Profiling consists of describing the preproject social conditions (i.e., the social description of Chapter 7). Shields (1974: 40–41) succinctly describes the rationale for profiling:

> The purpose of the profile is to establish a relevant planning setting and framework for analysis, and to assist the planner in identifying the needs and problems of the planning area as it currently exists [U.S. Army Corps of Engineers 1973: 5]. The task here is to accumulate and organize baseline data about the area in order to develop its sociological "portrait" prior to the start of project construction. At the least, this implies some dimensioning of the probable loci of impact . . . or delineation of the parameters of interaction between project and area. . . . A systematic approach which seeks to identify major groups and issues may prove highly useful in anticipating likely points of stress which might be created by project plans. That is, compiling baseline data to describe the area, the profiler should try to identify salient characteristics and problems in terms of their implications for impact assessment. One problem here is to delimit area (system) boundaries. . . . This necessarily involves giving careful thought to where indirect and longer-term impacts are likely to lodge so as not to be too narrow in boundary-setting.

Wolf (1974: 22) notes that two basic attitudes can be taken toward bounding the impact area [and describing the technology delivery system (TDS)]. One is "project related" and focuses on an existing project proposal; the other, and more difficult, is "area related" and focuses on accurate problem identification, rather than on specific project justification.

In any event, the profile features should cover a comprehensive set of the pertinent social-impact categories. The traits listed in Table 12.1 could serve as a checklist.

[1]Shields (1974) catalogs a wide range of social science methods that have been used in various SIA tasks. He also surveys a substantial number of SIAs, noting the extent of their coverage of the seven steps from the Section 122 Guidelines.

12.3.2 Data Gathering and Use

Data gathering and selection is a matter of maximum breadth of coverage balanced against convenience and cost. Social data are collected both to obtain the "before-project" profile and to record changes resulting from project planning and implementation. We distinguish between primary and secondary data-based analyses. *Primary analyses* rely on surveys, field visits, and public-participation processes to acquire new data bearing on social impacts. These are particularly pertinent to localized-area impact studies. *Secondary analyses* rely on existing data sources such as published reports, case studies, or relevant previous analyses. These are particularly salient in geographically diffuse TA /EIAs. Table 12.2 shows the use of both sorts in a study of social impacts of water-quality control legislation (Abt Associates 1975).

The heart of primary social data acquisition is the survey. This mode of inquiry is subject to easy abuse. One needs only to scan a recent newspaper to discover survey results biased because of improper sampling or loaded or unclear questions. Basic survey techniques are discussed in Section 6.4.4.1. Table 12.3 provides a useful checklist of some candidate sources of statistical and other data.

Social indicators are another form of social data which is becoming significant. Section 7.3.2 details this concept of using objective, time-series statistics to understand social states of well-being. One concern in the use of social indicators is to make sure that any data used represent the appropriate level of aggregation for the case being considered.

Another approach to primary data collection comes from the tradition of ethnographic and anthropologic field research. Each approach—survey or ethnographic [see Gold (1977) for a contrast]—has its strengths and weaknesses. Survey research is based on careful sampling procedures and is subject to detailed quantitative analysis using statistical techniques. Ethnographic research, as exemplified in SIA by the interactive approach of Vlachos et al. (1975), is much more qualitative and individualistic. It seeks to understand and appreciate the views and concerns of the parties at interest. Through sympathetic interaction, it tries to elicit their perception of the consequences of the project. The cameo entitled "An Interactive Approach to SIA" gives a systematic account of this approach. Notice the continuous tacking back and forth by the researcher between the particulars of the subcommunity being investigated and the context that is generating the technological intervention.

TABLE 12.2. Types and Selected Sources of Data at the Community Level with Illustrative Examples of Data Obtainable[a]

Type	Source	Illustrative example
Historical documents	Local newspaper files	Past controversies over water pollution abatement
	Local governmental records	Past local ordinance on water pollution abatement
	City or county public works department records	Local, often detailed, data on water use and planning
Secondary sources	U.S. Bureau of the Census[b]	
	1. City and county data book	Compilation of many demographic and social indicators
	2. First- and fourth-count census publications	Wide range of detailed population breakdowns
	3. Housing census	Detailed estimates of housing characteristics
	4. Agricultural and industrial census	Detailed estimates of elements of community's economic sector
	U.S. Department of Agriculture	Some data on crops and water consumption patterns available through county extension agents
Primary sources	Local residents	Subjective estimates of social impacts
	Local business and labor leaders	Estimates of past and future economic effects
	Local government officials	Impacts of changes on social programs (i.e., unemployment, campaigns to attract industry)
	Local water-related agency officials	Particulars of water-related activities

[a] From Abt Associates (1975).

[b] At the time of this submission, sufficient time had not elapsed for the development of a detailed list of data elements to be gathered. However, several of the key personnel involved in the study have developed such lists for other water-related studies.

TABLE 12.3. Selected Sources of Secondary Data[a]

U.S. Bureau of the Census
- Census of Governments
- Census of Transportation
- Directory of Federal Statistics for States
- Directory of Non-Federal Statistics for States and Local Areas
- Fourth-Count—Minnesota Analysis and Planning System (or other)
- General Housing Characteristics
- General Population Characteristics
- General Social and Economic Characteristics
- Other

State sources
- Governor's Office of Human Resources
- State Offices of Planning and Analysis
- Superintendent of Public Instruction
- State Department of Agriculture
- State Department of Business and Economic Development
- State Departments of Conservation
- State Departments of Finance
- State Department of Public Health
- State Department of Employment
- State Department of Public Works, Buildings, and Highways
- State Department of Highways
- State Department of Environmental Quality
- State Library
- Historical and Archaeological Commissions

Regional Planning Commissions
City (or Municipal) Planning Commissions
Departments of Public Health
Social Service Agencies
Boards of Education (or School Districts)
Social Science Research Laboratories (private and public)
University and College Libraries (theses and honors papers, as well as archives, journals, books, etc.)
Public Libraries
Local or Regional Water Resource Councils
Computer Centers
Agricultural Extension Centers (or Farm Labor Bureaus)
Others
- State Highway Maps
- U.S.G.S. Maps
- U.S. Department of Commerce
- Other

[a] From Vlachos et al. (1975).

AN INTERACTIVE APPROACH TO SIA
[from Vlachos et al. (1975)]

1. Identify the cultural composition and components of the community. Examples: Irish, Italian, Polish, Mormon, American Indian.
2. Learn their philosophy, religion, world views, beliefs, lifestyles, tastes, and other intangible background elements before studying the more material, tangible types of data: (a) if such information is available, learn as much of it as possible; (b) regardless of whether it is available the indispensable next step is to talk to some members of the cultural group(s) person-to-person, in order to learn further about these intangibles. (Caution: Observe and follow the modes of communication used in the cultural group. Do not rely on the so-called "leaders" recommended by outsiders. Go into the community yourself, and meet ordinary people whom you find there, not pre-selected or pre-arranged.) If sociological or anthropological books and articles on the cultural group are available, read them.
3. After obtaining a background as above, look through all the data first before sorting them out. It is important to spend as much time as needed in this step. Look at the data back and forth several times to see overall connections and patterns. Formulate a tentative pattern and several alternative patterns if possible. Go back to the data again to see if any pattern fits the data. If not, change it. [The pattern, that is.]
4. Go to the community again, and talk with people under step 2(b). Find out what categories are meaningful from their point of view. This is the endogenous relevance.

The incorporation of data gathered and analyzed for other purposes into an SIA is not a trivial task. However, secondary data sources are most important in TAs. Examples of the use of secondary data include the Organisation for Economic and Community Development (OECD) studies on noise pollution, the automobile, and natural resources. As described by Knezo (1974), these combined information from scholarly, "think tank," and OECD member government studies with that from newspaper articles and other data. Knezo also notes that more and

5. Choose what seem to be relevant considerations from the point of view of the relationship between the outside community (including the entire nation) and the project community. This is the exogenous relevance.
6. Organize the data around the endogenous relevance. This should be done in cooperation with someone from the community.
7. Interpret each item in the data in terms of the cultural context. If the data do not make sense, suspect that you are not sufficiently aware of the cultural context. Even if the data make sense, still suspect that the "making sense" may be an illusion due to consistent misinterpretation on your part. Always check the interpretations with people in the community.
8. Check whether the data and their interpretations depend on situational factors, and whether the "answer" may change if the situation changes.
9. Try to enter into the thinking of the people in the community; use their logic and frames of reference in describing and explaining the data.
10. Study the interrelations between the variables in the data. Study mutual causal relations, and identify mutually reinforcing causal loops as well as mutually counteracting causal loops.
11. Return to the exogenous relevance.
12. Interpret the data in terms of the context external to the community.
13. Check whether the "answer" may change if the situational factors external to the community change.
14. Study the interrelations between the variables in the community and the variables outside the community.

better social-science-oriented reports are available than have been used. These are reflected in the development of a considerable literature on social aspects of the environment and in reports that may be considered partial assessments (Knezo 1974).

Hamilton (1977) presents an attractive approach to the incorporation of secondary data into SIA (see cameo entitled "Use of Secondary Data in SIA"). Particularly note her inclusion of historical data series and analogous case studies.

USE OF SECONDARY DATA IN SIA
(Hamilton 1977)*

1. The technology and the context of the technology development are defined.
2. Based on those definitions, data sources are identified.
3. A questionnaire to "interview" the data sources is formulated and applied.
4. Based on the results of applying the questionnaire, a *standard development case* is compiled.
5. Notable exceptions to aspects of the standard development case are identified, described, documented, and the conditions under which they occurred are described.
6. If there are sufficient notable exceptions, one or more *alternative development cases* is developed.
7. A *standard baseline* (nondevelopment) *case* is formulated using the same questionnaire approach.

In closing this section we offer a general caution regarding the gathering of both primary and secondary data—sensitivity of selection. Data selection will color the impact assessment and reflect the biases of the assessor. Hence assessors should attempt to make explicit their use, if any, of social theories or models and their criteria for choice of social-impact indicators and measurement methods.

12.3.3 Projection

Shields' (1974: 41–43) description of "without project" and "with project" projection is helpful. The steps correspond to our forecasting tasks (see Chapters 6 and 7):

> *"Without Project Projection"*: This is "extended profiling": forecasting the real situation into the future under a "no project" assumption. Essentially this involves extrapolation of the baseline data in the profile to detect changes which might be anticipated in the absence of a project. This general forecasting phase should include some delineation of the time frame within which the predictions are cast as well as explicit statement of the assumptions grounding the predictions. . . . Predictions are, or at least should be better than "best guesses." But they are only as good as the assumptions upon which they rest: and if the assumptions change, as they often realistically do, then the predictions will err. One way of optimizing predictive validity is to hypothesize a set of alternative

> futures, assuming a range of parameter values for the designated project area. . . .This is particularly important at this step since the "without" project projection(s) are the baseline against which the "with" project projection(s) are compared.
>
> *"With Project Projection"*: This more specific project-related forecast involves speculation about possible futures of the area assuming one of the alternative plans is carried out. Causative factors and their effects are identified and comparisons are made against the "without" project projections. Previous impact studies of similar project types in similar project areas may be useful here. . . . In addition, effective sampling of the opinions of affected publics is one technique which could be more frequently used at this stage. . . . In principle and at least initially, a set of alternative project plans should be assessed as in the previous step. . . . The preparation of the "with project" projection is essentially the heart of effect identification. It can be best achieved by considering all factors going into a project and describing the effect that factor will have on the planning area [U.S. Army Corps of Engineers 1973]. In practice, however, it is probably inefficient, if not impossible, to consider *all* factors and their effects; rather, the analysis calls for some selective focus on a subset of the total according to some criteria of significance.

Miller (1976) adds an important caution, emphasizing that it is necessary to separate levels of impact in the selection of study areas. Mistaken comparisons can easily occur if "without" and "with" projections casually mix localized and wide-ranging effects. Offsetting local costs against national benefits is a prime example. Similarly, one must watch the *time basis* of comparisons. Balancing short-term versus long-term costs and benefits must be considered in light of the critical choice of discount rates. Possible changes in future societal values also enter into such comparisons.

Several technology assessments have used *scenario* construction (see Section 7.4.2) as their response to the uncertainties of projecting sociotechnical systems (The Futures Group 1975; International Research and Technology Corporation 1977). The future of such developments as geothermal energy or controlled environment agriculture is highly uncertain. Hence impact assessment can be served by *dissociating "projection" from impact assessment.* Rather than forecasting the most likely developments, an appropriate range of scenarios is postulated to capture the span of interesting impacts. The likelihood of these projections need not be particularly high; they serve as a vehicle to explore potential impacts. Extreme scenarios can serve to illuminate the impacts of near-term policy choices in a normative, rather than extrapolative, sense. Such a dissociation of forecasting and assessment represents an extension of the "with–without" scenario formulation. It is most appropriate when uncertainty is so great as to make any given projections very unlikely (i.e., for untried technologies, nonlocalized implementation area, and /or long time horizons), and when it does not violate guidelines (i.e., for EIS preparation).

12.4 IDENTIFICATION, DESCRIPTION, AND DISPLAY OF SOCIAL IMPACTS

12.4.1 Strategies

Identification of social impacts begins with comparison of the "projection with project" and the "projection without project"—the difference being the impacts. The cameo entitled "Social-Impact Questions" provides an incisive list of pertinent questions with which to pursue explicit impact identification. In addition, Table 12.1 offers a checklist of potential social impact areas that go beyond the generic treatment in Chapter 8.

We call attention to three significant problems in SIA.

1. *Selection.* Identification is colored by the categories used in profiling and the assessors' perspectives. Further, policy considerations will require focus on the significant impacts.
2. *Cause and effect.* A technology or project is only one of a complex of influences at work. Actual impacts will depend on other factors, such as parallel developments, policy actions, and social reactions.
3. *Anticipation.* It is not easy to grasp the implications of an unimplemented technology.

Approaches to this third problem involve the "description and display" of impact information. This can be a critical element in informing affected parties of potential implications of a development to, in turn, enable them to indicate their preferences.

Shields (1974: 43–44) recommends information displays in the form of charts, models, graphs, and cognitive maps to represent impacts in a way that facilitates public understanding and evaluation. The particular technique used will partly depend on the type of methodology employed in the impact study. For example, Gold (1974) used narration in approaching his study of the impact of coal-mining industrialization from an ethnographic perspective. Mack (1974) presented her findings about the impact of the North Springfield, Vermont dam using tabular displays of impacts and impacted groups.

Two additional examples indicate that obtaining public opinion can

SOCIAL-IMPACT QUESTIONS
[after Born and Besadny (1976)]

1. Were all impacted populations identified?
   a. Immediate property owners
   b. Community
   c. Regional
   d. Interest groups (local and nonlocal)
2. Were interactive impacts between impacted populations as well as direct impacts explored?
3. Was the question "Who benefits and who pays" clearly and hopefully quantitatively answered?
4. Were demographic impacts explored? Would project increase or decrease population (temporarily or permanently)? Would age, sex, or class mix change?
5. Were changes in institutions delineated? Would the project impact the ability of schools, churches, hospitals, town government, and so on to meet the needs of present and future citizens?
6. Were displacement and relocation impacts analyzed?
7. Was impact on community cohesion explored? Did or will the project divide the community? How deep would the divisions run? Would transient or new citizens dilute a stable (possibly unique) community?
8. Were changes in individual life-styles included as a social impact? Would the project detract or enhance personal satisfactions?
9. Was the development of new norms of behavior analyzed?

be important in TAs, as well as in EIAs. Martin Ernst, in the TA of electronic funds transfer (Arthur D. Little, Inc. 1975), recruited consumer advocates to attend a workshop to first learn about the technology before providing their perceptions of possible impacts. Haas and Mileti (1976), in studying the implications of earthquake prediction, followed a multistep process:

1. Discuss the technology with representatives of potentially concerned organizations.
2. Construct scenarios.
3. Present the scenarios to the representatives.

4. Obtain their comments on the scenarios.
5. Survey the representatives with structured questions regarding the impact of the technology on their organizations.

An analogous procedure may be used within the ethnographic approach (see Section 12.3.2). In this case the scenarios would be translated into the frame of reference of the community groups. No formal survey as such would be used.

We have stressed the importance of interaction between assessors and stakeholders regarding the nature of impacts and suitable policy actions (to which the treatment in Chapter 16 is also relevant). We now turn to a discussion of some of the more significant social impacts.

12.4.2 Some Specific Social Impacts

Shields (1974) presents an excellent review of analyses of various types of social impacts. We follow his treatment in discussing some important social impacts.

Displacement and relocation is probably the most widely treated social impact, especially in EISs. It is a pervasive impact with very high negative effects on people. Typically, it occurs in the case of large projects that lead to involuntary and irreversible movement of people. In most cases the persons being displaced, although they have no choice as to their displacement, do have a choice as to where they will relocate. Studies of displacement and relocation often note the presence of psychological and social stress.

Local–national divergences occur when the goals of national and/or regional development undermine or obscure the goals of the citizens in the most directly affected communities. Severe community disruptions have resulted from the relative neglect of these goals.

Distributional impacts refer to the question of equity, which involves economic as well as social impacts. They refer to situations in which some groups are, on balance, beneficiaries while others are losers because of a project or technology. For example, a new dam may benefit certain segments of a community, while those displaced by its construction suffer social and economic loss.

Effects on *community cohesion* include shifts in friendship networks, strains on existing friendships, intensification of class alignments and class awareness, changes in "neighborliness," and dissolution and reformation of coalitions. Such effects are associated with rapid development, the "boom-town" effect, which tends to lessen community cohesion (Gold 1974). However, other developments may sometimes enhance community cohesiveness [for example, roads and public transportation developments (Bigelow-Crain Associates 1976)].

In isolated communities, technological developments often lead to *changes in life-styles* by introducing persons whose background, goals, and values are different from those of the original residents. But beyond this project-by-project situation, many technologies, such as the contraceptive pill, have led to changes in life-styles on a wide-ranging scale.

Aesthetics refer to the visual impacts of technological developments. The National Environmental Policy Act of 1969 (NEPA) legitimates the consideration of aesthetics as it mandates preservation of "esthetically and culturally pleasing surroundings." A famous case of aesthetic impact is the Embarcadero Freeway in San Francisco. Information regarding this freeway was presented at public hearings and in news media to the community. However, it was only after the community saw that this elevated roadway interposed a wall between the city and its waterfront that the reaction halted freeway construction in San Francisco for over a decade. Harkness (1976) provides a discussion and an annotated bibliography on visual-impact assessment.

12.4.3 Other Considerations

Steps 6 and 7 in the U.S. Army Corps of Engineers (1973) Section 122 Guidelines deal with the evaluation of impacts and the policy options for mitigating possible adverse effects. These areas are covered in Chapters 14 and 15.

The main problem in addressing social impacts and policies is to relate them to the other types of impacts and policies. Attempts to do this rest, in part, on the use of good judgment for establishing the categories to be considered and ratings to be used. Burnham et al. (1974) present an approach to evaluate social with other impacts. In one phase of their approach, independent technical (expert) evaluations are made as to the magnitude of aesthetic, land use, water, air, economic, cultural/recreational, health/safety, and ecological impacts of alternative developments. Next, surveys of selected parties at interest first address their preferences among the alternative developments on each of the impact categories. For instance, prepared slides convey the land-use effects of the alternatives. Then the survey obtains a societal weighting as to the importance of the eight impact categories for the development alternatives at issue.

Before turning to a very special social impact area—human values—which deserves a separate treatment, we reiterate the point (Shields 1974: 31–32) that technologies and projects affect different people in different ways at different times. The diverse complexity of social impacts means that we are often faced with the problem of combining non-commensurables—the classic case of adding apples and oranges.

12.5 IMPACTS ON VALUES

We now discuss an important general issue in SIA—the determination of how a technology (project, policy, or whatever) is likely to alter societal values. This is different from placing a value on impacts (impact evaluation—see Chapter 14). However, both are related in a feedback loop of special importance for long-term impacts. Namely, assessors' attempts to evaluate the societal impacts require inferences as to the values held by future generations, assuming these are likely to undergo changes from presently held values. As Miller (1976: 22) notes, present generations act as trustees for future generations whose values we do not know.

Hornick and Enk (1978) illustrate the importance of realizing that both values and technologies are changing and affecting each other (Figure 12.3). In considering possible decentralized solar energy technologies, they note that we must be concerned about how values are involved in the assessment, and how values and the technology being considered relate to one another. They insist, furthermore, that we must also understand how values are changing, how these changes may affect possible development of the technology, and how use of decentralized solar technologies might affect changing values. Then, we must go another step to understand how the actual development and use of the technology will affect values as they are and values as they will have changed. Perceptions of how values and technologies may change will have effects as real as their actual changes.

For practical purposes we can consider two separate issues: (1) the forecasting of general changes in societal values and (2) the measurement of particular changes in values attributable to specific developments. The concept of general value change and suggestions on how to measure such change appear in Section 7.3.3. This discussion of how particular developments might alter societal values builds on that more general one. Indeed, due to the combination of the relatively long time periods in value change (e.g., decades at least) and little systematic study, we offer only suggestive illustrations.

Major value changes as a result of a technology are likely to be long range and indirect—attributable to the cumulation of more than one influence. Witness J. Coates' (1971) discussion of possible changes in family life due to television (cameo on p. 59). In contrast, individual projects may shift the values held by individuals as a result of reloca-

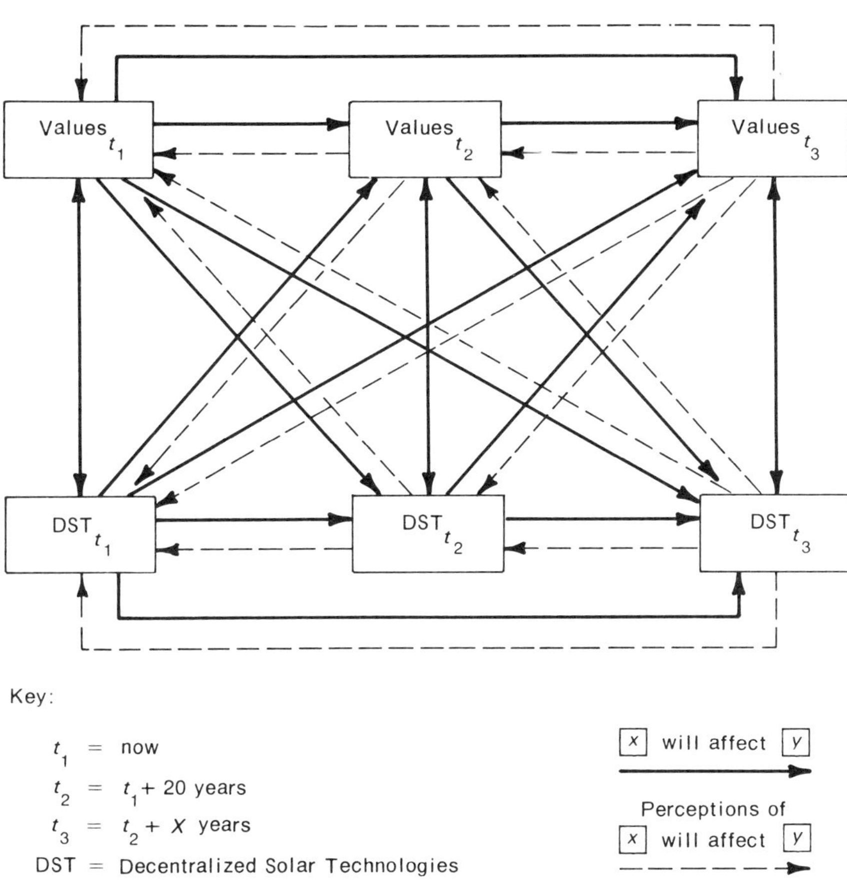

FIGURE 12.3. The complexity of interactions between values and technology. Source: Hornick and Enk (1978).

tion, exposure to new viewpoints, or whatever. Projects are less likely to affect broadly held values (e.g., those institutionalized in some fashion).

Toffler (1969: 3–4) provides a graphic illustration that a particular technology can alter a social value:

> Egbert de Vries, the Dutch sociologist, once described an African community whose inhabitants thought it necessary to start a new fire in the fireplace after each act of sexual intercourse. This bothersome belief

> meant that following each sex act someone had to go to a neighboring hut to bring back a burning stick with which to spark a fresh fire. In turn, this made adultery embarrassingly difficult, since each act of intercourse was, in effect, publicized.
>
> What changed all this was a simple bit of technological diffusion—the introduction of matches. Matches made it possible to light a new fire without first going to a neighbor's hut. The introduction of this technological artifact altered sexual behavior in the community. Did this bring with it a shift in values? Was adultery less or more frowned upon as a result? By facilitating the privacy of sex, did matches alter the valuation placed upon it? And what values led to the adoption of matches in the first place?
>
> Before we chortle superciliously about this incident, perhaps it would be well to ask how the increasing use of birth-control pills is altering our own sexual values today? And what about other kinds of technological change? What impact will widespread implantation of artificial organs have on our value system? Do the new communications technologies leave our values untouched? [Copyright © 1969 by Random House]

Major value changes due, at least in part, to a particular development are likely to be difficult to predict. This poses a problem for TA/EIA. Determining that a development may alter human values can be critically important in the assessment (e.g., consider value-change issues attendant to genetic engineering). On the other hand, building an objective case that a particular value will change in a certain fashion is likely to be difficult.

It may be useful to follow a procedure such as the following:

1. Identify value areas that the technology or project may possibly impact.
2. Determine, if possible, the trend of values in these areas (see Section 7.3.3).
3. Determine the direction and significance of any change due to the development under assessment. Note that the effects of technologies tend to pervade a society, whereas those of projects are relatively more localized.
4. Estimate a range of *possible* value impacts, possibly using scenarios as a vehicle.

This process depends on the good judgment of the assessors, including their ability to grasp the values of principal impacted groups. In some cases it may be desirable to include humanists on the study team, as was done in the TA on life-extending technologies (The Futures Group 1976).

As can be seen from our treatment, the study of impacts on values is an area where great uncertainties exist. This is due to the pervasiveness and subtlety of social values, as well as the difficulty in measuring both current values and value change. However, reason and human judgment will remain important elements in our understanding of values. This is an extremely important area in many assessments and one that has rarely received the treatment that it deserves.

12.6 PSYCHOLOGICAL IMPACTS

The impacts of technology on the individual have not been widely studied. A general framework for dealing with this area is provided by the concept of "quality of life" [U.S. Environmental Protection Agency (EPA) 1973b]. This concept can serve to provide a metric for the study of impacts on individuals, analogous to that of social indicators for social impacts. Not surprisingly, there is often considerable overlap when one compares lists of social and quality-of-life indicators (e.g., compare the cameo entitled "Quality-of-Life Factors" with the lists in "Suggested Indicator Groups" on p. 141).

One useful theoretical base for considering quality of life is to begin with identifiable, basic human needs. Maslow (1968) has developed the classic hierarchy of human needs. His five levels are (from lowest to highest priority):

1. physiological needs, such as food, clothing, sex, and health;
2. safety needs, such as stability, security, and order;
3. belongingness needs, such as affection, affiliation, and love;
4. esteem needs, such as prestige, success, and self-respect;
5. self-actualization needs, that is, the development of one's full potential.

The quality of an individual's life may be determined by how well these needs are met. The needs are founded on a basic cognitive and value structure. Changes due to technology may affect the basic structures, the needs they determine, and the means for meeting those needs. These are the impacts to be determined.

It is often convenient to think in terms of specific areas of quality of life. These have a bearing on fulfillment of the basic needs, but may be more directly estimated. The cameo entitled "Quality-of-Life Factors" provides a list of such areas. This can serve as a checklist to help identify impacts on individuals.

Some of these areas are suitable to measurement through the use of social indicators (Section 7.3.2). These especially include factors relating to the lower levels of the needs hierarchy (depending on the proper

QUALITY-OF-LIFE FACTORS

Physical well-being
Emotional stability
Food
Clothing
Housing (shelter, heat, light)
Weather
Personal income/wealth
Environmental quality (air, water, land, noise)
Safety of life and property
Equitable criminal justice
Stability (permanence, familiarity)
Love friendship (companionship)
Sexual satisfaction
Status (popularity, acceptance)
Self-confidence (self-respect)
Opportunity for employment
Neighborhood quality
Consumer product availability
Time available for leisure
Leisure activity opportunities
Cultural opportunities (art, music, aesthetics)
Mobility (transportation)
Family relationships
Novelty (change)
Privacy
Freedom
Sense of humor
Individuality
Job satisfaction
Opportunity for learning
Power (political, social)
Self-actualization (self-fulfillment)
Faith (philosophical rootedness)

level of aggregation for the TA/EIA). For example, indicators of the health of a population should be available, as should crime rates and safety statistics. For the higher need levels, indicators developed for the individual situation are likely to be necessary. Such indicators might include status, opportunity for leisure and culture, availability of privacy, and job satisfaction.

In considering impacts on the quality of life, it is useful to think in terms of ends and means. Fulfillment of the basic human needs is the ultimate end. The quality of life factors are means to meet these needs. Technology can directly affect these factors (e.g., sufficient food, personal mobility) or the means to attain them. While the relationships between the ends and the means are not always clear-cut, ends–means analysis provides a useful framework for studying impacts on individuals.

An approach to the problem of psychological impacts in line with these considerations was undertaken by Clippinger (1977). He studied an isolated group impacted by technological change, the loggers of northern New England, who were faced with changes in the physical technology of logging and the organization of the logging industry. The changing situation brought with it changes in the underlying values and life-style of the loggers.

Clippinger's objective was to develop and test a framework for describing the effects of different technological configurations on the beliefs and value systems of the users of the technology. He also intended to develop a method for describing and measuring value and belief processes, in such a way that changes could be empirically related to certain properties of the technology.

His approach distinguishes between needs (goals) and means. He selected appropriate sets of goals (along the lines we suggest) and means, and developed scales for both specific to the loggers. The "means" to individual fulfillment potentially influenced by the changed technology were (1) activities (e.g., work, recreation), (2) institutions/organizations/settings (e.g., education, local government), (3) people (e.g., immediate family, co-workers), and (4) technologies (e.g., logging technology, work organization). Among Clippinger's preliminary observations were the following:

- The new harvesting technologies in the logging industry favor traits at variance with traditional values such as independence and self-reliance.
- The high-technology loggers think in terms of free time and vacations; traditional loggers do not.
- High-technology loggers appear to "compartmentalize" their personal, as well as professional, lives to a greater degree than do traditional loggers.

This work suggests a potentially useful approach to the study of psychological impacts in TA/EIA.

12.7 CONCLUDING OBSERVATIONS

The analysis of impacts on human beings is one of the most unique and important promises of TA/EIA. However, serious study is still in its infancy, and the field suffers from the enormous uncertainties of most new areas. We divide impacts on human beings into three partially overlapping categories: (1) social impacts, (2) impacts on values, and (3) psychological impacts.

For the analysis of social impacts, we follow the guidelines of the Corps of Engineers, a leader in the field. The logic of their analysis is that the social impacts due to a technology or project are the difference between the projected state of the world with the intervention and the state without the intervention.

Great uncertainties are involved in estimating these two conditions. Primary data gathering, in either the tradition of survey research or that of ethnographic research, is usually necessary. The use of secondary social data, including social indicators, is essential.

Important social impacts include displacement and relocation, local–national divergences, distributional impacts, community cohesion, changes in life-styles, and aesthetics.

Values can be affected by technological developments. Value changes due to technologies are more pervasive and long range than those due to projects.

The study of psychological impacts may be approached by considering the differences in individual goal states and the means to achieve them, relating to the basic human needs—physiological, safety, belongingness, esteem, and self-actualization. Procedures for this inquiry are not well established at present, but a sample approach and quality-of-life factors checklist are provided.

More so than in the analysis of environmental and economic impacts, uncertainties abound, and methodological development is in the preliminary stage. However, the centrality and importance of the study of impacts on human beings make it a priority area for substantive and methodological development. At the same time, the interrelationships between impacts on human beings and other classes of impacts need to be better understood.

RECOMMENDED READINGS

Clippinger, J.H. (1979) *Assessing Sociotechnological Change: A Case Study and Methods.* Cambridge, MA: Kalba-Bowen Associates.

Develops and applies an approach to the study of psychological impacts.

Finsterbusch, K., and Wolf, C. P. (eds.) (1977) *Methodology of Social Impact Assessment.* Stroudsburg, PA: Dowden, Hutchinson and Ross.

The most up-to-date and comprehensive source of information on SIA.

Shields, M. A. (1974) *Social Impact Assessment: An Analytic Bibliography.* Prepared for the U.S. Army Engineer Institute for Water Resources, Kingman Building, Fort Belvior, VA.

An expository analysis of social impact studies to 1974 with an annotated bibliography.

Wolf, C. P. (ed.) (1974) *Social Impact Assessment.* Milwaukee, WI: Environmental Design Research Association.

A collection of papers dealing with various aspects of SIA.

Wolf, C. P. (ed.) (1976—) *Social Impact Assessment.* New York: Environmental Psychology Program, CUNY Graduate Center.

A regular newsletter that keeps its readers abreast of the latest developments in SIA.

EXERCISES

12.1. (Project-suitable) Select a topic for assessment.

1. Outline how you would conduct an SIA in this instance. Specifically consider the possibility of impacts on values and psychological impacts.
2. What methods appear appropriate for use in this SIA?
3. What data would you seek? How would the data differ for analyses of social impacts, value impacts, and psychological impacts?
4. If several people have done the exercise, compare the findings.

12-2. Preparation of an SIA proposal. Select a topic for a social-impact assessment. Prepare an informal proposal (a preproposal) to submit to a particular agency for support to perform an SIA. Consider yourselves to be a consulting firm and thoughtfully imagine a suitable sponsor agency. Focus only on the SIA—you may assume that any needed companion impact analyses will be done by others.
In a total of about three to five pages (double spaced), describe:

1. the need for an SIA (the situation, the issues, the uncertainties);
2. what will be covered in your proposed SIA;
3. what methods and what data sources you will use;
4. who will be on your assessment team and how interested parties to the issues will be involved; and
5. a management plan to accomplish the SIA (who will do what, when, total time needed, and costs—see Chapter 17 if necessary).

12-3. Use of analogous cases. Obtain two EISs on similar types of project (e.g., two Interstate highway segments or two reservoir-construction projects). Designate one (possibly the first-dated one) as a previous EIS and imagine that you are now preparing the second one.

1. After familiarizing yourself with the second situation, "interview" the first EIS for relevant ideas and information with respect to social impacts. (The cameo entitled "Use of Secondary Data in SIA" may help.)

2. Now consider the second EIS. Determine which ideas or information from the analogous first EIS
   a. would have improved the second one;
   b. are equivalently considered in the second one; and
   c. do not properly apply to the second one.

12-4. Critique of SIA. Select a particular TA or EIA with which you are now somewhat familiar. Focus on the SIA conducted in that study.

1. What potentially significant effects appear to be left out? Which are inadequately treated? (Value impacts? Psychological impacts?)
2. Have all of the relevant stakeholders been considered? How? What forms of participation in the assessment were used? How satisfactory were they?
3. Critique the SIA methods used. Critique the coverage of impacts on values and psychological impacts.
4. If several people complete the exercise, compare and discuss results.

12-5. Impacts on values. Select a TA with long time horizon (at least 20 years).

1. Refer to the cameo entitled "Modes of Value Change" (p. 145) and Table 7.1 (p. 144) for a consideration of value changes. Did the TA address such issues? How? How effectively?
2. What additional changes in societal values could you foresee resulting from implementation of the technology in question?
3. What means (methods) would you suggest to better incorporate analysis of value impacts into this TA? Be sure to note what these would cost in time, manpower, and money.

12-6. Taking a broad and long-range perspective, consider the possibility of human cloning (i.e., asexual reproduction of genetically identical humans).

1. As a group, "brainstorm" possible implications of routine availability of this technology for human values.
2. Consider possible implications for individuals. Use the "Quality-of-Life Factors" cameo as a checklist.
3. Discuss the feasibility of performing a formal TA incorporating these impact areas at this time.

13

Technological, Legal, and Institutional/Political Analyses

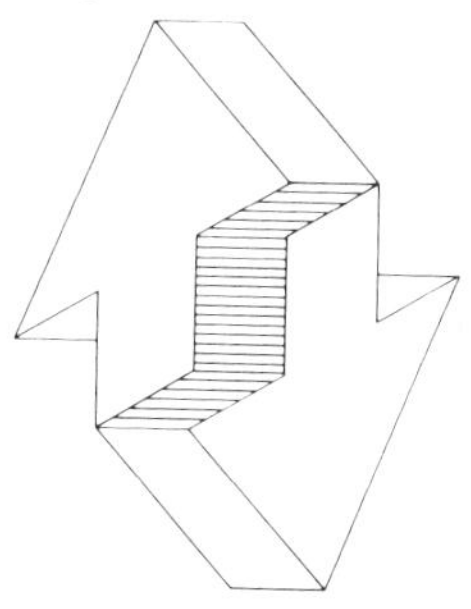

This chapter completes the EPISTLE elements by introducing technological, legal, and institutional/political factors into the assessment. It discusses the important concepts underlying each of these classes of influences on, and impacts of, technology. Most importantly, the chapter provides analytical strategies to undertake technological, legal, and institutional/political analyses.

13.1 INTRODUCTION

This chapter treats the remaining elements of EPISTLE—technological, legal, and institutional/political analyses. The rationale for this aggregation is twofold. First, our treatment of these areas is briefer than that of environmental, economic, and social/psychological impacts. In particular, consideration of technological interactions leans heavily on Chapter 6. Second, and more significantly, these areas are tightly interlocked, especially the legal, institutional, and political ones.

FIGURE 13.1. Technological, legal, institutional, and political interlinkages.

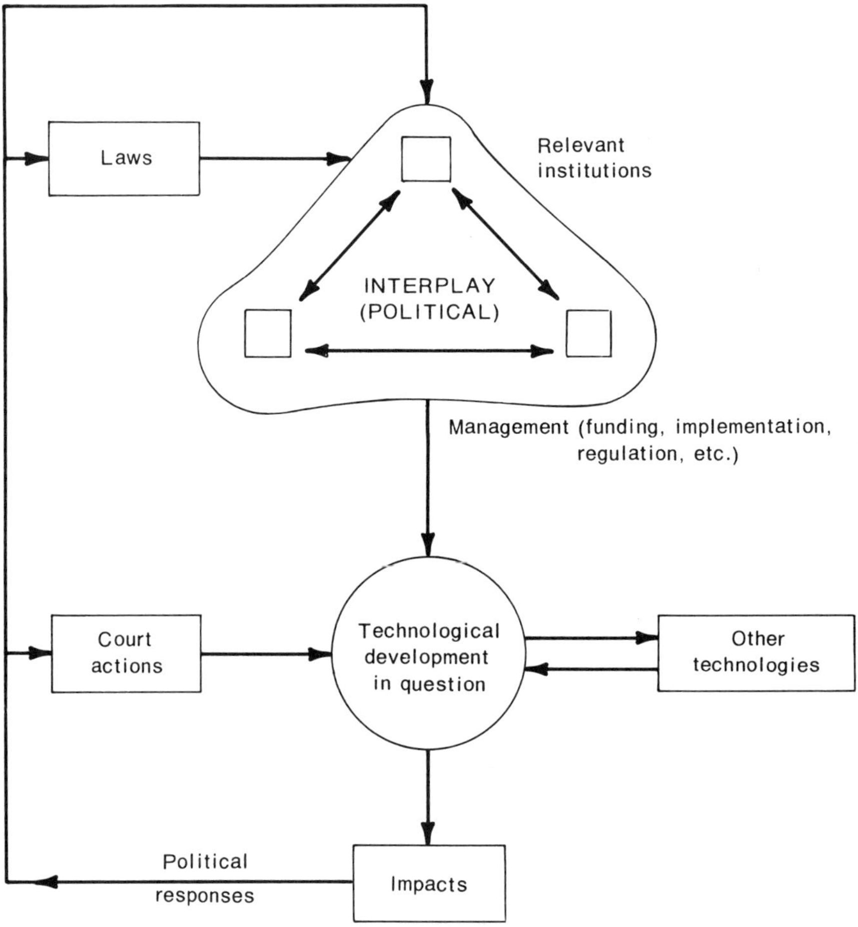

Figure 13.1 expands on the technological delivery system (TDS), shown in Figure 4.1 (p. 44), to emphasize the interrelationships most pertinent to the present discussion. *Laws* (statutes) set forth *institutional* authorities to develop, regulate, and manage the technologies in question. Procedural requirements (e.g., public hearings and EIS preparation) may also be mandated. Institutions, such as federal operating agencies and private firms, interrelate as participants in the TDS. The technological development in question, itself, interacts with other technologies. The technology causes "impacts" on society as it is implemented. In response to perceived impacts, parties at interest may take *political* action (e.g., press for legislative revision or pressure executive institutions) or *legal* (court) action at some level of government.

The reader may have noticed that "impact" has been omitted from the chapter title. The reason is that we treat both "influences on" and "impacts of" the technology in considering technological, legal, and institutional/political factors. In current practice, institutional analysis emphasizes issues in the implementation of the technology. The persons responsible for assessing how the technology is likely to influence other technologies, affect institutions, and alter laws are also likely to be addressing issues of implementation, thus completing the loop. The resultant composite "institutional" analysis may prove a most influential avenue to affect the planning and policy processes. Thus this chapter is closely linked with policy analysis (Chapter 15).

The next section discusses the general methodological characteristics of technological, legal, and institutional/political analyses. The "influences on" and "impacts of" each of these areas are then addressed, in turn, in Sections 13.3–13.5.

13.2 GENERAL METHODOLOGICAL CONSIDERATIONS

The methodology for addressing technological, legal, and institutional/political factors can be characterized as qualitative but systematic. With the exception of certain techniques for technology forecasting, there does not appear to be a major role for quantitative analyses. Assessors need to understand the appropriate TDS (see Figure 4.1, p. 44). Systematic collection of basic documents such as laws, legal cases, and agency regulations, augmented by discussions with informed people, is vital. This information should be synthesized to generate insights into critical factors for implementing the technology in question, the likely effects on existing institutions, and possibilities for new laws and institutions.

Several specific techniques may be useful for these tasks. These include *brainstorming, opinion measurement, scenario construction, historical analogy*, and *qualitative modeling*.

Brainstorming (see Chapter 1) can generate a wide assortment of potential institutional actors, actions, and impacts for consideration. Systematic procedures, such as relevance trees (Chapter 8), could prove useful to assure coverage of all relevant factors. However, the main thrust should be informal idea generation.

Opinion measurement (see Chapter 6), especially informal discussions with informed parties, is probably the most significant source of institutional information. Identification of potentially informative sources is important, and several techniques, such as "snowballing" (in which one continues to follow the leads of the last informant until there are no new leads) can help. Public-opinion measurement and citizen participation in the study may also yield insights into the acceptability of various technological and institutional options under consideration.

Scenario construction (Chapter 7) can aid in focusing alternatives into coherent packages. These, in turn, may provide a better understanding of the likely interactions among legal, institutional, and political forces, and thereby distinguish workable options. For instance, it may be difficult to grasp the political ramifications of producing a large additional amount of geothermal energy nationwide. A scenario that portrays a program of tax incentives, involvements of electric utility companies, and the regional distribution of the resultant additional energy may usefully suggest legal and institutional courses of action, as well as political implications. A variation in which one sets forth desirable future states, and then figures out how to get there from "here" (normative forecasting), can be especially fruitful in the generation of institutional options.

Historical analogy examines relevant prior developments to suggest how particular legal, institutional, and political strategies have fared and how they might work in the present case. Obviously, one must avoid hasty generalization from one situation to another. Yet there may be rich lessons here, for instance, from consideration of prior nuclear power plant sitings in preparing an EIA of a new plant.

Qualitative modeling (Chapter 9), especially constructing influence diagrams, such as those produced by interpretative structural modeling (Chapter 5), is likely to be quite useful in synthesizing information on the complex of "institutional" issues. Specific study strategies used to deal with various types of "impacts of" and "influences on" can be found in the following sections.

13.3 TECHNOLOGICAL FACTORS

Both the "influences on" the new development by existing technologies and the "impacts of" the new development on existing technologies may be important in a TA or a program EIA. The "impacts of" are usually not significant in project EIAs. The general problem is to consider the mutual causal influences of a set of related technologies. "Influences on" and "impacts of" represent opposite directions of causality. The whole process changes in time. For example, the existing state of technology may influence the development of a new technology, which in turn may lead to the modification of existing technologies, creating further technological needs and opportunities, and so on.

To illustrate these processes, we discuss some typical configurations based on the Kelly et al. (1978: Chapter 2) study of technological innovation.

"Technological readiness" refers to the *influence* of supporting technologies *on* a new technology. In the extreme case, introduction of the new technology will be impossible if the required supporting technologies are lacking. The state of their existence will determine the performance level and characteristics of the innovation introduced. For example, the introduction of a vertical or short takeoff and landing (V/STOL) aircraft system requires, in addition to the aircraft, avionics, pollution abatement, and ground traffic-flow technologies. The potential performance level of the aircraft themselves is limited by the current state of these technologies.

Competing technologies are also "influences on." In the case of the V/STOL aircraft system, the status of potential competing technologies, such as conventional aircraft and buses, would influence the decision to develop and adopt the V/STOL system. The availability of a technology without necessary related and supporting technologies creates a "technological imbalance." This "impact of" V/STOL technology may serve as a stimulus to further development in supporting technologies. A sequence of such imbalances, which illustrates the complex mutual causality involved, occurred in the 19th-century textile industry. Kay's invention of the flying shuttle sped up weaving, upsetting the ratio of spinners to weavers. At that point, either the number of spinners had to be increased or else innovations to quicken the spinning process were required. A number of innovations by Hargreaves, Cartwright, and Crompton sped up spinning. Then Cartwright mechanized weaving by the invention of the power loom. These machines lowered the price of cotton textiles, leading to a corresponding increase in demand. The bottleneck now lay in the supply of raw cotton, where the main difficulty was the labor involved in picking the seeds from the bolls. This bottleneck was broken by Whitney's invention of the cotton gin, which more than tripled the rate of picking seed-free cotton (Mantoux 1961).

Understanding technological impacts depends on understanding the connections among related technologies. Information can be obtained by literature review and by contact with experts in the technologies. The question of the future of these technologies and how, if at all, their developments will interact, must follow. This sort of analysis is covered in technology forecasting (Chapter 6). Appropriate techniques include trend extrapolation, with its several variants, and expert opinion. Quantitative techniques are more appropriate in dealing with technological parameters than with any other topic in this chapter.

A procedure for dealing with technological impacts is the following:

1. Identify technologies related to the technological development of interest.
2. Determine how these technologies are related. Specifically, emphasize relationships of dependence among particular technical parameters (e.g., metal strength and aircraft speed).
3. Consider any nontechnological influences that may affect these relationships.
4. Forecast the constellation of these parameters.
5. Interpret the causal relations between the parameters pertaining to the technology of interest and the related technologies, considering each as both cause and effect.

To a large degree, determining interactions of a technology or program with other technologies reduces to technology forecasting. The principal difference between "influences on" and "impacts of" is the point in the causal chain from which they are considered. The policy considerations relating to technological factors cluster around: (1) incentives to innovate, including public funding of research and development, (2) protection, through patenting, and (3) the public regulation of technologies, e.g., through antitrust laws.

13.4 LEGAL FACTORS

13.4.1 Implications of the Law for Technological Development

Legal analysis is considered first in terms of the effects of the law on the technology and then in terms of the effects of the technology on the law. Let us begin with some basic reflections on the law and how it operates.

The law is best seen as the active embodiment of a society's values (Tribe 1973a: 47). It furnishes the architecture for social edifices ranging from marriage to representative government to economics. The law can operate negatively by constraining interactions (e.g., criminal law); it can operate positively by facilitating human relations and aspirations, or by establishing institutions and providing authority to perform designated functions.

Legal norms can be created when interested persons induce legislators, administrators, or judges to pass a law, promulgate an administrative rule or decision, or create a judicial precedent during the course of litigation (Changnon et al. 1977). Technological development can be influenced by such legal actions in three basic modes (Tribe 1973a: 52): (1) *specific directives*, (2) *modifications of market incentives*, and (3) *changes in decision-making structures*.

Specific directives can prevent development, such as by declaring an area a national wilderness. They can also permit it, as when Congress mandated construction of the Alaska pipeline to close off court actions on the adequacy of the EIS.

Modifications of market incentives include alterations in taxes and credits to encourage certain forms of technology (e.g., the oil-depletion allowance). Tort actions are a prime means to internalize social costs by making a firm liable for the effects of its processes and products (Katz 1969). Recently, bases for liability claims have broadened, placing a greater responsibility on firms to assure the safety of their actions and products (e.g., lawsuits on the effects of chemicals released into rivers). On the other hand, it is desirable to protect the firm sufficiently from prohibitive financial risk to encourage innovation. For instance, the Price–Anderson Act provides for a system of private insurance and government indemnity totaling $560 million to pay public liability claims in case of a nuclear accident.

The notion of liability as a mechanism for controlling technological development deserves additional attention. The jeopardy of being sued can be a major consideration in the implementation of a technology. By inducing "cost internalizing" (see Chapter 11), liability can improve society's benefit:cost ratio, but it is by no means obvious which rules of liability would maximize human satisfaction (Tribe 1973a: 96). Furthermore, liability can be difficult to prove. The burdens of suit may outweigh the benefits when the disturbance is diffuse, and the courts usually lack expertise to address sophisticated technical issues adequately.

Figure 13.2 denotes four bases for legal liability, for the case of hail suppression. Successful suit requires proof on one of these grounds. As Davis points out in analyzing the hail suppression case, different political jurisdictions may view matters quite differently (Changnon et al. 1977: 154). Pennsylvania considers weather modification to be abnor-

FIGURE 13.2. Basis for legal liability.
Source: Chagnon et al. (1977: 152).

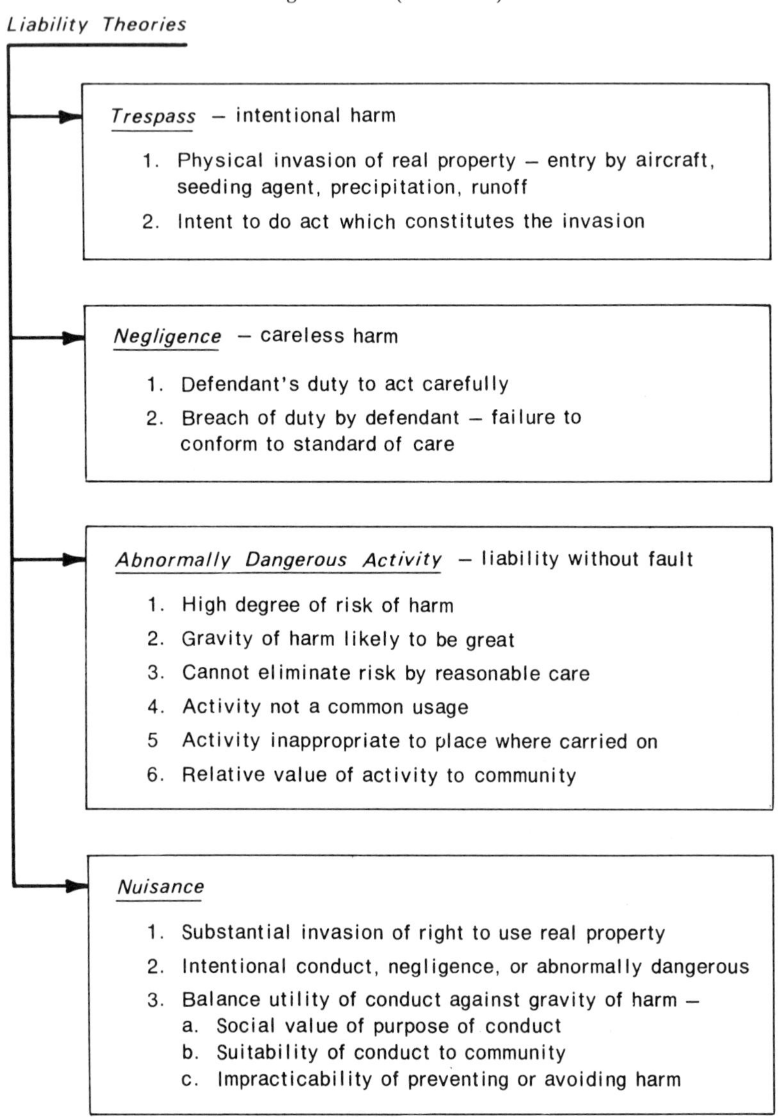

Liability Theories
Trespass – intentional harm
1. Physical invasion of real property – entry by aircraft, seeding agent, precipitation, runoff
2. Intent to do act which constitutes the invasion
Negligence – careless harm
1. Defendant's duty to act carefully
2. Breach of duty by defendant – failure to conform to standard of care
Abnormally Dangerous Activity – liability without fault
1. High degree of risk of harm
2. Gravity of harm likely to be great
3. Cannot eliminate risk by reasonable care
4. Activity not a common usage
5 Activity inappropriate to place where carried on
6. Relative value of activity to community
Nuisance
1. Substantial invasion of right to use real property
2. Intentional conduct, negligence, or abnormally dangerous
3. Balance utility of conduct against gravity of harm –
a. Social value of purpose of conduct
b. Suitability of conduct to community
c. Impracticability of preventing or avoiding harm

mally dangerous so that proof of fault is not necessary to recover for harm, whereas Illinois statutes declare that it is not abnormally dangerous. In the event a plaintiff should be able to establish a liability theory, harm, and a causal linkage, defendants can still escape liability by establishing a defense. Davis notes several possible lines of defense: contributory negligence by the plaintiff, consent to the risk, the legal right to protect one's property (e.g., stop hail damage), and sovereign immunity (some states are immune from liability claims). Legal analysis requires careful attention to liability issues affecting the technology in question.

Changes in decision-making structures are obviously a powerful lever on development decisions. Institutional authority may be changed, as in the partition of the Atomic Energy Commission's regulatory functions to the Nuclear Regulatory Commission and its promotional functions to the Department of Energy. One intended result of this separation was the enhancement of the regulatory aspect.

13.4.2 Legal Analysis

This discussion of the law suggests its role in the social management of technology. In considering a particular technological development, each of the three modes of legal control (directives, market modifications, and structural changes) should be considered in regard to existing and potential laws, administrative regulations, and court cases. Table 13.1 provides a general framework to assure systematic consideration of these factors. The TAs of hail suppression (Changnon et al. 1977) and remote sensing (of Earth resources by satellite, etc.) (Zissis et al. 1977) provide two rich examples of legal analyses. The cameo entitled "International Legal Considerations in Remote Sensing" illustrates the sorts of issues that can arise.

13.4.3 Legal Impacts

The legal means to control technology and the impacts of the technology on the law are strongly linked. For instance, definition of sovereignty limits certainly affects the use of remote sensing (see cameo). Conversely, advancements in satellite technology have spurred concern and consequent development of relevant legal principles. Accordingly, the *consideration of legal impacts should take place in conjunction with the legal analysis* (Table 13.1). In discerning legal issues and options, one is also identifying legal impacts.

Major new technologies can threaten formerly secure human values and thus involve serious and broad-ranging legal issues. Several examples illustrate this.

Photocopying has drastically eased the reproduction of printed material, undermining copyright protection. Recently a revised copyright law prescribed new provisions to cope with the photocopy problem.

Electronic surveillance devices have provided new capabilities, requiring new legal determinations of rights of privacy.

The efficacy of blood transfusions has induced the courts to rule against religious tenets of the Jehovah's Witnesses to protect human lives by providing transfusions.

New biomedical technology for life extension through artificial maintenance has led to legal dilemmas on "pulling the plug" and the definitions of life and death.

It is desirable that the TA/EIA probe the possibility of such profound

TABLE 13.1. Legal Analysis

Task	Suggestions
1. List the *legal issues* affecting the technology under consideration. These can be categorized by the *functions served:* Research (e.g., funding, ethics) Development (e.g., authority) Planning (e.g., authority) Implementation (e.g., operations) Financing (e.g., sources, constraints) Regulation Liability	1. Begin with the obvious issues, identify additional issues by brainstorming and informal interviews with knowledgeable persons. Extend these through review of legal periodicals and topic-focused sources (e.g., articles and books on weather modification are likely to touch on legal issues). A broad range of viewpoints should be obtained. For new technologies lacking in precedents, it will be necessary to draw on related technologies and analogous cases.
2. Identify and analyze the *legislative statutes* and *administrative rules* that pertain. These can be categorized by the functions served (task 1). Note the three basic modes: Specific directives Market modifications Decision-making authority granted and constraints imposed	2. Research applicable codes at all levels of government. Agencies may have gathered and summarized pertinent statutes and administrative provisions. Consult with administrators and planners, including comprehensive planning bodies as well as technology-specific offices.
3. Establish the pertinent *case law*. This could well follow the classification of legal issues; also note the three basic modes (task 2).	3. Focus legal research on the technology when rich precedent exists. When this is not the case, one will have to consider related topics, pending cases,

legal /value issues (see Section 12.5). Identification of critical impacts can be aided by a legal analysis as in Table 13.1.

13.5 INSTITUTIONAL/POLITICAL FACTORS

13.5.1 General Considerations

Institutions and politics are vital factors in the implementation of a technology. In the process of such technological implementation, they may be themselves drastically affected. Section 13.5 discusses some basic institutional /political characteristics, presents a strategy to analyze how these factors can impinge on a development, and finally highlights the sorts of impact on institutions and political functions that can arise from a technological change.

TABLE 13.1 *(continued)*

Task	Suggestions
	miscellaneous policy statements, and current practices. Computer retrieval may help.
4. Identify *legal options* for the social management of the technology under consideration: Available authorities and constraints Possibilities presently under consideration (e.g., bills in process) Additional options generated on the basis of analysis of the technological and legal options involved.	4. Discuss possible options with a variety of concerned parties. Determine what legislative bodies have an interest, examine legislative calendars, discuss pending legislation with appropriate staff. Generate novel, but not unrealistic, options to span the range of possibilities.
5. Evaluate the combinations of technological and legal options: Include the baseline situation Identify the steps required for implementation of the legal options, taking into consideration the political factors involved Compare the implementation of the technological options under alternative legal options (take into account the institutional / political analysis).	5. Evaluation should be done hand in hand with the institutional analysis as part of the overall policy analysis.

INTERNATIONAL LEGAL CONSIDERATIONS IN REMOTE SENSING
[based on Zissis et al. (1977)]

The use of satellites for surveillance of various Earth features is a relatively new technology that involves equally new areas of the law as well. Some of the international legal issues are the following:

"Space law" principles are largely uncodified; they must be inferred from treaties, national law, and customary usage.

Airspace vs outer space boundaries are murky; how high does national sovereignty extend? (The Soviets favor regulation according to the nature of the activity rather than the altitude.)

What constitutes legitimate observational activity? What constitutes espionage?

For the purposes of TA/EIA, *institutions* refer to groups of individuals (i.e., organizations) whose collective activities have a bearing on the TDS in question. Institutions may be formal (i.e., formed by law or contract) or informal. Governmental agencies, interest groups, and cultural institutions function so as to further certain social values. In pursuit of their goals, institutions become involved with particular technologies.

Politics can be considered as the domain of power relationships. Presumably, the parties with power in a situation are those who obtain the most benefits with the fewest costs. Politics is tightly linked with the policy process to which TA/EIA aims to contribute.

Politics and institutions are so closely intertwined that we choose to present a combined analysis. Our analytical approach (Table 13.2) considers both power dynamics within institutions and the power relationships among institutions.

From another perspective, institutions and political interactions can be recognized as, respectively, *structures* and *processes.* Structures are a formalization of patterns of relationships. Processes are the behaviors that take place within a structural context. Together, structures and processes combine to determine the performance of a TDS (Gross 1966). Figure 13.3 illustrates a TDS for the private housing market for solar energy. The critical role of institutional structures and their functioning in determining the success or failure of this technology stands out. Our institutional analysis strategy (Table 13.2) attempts to capture the important elements of both structures and processes.

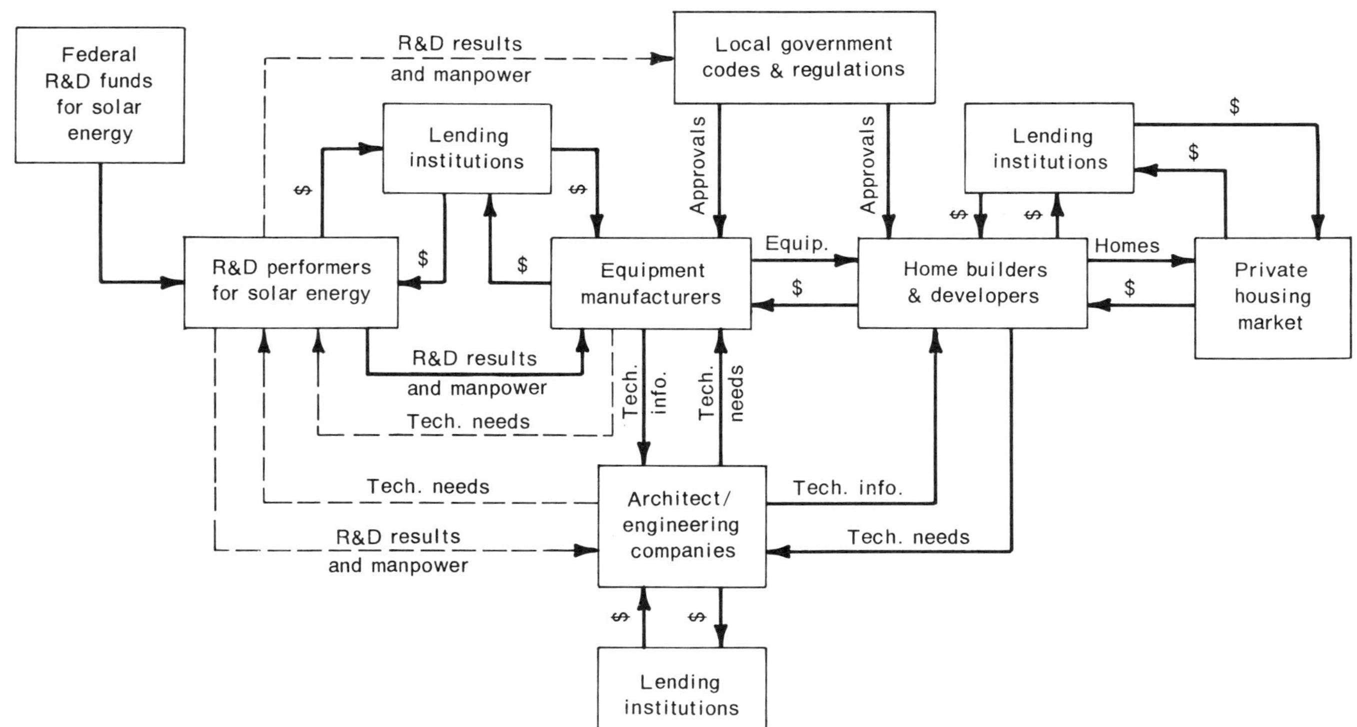

FIGURE 13.3. The TDS for the private housing market showing required interactions between solar-energy R&D performers and other components. Broken lines indicate linkages to be established or strengthened.
Source: Ezra (1975). Copyright © by the American Association for the Advancement of Science.

TABLE 13.2 Institutional/Political Analysis

Task	Suggestions
1. Identify *formal organizations* [public (at all levels of government) and private] involved and their: Statutory authority with respect to the technology of interest General characteristics Mode of operation Interfaces with other institutions Where they get their funds Who has authority over them in various functions Real and perceived constraints on their operation Functions served for the technology in question Facilitation (information exchange, planning, coordination) Financing Rule making (establishment of standards) Enforcement (monitoring, arbitration, application of sanctions) Operation Perceived and potential interests in the technology in question Political concerns (power relationships within the organization)	1. Begin with the obvious important organizations; identify additional ones through discussions and document review. Obtain and review procedural manuals. Interview those in policy making positions and staff involved in implementing policies. Consider actual practices (as contrasted with official descriptions) and understand any indirect ways to get things done. Identify hierarchical levels and their respective roles. Explore interaction patterns by questioning acquaintanceships, information exchange practices, and joint programs. Compose images of actual practice based on multiple views. Consider institutional innovativeness (record of adopting new programs), centralization, and formalization.
2. Identify *informal organizations* involved, i.e., parties at interest (consider parallel points to those listed	2. As above; also, "chaining" contacts may help in identifying persons for interviews.

To amplify the notion of process, it is worth emphasizing that institutional analysis attempts to play a distinctly *active* role in TA/EIA. More than a passive enumeration of the impacts likely to follow from a particular development, the intent is to *design* a "successful" implementation. Careful analysis of institutional and political functions can suggest ways to mitigate negative aspects of a development.

In a TA or program EIA, impact analysis is likely to take the form of large-scale (e.g., national) strategies. In a project EIA, it will be much closer to the day-to-day workings of the local institutions involved. In

TABLE 13.2 *(continued)*

Task	Suggestions
for formal organizations, and links with various formal organizations).	
3. Analyze the *effectiveness* of the various institutions *with respect to implementation of the technological options* under consideration. Consider both on an individual basis and as a system.	3. Consider necessary functions to be performed by the institutional structures. These can generate criteria for judging institutional effectiveness. A matrix format can be used to summarize how well institutions satisfy these criteria under different policy options. A system process flow diagram may help to assess the importance of particular institutions' roles.
4. Generate *new institutional options* potentially more conducive to implementation of certain of the technological options. Consider the political realities with care.	4. Draw upon task 3 to indicate weak functional areas of existing institutions, especially critical leverage points for the performance of the overall TDS. Generate institutional possibilities to improve this situation, seeking ideas in current proposals and pending actions, plus suggestions from a wide network of parties at interest.
5. *Evaluate* the combinations of technological and institutional options: Include the baseline institutional case and the no-action technological option Identify the sequence of institutional actions and the time needed for implementation of the technological options (take into account the legal analysis).	5. Systematically apply the functional criteria to various institutional arrangements. Preparation of a few selected institutional scenarios may convey the flavor of the evaluation better than a comprehensive exercise to evaluate most possible variations. Consider institutional flexibility in responding to various contingencies.

both, there is a real opportunity for institutional analysis to contribute to improved functioning of the TDS under study. Institutional manipulation is often the key ingredient in policy making. It offers a major lever to direct the course of development of a technology.

13.5.2 Institutional Analysis

Having discussed institutions and politics, we turn to the issue of how to perform institutional analysis. Table 13.2 embodies the analytical framework based on the work of Sarah Taylor (see especially U.S. Army Corps of Engineers 1976).

Several points can be made about the tasks outlined. First, tasks 1 and 2 should proceed together, drawing upon documents and informal interviews. Literature on the technological topic can help identify institutional actors in the TDS. Determination of statutory authority is involved in legal analysis. Table 13.3 provides a sample entry in an institutional analysis that juxtaposes statutory authority with functions served and general characteristics of the formal organizations identified. Analysts should be familiar with organization theory to be aware of typical behavior patterns. For instance, staff personnel with personal loyalties to their boss ("courtiers") may act contrary to bureaucratic norms (Dexter 1977). There are four categories of interaction between informal and formal institutions: (1) legitimate, (2) clientela (direct relations), (3) parantela (fraternal, indirect ties, as between political leaders and the bureaucracy), and (4) illegitimate (bribery) (Peters 1977). The use of scheduling procedures (Chapter 17) to determine the progression of actions in implementing the technology in question can be informative (U.S. Army Corps of Engineers 1976). Political sensitivity is needed throughout the analysis, especially to possible institutional changes.

13.5.3 Institutional/Political Impacts

To provide some flavor to this short discussion, Table 13.4 distinguishes institutional /political influences and impacts. The list could be greatly extended, and, in many cases, particular incidents could be fit into two or more of the four categories.

Institutional impacts are troublesome to document and also highly sensitive because institutions act to protect their interests. This portion of a TA /EIA needs to be considered in an active, feedback perspective. As potential impacts are identified, they will suggest interventions in the course of the development. Although this makes prediction difficult, it can offer better-informed participation in the policy process—a prime objective of TA /EIA. We recommend that the identification and analysis of institutional and political impacts take place in conjunction with the complementary implementation analysis (Table 13.2).

13.6 CONCLUDING OBSERVATIONS

The remaining elements of EPISTLE (technological, legal, and institutional /political) are introduced. The latter two are tightly interlocked. In practice, institutional analysis emphasizes implementation issues. We separate institutional influences on technological development from technological impacts on institutional factors. However, the analytical strategies emphasizing the former also allow us to deal with the latter.

The useful methodology is systematic but qualitative, rather than quantitative, in character. Qualitative modeling, brainstorming, opinion measurement, scenario construction, and use of analogous cases are pertinent techniques. Each area (technology, law, organizations, and politics) adds its own professional insights. Synthetic understanding is the key to successful institutional study.

The development of a technology affects that of other technologies and, in turn, is affected by their development. This complex mutual causality produces both "influences on" a technology and "impacts of" a technology. Technology forecasting techniques are most important for analyzing this development, once the relationships among the technologies have been established.

The law embodies society's values through legislative statutes, executive regulations, and judicial precedents. The chapter discusses basic ways in which the law can influence technology, with some emphasis on establishment of liability.

Institutional structures and processes go far toward determining the viability of technological options.

Tables 13.1 and 13.2 present complementary analytical strategies for legal and institutional/political analysis. These analyses are highly interactive—both between one another and between assessors and parties at interest. We seek to determine better "institutional" arrangements to mitigate negative aspects of technologies and, in the process, significant alterations of institutions are liable to result.

Examples of legal and institutional/political impacts indicate their often indirect but vitally important nature.

RECOMMENDED READINGS

Changnon, S. A., Jr., Davis, R. J., Farhar, B., Haas, J. E., Ivens, J. L., Jones, M. V., Klein, D. A., Mann, D., Morgan, G. M., Jr., Sonka, S. T., Swanson, E. R., Taylor, C. R., and Van Blokland, J. (1977) *Hail Suppression: Impacts and Issues*. Urbana, IL: Illinois State Water Survey. Prepared for the National Science Foundation, Office of Exploratory Research and Problem Assessment.

A fine example of legal analysis in a TA. [Zissis et al. (1977) is another.]

U.S. Army Corps of Engineers, (1976) Baltimore District. "Binghamton Wastewater Management Study: Institutional Analysis Appendix."

A stellar example of a detailed institutional analysis.

TABLE 13.3. A Partial Sample of Institutional Identification—Wastewater Management on a National Scale[a]

Federal agencies			Comprehensive wastewater interests								
Appropriate act/statute	Designated institution	Administrator	Planning, policies, related services	Financing, policies, grants, related services	Property-acquisition policies	Construction, maintenance, related services	Wastewater management administrative policies, budgetary considerations	System operation, policies	Water-supply services, management	Solid-waste management	Is concerned with
Water Resources Plng. Act of 1965. NEPA of 1968 (P. L. 91-190) (P.L. 89-298 approved 27 Oct 65; Section 206). Rivers & Harbors Act of 1966 (P.L. 89-789, 7 Nov 66; Sect. 102). Federal Water Pollution Contr. Act Amend. of 1972 (P.L. 92-500) approved 18 Oct. 72; Sec. 404	Department of Defense	U.S. Army Corps of Engineers	X		X	X	X	X	X		In addition to its comprehensive water-resources management program (including flood control, storm-water runoff, disposal areas for dredge material) the Corps is involved with a study of alternative means to manage wastewater on a regional basis, in various portions of the country.

											In a field somewhat related to comprehensive wastewater management, the Corps issues permits for the discharge of wastes, other than municipal, into navigable waters of the United States.
Section 204 of the Demonstration Cities & Metrop. Dev. Act of 1966. Title IV of the Intergovernmental Cooperation Act of 1968. Housing Act of 1954	Department of Housing and Urban Development (HUD)	Office of Community Devel.	X	X		X					HUD has a number of grant programs available for community planning, loans for public works construction—including sewer and water-facility construction. Both agencies administer the 701 Planning programs under the Housing Act of 1954.
		Office of Community Plng. (Both under the Office of Community Plng. & Mgt.)	X	X		X					

[a]From U.S. Army Corps of Engineers (1976: 18).

TABLE 13.4. Sample Institutional and Political Influences and Impacts

Influences of	Impacts on
Institutional factors	
Automated guideway transit technology is available and needed; transit-authority institutional difficulties are the main barrier to deployment (U.S. Congress, OTA 1975)	Transition to the metric system could provide impetus for nationwide building codes and consequent industry standardization (Wertz 1977)
Dissolution of the Congressional Joint Committee on Atomic Energy makes nuclear energy more vulnerable to its opponents	Movement to electronic funds transfer could cause a relative disadvantage to smaller financial institutions and retailers, and pressure toward a national banking system (A. D. Little, Inc. 1975)
Multinational corporations have rationale to promote U.S. adoption of the metric system to facilitate their cross-national operations (Wertz 1977)	Movement to a hydrogen economy would necessitate a new international infrastructure for fueling (Dickson et al. 1976a)
Political factors	
Energy conservation initiatives have encountered a variety of political forces, ranging from oil companies to cost-conscious consumers, tending to stymie coherent planning	School bussing to achieve desegregation affected local, state, and federal elections
Political factors influenced the U.S. decision to forego building the supersonic transport (SST)	The British–French supersonic transport, the Concorde, influences international politics, e.g., in landing rights for it and plane orders from the United States
Environmental groups, given a basis for intervention by the National Environmental Policy Act of 1969, carry weight in most development decisions, e.g., on power-plant siting	Nuclear power, with plutonium recycling, presents increased possibilities for political terrorism (U.S. Nuclear Regulatory Commission 1976)

Kelly, P., Kranzberg, M., Rossini, F. A., Baker, N. R., Tarpley, F. A., Jr., and Mitzner, M. (1978) *Technological Innovation: A Critical Review of Current Knowledge*. San Francisco: The San Francisco Press.

Synthesizes the literature on technological innovation.

Tribe, L. H. (1973) *Channeling Technology Through Law*. Chicago: Bracton Press, Ltd.

A meaty discussion of the interaction between law and technology, illustrating legal research applied to technological issues.

EXERCISES

13-1. (Project-suitable) Choose a technology. Work through the five-step procedure for dealing with technological impacts in Section 13.3 without going into excessive depth or detail. Use technology-forecasting techniques as appropriate. List the parties at interest in the technological impacts and some policy sectors that may be involved.

13-2. Argue through cases for the plaintiff and for the defense as to liability concerning increased air-pollution levels from a new coal-fired power plant. Assume the plant has been in operation for 1 year and that air-pollutant levels have increased in an adjacent urbanized area. In particular, identify the important contingencies in establishing who might be liable for what.

13-3. (Project-suitable) Choose a technology. Step through a legal analysis to an appropriate level of detail following the strategy of Table 13.1

13-4. (Project-suitable) Choose a technology. Step through an institutional/political analysis to an appropriate level of detail following Table 13.2.

13-5. (Project-suitable) An exploratory comparative synthesis.

1. Focus on a particular impact assessment topic, preferably one with which you are already familiar.
2. Specify two significant alternatives (you may wish to expand these into brief scenarios—see Section 7.4.2)
3. Identify significant institutional, political, legal, and technological *influences* that *differentially* favor one of these alternatives (e.g., present law prohibits one alternative.)
4. Identify significantly *different* institutional, political, legal, and technological *impacts* of the alternative developments.
5. Discuss how you would pursue analysis of these sensitive differential factors to help resolve the choice between the alternatives. Would any specific techniques be useful, and if so, how should they be applied?

14
Impact Evaluation

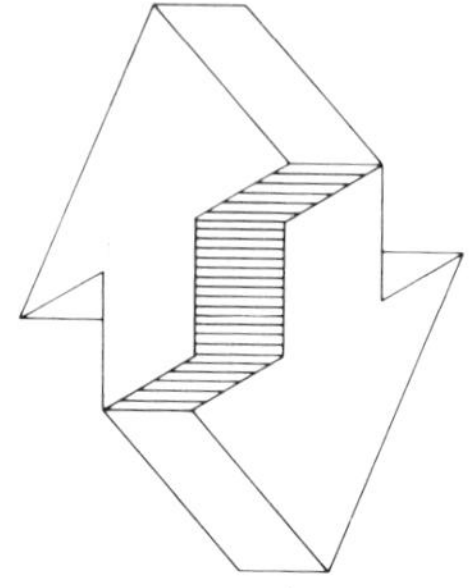

This chapter is designed to assist the reader in the evaluation of the impacts whose identification, description, and analysis have been discussed in the preceding several chapters. Impact evaluation seeks to assess and compare the impacts that proceed from a technological activity, so as to provide a basis for policy formulation. Evaluation is, by definition, the process of determining value. Thus the value dimensions involved in its performance are considered in the opening sections of the chapter. Both value issues internal to the assessment process and those arising from the larger societal context of TA/EIA are developed. Strategy for carrying out impact evaluation is next described and discussed. Three methods of varying complexity and sophistication—dimensionless scaling, decision analysis, and policy capture—are explained in the last section.

14.1 INTRODUCTION

Evaluation is the process of assigning value. The value of something is usually assigned relative to that of something comparable with which the evaluator is familiar. For example, the currencies of the various countries are constantly valued and revalued relative to one another on world money markets. Ultimately, however, evaluation must rest upon some stated or unstated criteria that provide the yardstick by which judgment is made. Thus a currency's value rests upon the economic health and prospects of a country as indicated by parameters such as attractiveness of exports, opportunities for investment, and domestic rate of inflation. Underlying these judgment criteria are goals and aspirations that themselves are the external manifestations of individual or group values.

Evaluation thus requires two elements—criteria and measures. Criteria reflect the values held by the evaluators or the parties whose judgment they are trying to reflect. Measures reflect the degree to which the criteria are met. For instance, if one were concerned about sufficient food to maintain a nation's people, appropriate measures could include malnutrition and starvation rates.

Impact evaluation involves the assessment of the effects of a technological development in the light of societal values. It is related to the treatment of values elsewhere in this book. In particular:

Chapter 2 poses broad societal value considerations affecting technological development;

Chapter 7 discusses the concept of value and ways to measure value change; and

Chapter 12 considers the impacts of technological development upon societal values.

Here we emphasize the way in which values translate into criteria for evaluating impacts.

The preceding several chapters have discussed ways to identify and analyze various impacts. These analyses provide the basic information on which to base impact evaluation.

Evaluation takes place throughout the TA/EIA process. Taken broadly, evaluation plays a role in most steps in TA/EIA. The assessor makes repeated decisions (preceded by conscious or unconscious evaluation) during the assessment, beginning with selection of the technology to be assessed. These decisions, both individually and collectively, shape the character and results of the study.

Taken more narrowly, evaluation can be restricted to the techniques by which impacts are consolidated to provide input to the policy analysis process. Since each TA/EIA step contributes to policy analysis, impact evaluation is no exception. Thus impact evaluation should be considered in light of how it clarifies the choice of policy options, and not how impressively it marshalls and displays detail.

The major sections of this chapter address two themes: (1) a discussion of the role of values in impact evaluation and (2) the process of doing impact evaluation. The latter concerns both general strategic choices and specific techniques that may be useful.

14.2 THE ROLE OF VALUES IN IMPACT EVALUATION

Values determine which of many feasible goals or futures is preferable. Policies, or means, are then evaluated on the basis of their perceived capacity to achieve those goals. Thus the selection of means, no less than ends, involves the application of a set of values. In the context of TA/EIA, the assessor generally desires to produce a valid, useful study (a goal). In each step of the assessment, therefore, strategies and techniques (means) are chosen that are believed to lead to that goal.

Values enter into impact evaluation from two sources—factors external to the assessment and preferences internal to the assessment team. External factors consist of the societal concerns that generate the TA/EIA. In fact, societal values advance the concept of TA/EIA as a useful means for anticipating the achievement of societal goals. Other value issues arise during the conduct of the study and are manifest in decisions about how the study will be executed. It is a mistake to assume that internal and external value sets are necessarily identical or even compatible (see the cameo entitled "Internal- and External-Value Issues"). We now briefly discuss the topics raised in the cameo.

14.2.1 Value Issues Arising from the Societal Context for TA/EIA

TA/EIAs are intended as tools to analyze costs and benefits of the application of a technology—but which impacts should be accounted as costs and which as benefits? Costs and benefits are determined in the light of values, and societal values are not monolithic. In today's society, for example, the values that produced the "Pepsi® generation" coexist with those that support the Amish life-style, and thus automobiles still share the road with the horse and buggy, in some areas of the United States.

Which values should the assessor apply to impact evaluation? Sometimes the values to be used are embodied in specific laws and/or regu-

INTERNAL- AND EXTERNAL-VALUE ISSUES IN IMPACT ASSESSMENT

Value issues arising from the societal context for the TA/EIA

1. Selection of value emphases from among conflicting value sets
2. Acceptance of current, rather than future, societal values as the basis for evaluation
3. Acceptance of current technologies

Value issues internal to the TA/EIA

1. Selection of evaluation criteria
2. Determination of scope of the assessment
3. Implications of methodological choices
4. Disciplinary biases within the assessment team
5. Compatibility of assessment team and societal values

lations, as with EIA. In general, however, it would seem advisable to consider alternative value sets representative of the different primary stakeholder groups (Carpenter and Rossini 1977). Failure to do so would weaken the capacity of the assessment to anticipate stakeholder reactions and thus diminish its contribution to policy formulation.

Early in the development of TA, Daddario (1968: 12) wrote that changes in values, attitudes or institutions are important, but that "because of their slow evolution, present human values and political institutions will serve as the frame of reference for purposes of measurement and appraisal."

Thus Daddario suggested, in effect, that future impacts be evaluated within the framework of present values. Because values do change slowly, this seems a legitimate approach for assessments with short time horizons, but its validity for long-time-frame studies appears doubtful. The relatively long gestation period of major value changes, however, makes it likely that value sets that will dominate the future are already held by significant segments of present society (see Section 7.3.3). Thus, incorporation of alternative value sets in impact evaluation is doubly important as it provides partial accounting for, not only present, but future value diversity as well.

The third sort of external value issue suggested in the cameo concerns the acceptance of current technologies. By and large, societies tend to be conservative—favoring what already exists in preference to unfamiliar alternatives. Assessors should beware of overstatements of the case for present technologies. Stakeholders' perspectives thus may need to be somewhat discounted, and the negative impacts of novel technologies should be considered together with possible ameliorative policy actions.

14.2.2 Value Issues Internal to TA/EIA

Problem-focused assessments address the suitability of various technologies for reaching a single desired end-state (goal), an end-state generally specified by the project sponsor. In a technology-focused assessment, the suitability of a technology for multiple end-states is addressed [e.g., potential uses for the videophone—Dickson (1974)]. In this latter type of assessment, the choice of end-states to be considered is normally left to the assessor. In most TA/EIAs, the assessors will have great leeway in the operationalization of evaluation criteria and the specification of measures. Some inherent value-laden choices that might be left to assessors include the following.

The importance of neighborhood integrity in assessing a proposed highway location. Choices include whether to consider the issue at all, how to define neighborhood, and how to measure integrity.

The valuation of a new energy source (e.g., a power plant) for a region. Choices include definition of future energy need (a function of emphasis on conservation), ways to define the region, and assumptions about alternative energy supplies from within or without the region.

The interpretation of environmental changes associated with alternative urban waste-disposal schemes. Choices include the definition of significant changes and the choice of criteria on which to gauge changes (e.g., a landfill might be seen as beneficial or not).

Selection of study scope can also have major effects on the value choices reflected by an assessment. The TA on the automobile produced by Grad et al. (1975) illustrates how selection of study scope can eliminate end-states from consideration. As mentioned previously, the study team narrowed its focus to the analysis of automotive air-pollution problems using current technologies, thereby excluding energy-consumption issues.

Similarly, methodological choices can affect impact evaluation. For example, were assessors to opt for quantifiable impacts, the assessment would likely weigh technological and economic impacts most highly. This could underemphasize social impacts in the evaluation process.

Yet another internal value issue arises from disciplinary biases within the assessment team. TA/EIAs are interdisciplinary activities that must be executed by individuals trained in different disciplinary contexts. Thus different models of explanation, experimental designs, and conceptual maps lead to disagreements, not only over the criteria with which to govern the study, but also over the standards that should define a bona fide TA methodology. One might readily imagine an economist, an engineer, and a sociologist differing over the evaluation of a series of impacts.[1]

Although methodological and disciplinary value differences develop, differences in background cultural values among TA/EIA team members tend to be slight. Rokeach (1973) documented a ranking of 36 values upon which such researchers were in essential agreement. For example: Action should be legitimated by reason; problem solving involves breaking down of a complex problem into manageable components; quantification is valuable; and intuition is suspect. Given this value and cultural homogeneity, one might reasonably ask whether a TA/EIA can be expected to accurately reflect the values of the wider society. The answer, of course, is that it probably cannot. However, approaches to deal with this problem are discussed in the next section, as we turn to strategic choices in impact evaluation.

14.3 IMPACT EVALUATION STRATEGY

The first strategic issue is whether impact evaluation procedures are to be explicit or implicit. Considering the numerous opportunities for value judgment in any impact assessment (as just discussed), it appears naive to assume that an assessment can be neutral—no matter how earnestly that neutrality is sought! The most responsible approach is thus to describe, as completely as possible, the assumptions, evaluation procedures, and analytical methods employed by the team. The team members and their backgrounds should also be identified, along with major providers of information, and members of advisory or oversight committees involved in the assessment. Through such a *full-disclosure* approach, the user is given a chance to determine distortions that have been introduced through the use of value sets different from his or her own. As much as possible throughout the study, specific value intrusions and the decisions that proceed from them should be noted. Asses-

[1]To refine an old tale, the three were shipwrecked on a desert island with only canned food. Their approaches to opening the cans differed. While the engineer banged cans open on a rock, the sociologist waited to survey his two colleagues. The economist had a more perfect methodology: He simply assumed the existence of a can opener.

sors should also make certain that team members are aware of value implications in the assessment.

We now address three additional strategic concerns in impact evaluation:

Who is to be involved in it?

What criteria are to be used?

What impact measures are to be considered?

These will lead us toward a fourth issue: How can impacts be evaluated? That issue is addressed in Section 14.4 in terms of possible techniques for impact evaluation.

14.3.1 Who Is to Be Involved?

The first option for a TA/EIA team is to perform impact evaluation *without outside involvement*. Outside perspectives could be used in defining the technology and understanding the impacts, but evaluation would be internal. This can reduce the appearance of bias, as in the Jet Propulsion Lab (JPL) study of automobile propulsion alternatives, which was sponsored by Ford Motor Company (JPL 1975).

A second approach ensures that a full range of values is reflected by providing for participation by stakeholder representatives. This concept, *participatory assessment*, is not new. The report of the U.S. National Academy of Sciences (1969), for example, places great emphasis on citizen participation. Further, EIS procedural requirements specify hearings and reviews to provide the opportunity for such participation. In fact, in various EIAs, citizen commentary or court action (a post hoc participation) have markedly altered the scope and findings of final EIS reports. We consider participatory possibilities in TA/EIA in general in Section 16.4.2. Here our concern is with stakeholder representation in impact evaluation.

Several levels of stakeholder involvement are conceivable. A low level would involve *role playing* by the assessment team, with particular members attempting to represent a constituency's viewpoint. An opportunity for *interested parties to comment* on an assessment (and hence on the impact evaluation) is another possibility. This is, of course, mandated for EIS preparation and has been used as a basis for follow-up revision of a TA (JPL 1975). *Conferences* held during the course of the TA/EIA, at the point at which impacts have been analyzed, can provide an opportunity for participation in the evaluation phase. Alternatively, *interviews* with selected parties at this stage of the assessment process can be used (The Futures Group 1975, 1976). A more involved level includes *direct stakeholder* participation in the

TA/EIA. Among other responsibilities, the six-member public interest group associated with the Arthur D. Little, Inc. assessment of solar energy (Berkowitz and Horne 1976) was to "provide non-technical judgments on relevant political and socioeconomic consequences" (Arnstein 1975: 71).

The substantive and procedural problems associated with participatory impact evaluation are formidable. For example, Carroll (1971: 652) notes that manipulation of a concerned but technically unsophisticated citizenry is a strong temptation for technical and administrative people. Citizen groups, on the other hand, can at times resort to sheer obstructionism to obtain their ends. In the Arthur D. Little, Inc. experience, the technical team found it difficult at times to work with the public-interest group due to severe value differences and consequent lack of trust.

Finally, the presumption that TA is unavoidably biased has led some to propose that it should be employed as an *advocacy* tool.[2] Advocacy is not intended to imply that basic data gathering should be distorted, but merely that the value set used in the study should be openly that of the sponsor. Advocacy TA transforms conflict over goals and values into an external clash between conflicting TAs. Hence there is an option for, say, General Motors and the Friends of the Earth to each prepare an assessment of the development in question. Conflict between advocacy TAs might occur in the court, on the floor of Congress, or before the bar of public opinion. Opponents of advocacy TA claim that the process would involve unnecessary duplication of effort. Supporters argue that the results of such duplication would serve to establish and highlight value differences between advocacy groups, thus providing a substantive focus for debate.

14.3.2 What Criteria Are to Be Used? The Concept of Utility and Alternatives to It

We have emphasized that values underlie the criteria for impact evaluation. When different values cause conflict over the issues, what general criteria are to be used to resolve them? For instance, suppose a proposed highway yields a favorable societal benefit:cost ratio but is decidely negative for a displaced resident. These are typically weighed against each other on the basis of social utility.

Utility is the state or quality of being useful. *Utilitarianism* is "the doctrine that the greatest happiness of the greatest number should be

[2]This position has been advanced in works by Green (1970), Arnstein and Christakis (1975: 171), Rossini et al. (1976), and Roper (1976).

the end and aim of all social and political institutions; or the doctrine that utility is the standard of morality, that actions are right in proportion as they tend to promote happiness" (Thatcher 1971). This is the principle that in one form or another underlies procedures such as decision and cost–benefit analyses, in which one tallies the sum of all the positive and negative units of utility for a nation or group.

However, whereas most of the impact-evaluation approaches rely on the notion of utility, it is neither a panacea for evaluation nor the only viable perspective. First, the calculation of *utility functions* (i.e., composite indicators of utility, such as benefit:cost ratios) is fraught with hazards. Different parties at interest hold different values to be important. Thus, even if all parties agree on the magnitude of the impacts, they will experience and evaluate them differently. Typically, TA/EIAs address complex situations on behalf of a pluralistic society for whom no single utility function can suffice.[3]

Utility is not the sole criterion for evaluating impacts. Gastil (1977) proposes a broader spectrum of grounds for decisions. This serves to place utility in a better perspective—still prominent but not the sole basis for evaluation. We endorse an evaluation perspective that adds, to utility, considerations of *equity* in distribution of goods and services and *nonmaterialistic* ends. This can serve to alert the assessor to alternative values and potential points of conflict concerning technological impacts. The following discussion is based on the work of Gastil (1977):

> While the utilitarian principle is based on enhancing the good of the whole society, a *distributive* ethic concentrates attention on the way in which social goods are allocated. One distributive ethic suggests equity as the highest principle. While a formulation of utility may lead to the computation of an overall benefit–cost ratio, distributional justice demands to know *who gets the benefits and who pays the costs.* It may be considered preferable to have fewer goods, more equitably distributed.
>
> Parallel to the material areas of concern are the *non-material* or spiritual areas. At all times many men have devoted their lives and their efforts to non-utilitarian goals. The Australian Aborigines traditionally devoted an astonishing part of their meager resources to ritual behavior; the Greeks, to learning, sculpture, and architecture; medieval Europe, to monasteries;

[3]Arrow's theorem explains the logical impossibility of a social-utility function—there is no rationally defensible basis for aggregating individual preferences (Arrow 1963).

modern America, to moon flights. Implicit in such efforts is a conception of the good society embracing non-utilitarian creation and achievement. In such a conception of society, "utility" and "equity" are simply means to higher ends which transcend day-to-day human concerns.

Many human groups have established a distinction between the sacred and the profane that requires not only a sense of transcendence of the human situation by action, but also the curtailment of action through a sense of *reverence*. To the new breed of ecologist, reverence may mean denying man the right to eliminate a species by his actions.[4]

Impact assessments of developments with implications for life-styles should consider spiritual values. Examples are not hard to find—for instance, genetic engineering, alternative work schedules (affecting religious observances), or even highway projects passing near grave sites. Specifically, the Futures Group's (1976) assessment of life extending technologies raised concerns about the desirability of life extension which surely fall beyond the horizons of a utilitarian "more life is better" approach.

Another facet of the definition of the good society is the range of concern for others. This may extend beyond the self to those for whom one has immediate concern (such as the family) to the nation state, to all humanity, into the future, and even beyond. This acceptance of responsibility for others is a basis of social ethics which often goes against immediate self-interest. Concern for the welfare of others is important in TA/EIA when one determines the range of stakeholders to consider, and when assessors act as stand-ins for potentially impacted parties. In this sense the assessor's responsibility to others is awesome.

14.3.3 What Is to Be Evaluated?

Impacts have been identified and analyzed in the steps of the assessment that precede evaluation. Most impact-identification techniques, as noted in Chapter 8, uncover far more items than the assessor can hope to analyze. Therefore, a subset of significant impacts must be selected for analysis. Analysis itself further reduces and alters this subset. The assessor thus must be aware that "evaluations" have been

[4]Ehrenfeld (1976) makes a strong case that utility is an inadequate criterion to support conservation of environmental attributes that are "non-resources." The argument that preservation of certain exotic or trivial species has human survival value can become quite weak on occasion. Ecological reverence is a more appropriate source of value to argue for preservation of a 3-inch noncommercial fish, the snail darter, in lieu of a Tennessee Valley Authority dam to provide energy.

made before formal evaluation begins. It is suggested that before proceeding, the effects of this "pre-evaluation" be reviewed.

When impact information emerges from analysis it will naturally be categorized along whatever lines have been chosen to subdivide the impact field. If, for instance, a disciplinary subdivision is employed, all economic-impact information may be grouped together, as will be technological, social, and environmental impacts, and so forth. Stakeholder, logical categorizations, and arrangements according to functional area of technology are also possible. The information developed in the various areas of analysis will likely be produced in different formats and involve different degrees of quantification.

For project-based EIS and narrow-focus TA studies, categorization along disciplinary lines will probably prove adequate. Broader-focus TAs, in which a wider range of impacts emerge, may require imaginative recategorization for effective evaluation. For instance, the assessors involved in the TA of life-extending technologies (The Futures Group 1976) found that separating technologies into those that extend maximum life-expectancy and those that reduce early deaths allowed effective comparative evaluation.

Integration of the discrete impact analyses may be necessary to properly gauge the net implications (see Section 17.4.3). This may be possible within a single framework (e.g., the cost–benefit scheme in Figure 11.1, p. 262). Changnon et al. (1977) used a somewhat different framework to draw impacts of hail suppression together. For each "system level" (individual, community, state, nation, and world), a contrast was drawn between those adopting the technology and those not. All impacts were listed (under alternative assumptions of implementation of the technology) for each, along with an estimate of their likelihood, whether they were advantageous, their importance, and the order of the impact (first, second, or higher) (see Table 14.1). In this manner classes of impact (like EPISTLE) were brought together for useful review and reflection.

14.4 EVALUATION TECHNIQUES

Vlachos et al. (1975) and Paschen et al. (1975: 125), among others, have considered specific requirements that must be fulfilled by evaluation techniques. These include that the technique must

be capable of dealing with both quantitative and qualitative criteria;

include both economic and non-economic aspects of each alternative;

TABLE 14.1. Evaluation of Selected Impacts of Hail Suppression on Communities Either Adopting the Technology or Not[a]

Impact	Likelihood	Benefit, or not	Importance[b]: Crop area	Importance[b]: Nation	Sequence order
Adopting					
Increase in farmland values	L	+	Mi	Mi	3
Local tax revenues increase	L	+	Mi	Mi	3
Improved public services	L	+	Mi	Mi	3
New, local government administration units	VL	+	Mi	Mi	3
Some local revenues diverted to hail suppression	VL	0	Mi	Mi	3
Controversy: opponents vs supporters	VL	−	Mi	Mi	3
Population outmigration slowed	L	+	Mi	Mi	3
Nonadopting[c]					
Slight dampening of local business activity	L	−	Mi	Mi	3
Local tax revenues decline	L	−	Mi	Mi	3
Lower farm land values	L	−	Mi	Mi	3

[a] Abstracted from Changnon et al. (1977: 357).

[b] This assumes the most optimistic of three projections of hail suppression capability in 1995—namely, 80% reduction of hail damage in the western United States. Code: VL = very likely, L = likely; (+) = primarily beneficial, (−) = mostly a cost, (0) = no significant net cost or benefit; Mi = minor; (2) = second-order, (3) = third-order impact.

[c] "Nonadopting" refers to agricultural areas with crops similar to those in adopting areas.

provide an objective framework that displays the underlying logic and process of choice in a fashion understandable to third parties;

allow for consideration of trade-offs and mitigating circumstances;

provide flexibility and monitoring mechanisms for considering new goals and criteria as values change and insight grows; and

provide not only an *overall* evaluation of all consequences, but *individual* evaluations of all consequences as well.

Evaluation results should be presented so as to support the final judgment in a fashion that portrays both the end result and the procedures by which it was reached. Further, the method used to consolidate individual evaluations into an overall decision must be substantiated. Finally, conflicting evaluations, based on the stance of different stakeholder groups, should be presented.

When the TA/EIA is to be used by diverse audiences, it is particularly advisable to show the "raw"-impact information. When users' value preferences differ, the opportunity to evaluate for themselves is a useful supplement to the assessors' evaluations.

One of the formidable tasks in impact evaluation is the combination of quantitative and qualitative impacts. Both balance and style are involved in evaluation and summary comparison. Numerical effects can be captured in charts and graphs; qualitative ones require a descriptive style. The dimensionless scaling variations described in the next section provide one mechanism for consolidation of both sorts in certain situations.

It is noteworthy that the three techniques we highlight are based on utilitarian criteria. However, they allow room for consideration of the other basic criteria (distributional equity, transcendence, and reverence) we discussed. Matrix representations using dimensionless scaling can accommodate diverse impacts and weigh them as the evaluators see fit. Policy capture can potentially stretch to decision criteria including equity and nonmaterial factors. These nonutilitarian criteria deserve full consideration in TA/EIA.

14.4.1 Dimensionless Scaling

In this simple approach, impacts are rated on a common "dimensionless" scale and displayed in a *matrix* format. Matrix formulations have been discussed extensively in previous chapters (cf. Sections 8.3 and 9.3). The formats most appropriate to impact evaluation appear to be impact × activity and impact × stakeholder group. An entry in a cell of the matrix is used to represent the direction, or direction and intensity, of the impact.

Possible scales range from qualitative to quantitative; for instance,

(−) adverse, (0) no significant net effect, (+) favorable;

(0) none, (1) minor, (2) significant, (3) major; and

(1) minimal, (2) minor, (4) significant, (8) major.

Cetron and Bartocha (1973) suggest that a nonlinear scale, such as the last one noted, better captures human judgments (see Figure 14.1A,B).

If a quantitative scale is used, impacts can be summed in various fashions:

total intensity and direction; for example, a column sum in Figure 14.1C giving a net result on a particular group;

summation of absolute values to indicate intensity without canceling positive and negative impacts against each other (Figure 14.1D); and

summation over weighted categories (Figure 14.1E).

For qualitative scales, one may dispense with any sort of summation or merely indicate number of positive and negative impacts. An interesting nonquantitative display is achieved when the differences (positive less negative effects) are presented as shaded areas of varying darkness; that is, the darker the area, the more positive the impacts (Roper and Dekker, 1978).

Figure 14.2 illustrates an interesting variant matrix presentation. Here the scaling is presented on the axes, with specific cases represented in the cells. The particular version shown displays surveyed stakeholder perceptions of the likelihood × desirability of a particular policy action.

Matrix entries are generally determined by team members working together or in cooperation with consultants. They are thus subject to distortions caused by differences between team and general societal values. Stakeholder groups can also be employed for this task, although there is an obvious temptation for them to overstate their case (recall the discussion of Section 14.3.1). In fact, this approach could be adapted to distinguish the biases and viewpoints of stakeholder groups.

The chief advantages of dimensionless scaling are that it is simple, takes relatively little time to perform, and produces an output that is clear and easy for users to comprehend. The primary disadvantage is that the information on which it is based is generally subjective, and any quantification is open to criticism. Further, results are sensitive to the number of impact areas, actions, or stakeholder groups chosen for representation in the matrix. However, the subjectivity of dimensionless scaling is obvious and thus easily evaluated by stakeholders. More sophisticated evaluation techniques generally use equally subjective information.

14.4.2 Decision Analysis

"Decision analysis" is the generic term for a number of techniques intended to quantify and systematize decision making, particularly under conditions of uncertainty or risk. Decision analysis can be subdivided into four basic categories (Sage 1977: 309), reflecting the conditions under which decisions are made:

1. *certainty*—an action results in one and only one outcome;

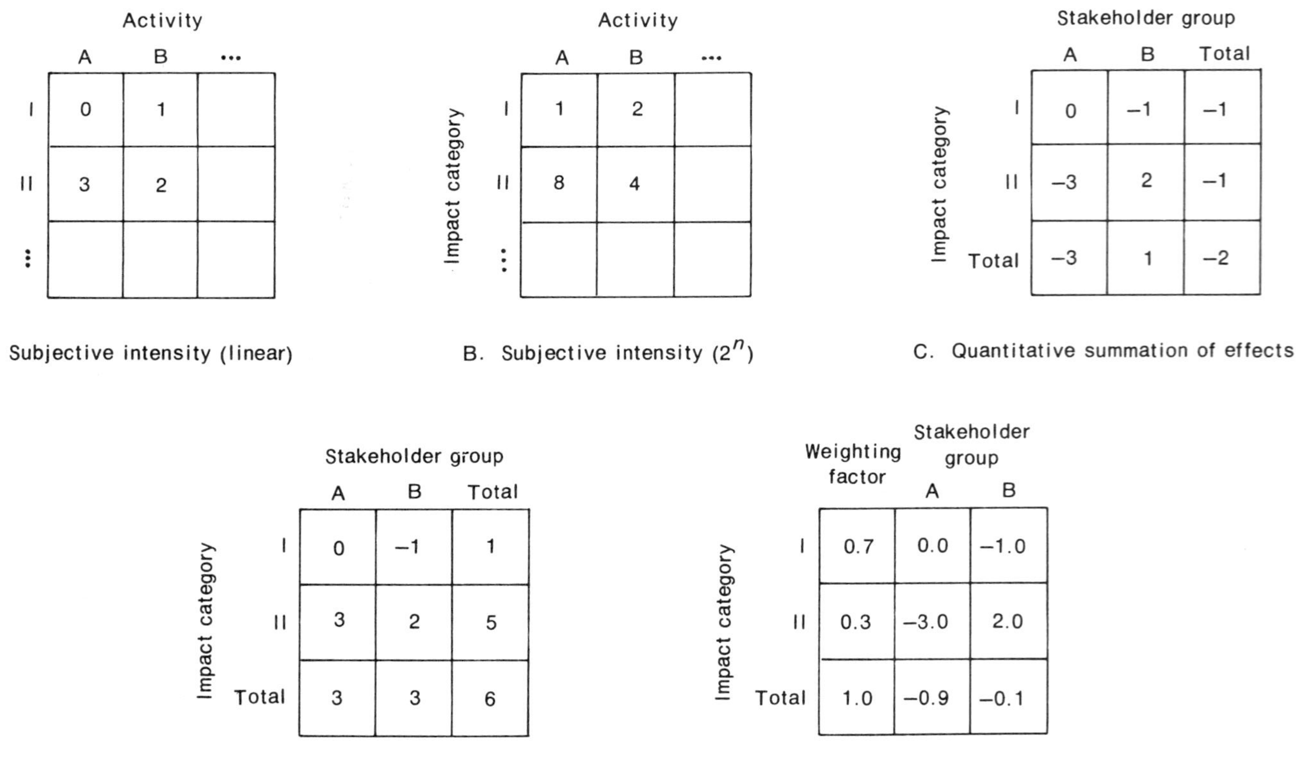

FIGURE 14.1. Some examples of dimensionless scaling.

	Very desirable	Desirable	Neutral	Undesirable	Very undesirable	No opinion
Almost certain				3		
Very likely		7 8	16	9	13	
As probable as not		12	10		4	
Very unlikely		1 2 12				
Almost impossible						6
No opinion						

Participant codes

1. Private industry
2. Private industry
3. Government – administrative
4. Environment and research
5. Public utility
6. Government – administrative
7. Government – legislative
8. Environment and research
9. Public utility
10. Government – legislative
11. Public utility
12. Private industry
13. Environment and research
14. Private industry
15. Private industry
16. Government – administrative

FIGURE 14.2. Impact likelihood vs desirability matrix: individual responses (by participant code) to the notion of creation of tax shelters for investment in geothermal development. [Comments (by participant code): (1) tax shelter not the way to go; (3) big companies don't use them much (it means more regulations and might encourage smaller companies—encourage drilling, not necessarily discovery, since indiscriminate activity might result); (6) subsidies are generally given to "wrong" people—they are not really productive; (10) in normal times this would be a natural development—but now public attitudes are against it; (12) tax incentives are needed for development.]
Source: The Futures Group (1975: 369).

2. *risk*—an action can result in more than one outcome depending on external conditions that have *known* probabilities of occurrence;
3. *uncertainty*—an action can result in more than one outcome depending on external conditions of *unknown* probability; and
4. *conflict*—external conditions are replaced by a competitor.

Decision analysis techniques dealing with conditions of risk and uncertainty are of the most interest in impact evaluation, although conflict conditions are appropriate when synthesizing evaluations by competing stakeholder groups. Decision under certainty seldom enters into the evaluation of the complex net of impacts developed in TA/EIA.

This section first deals with the basic premises of decision analysis, and then develops a simple example of analysis under conditions of risk. We discuss decision under risk, rather than uncertainty, since estimates of the probability distribution of the various development factors can be generated for a TA/EIA, if they are not available. Several useful approaches to decision under certainty (e.g., the Laplace, pessimist, optimist, min–max regret criteria, and optimist index) are described by Hadley (1967). An extensive discussion of decision analysis is presented by Sage (1977: 307–406).

The basic tasks of the decision analysis process are to

1. define the set of possible alternative actions;
2. define the set of external condition states;
3. define the set of possible decision outcomes, each of which corresponds to a discrete combination of an action and an external condition state;
4. evaluate each outcome according to the values and objectives of the decision maker, thereby arriving at its utility.

The utility function is an attempt to resolve questions of preference, particularly preference among outcomes, such as changes in employment and pollution level, that are not directly comparable.

The calculation of utility assumes that all outcomes can be assigned units on a common scale that indicates their value to the decision maker. The most common units to measure utility are dollars. Although many impacts will not be economic, cost–benefit analysis, for example, has developed means for sometimes expressing these impacts on a monetary basis (Chapter 11). Other bases, such as the time that an individual would be willing to spend to assure or avoid a given outcome, can be used instead of dollars.

In decision analysis, the decision process is represented schematically as a *tree*. Forks (or nodes) of the tree represent points at which decisions are made or chance outcomes determined. Branches emanating from each fork must depict mutually exclusive actions or outcomes and be exhaustive of all possibilities. The utility, or payoff, associated with each outcome is displayed at the tips of the tree.

To apply decision analysis the evaluator must execute four steps:

1. Construct the decision tree.
2. Assign payoffs to the tips of each branch.
3. Determine probabilities for each branch emanating from a chance fork.
4. Start at the tips and compute backward to determine the payoff associated with each decision fork.

The cameo entitled "A Decision Analysis Under Risk" illustrates a simple decision analysis. It assumes that all probabilities are known or can be computed from known probabilities. Computations proceed from the tips of the tree backward toward the trunk. If some probabilities are unknown, computations normally proceed forward through the tree, delaying estimations of unknown probabilities until the end of the process so that the sensitivity of conclusions to those estimates can be determined.

The major advantage of decision analysis in the evaluation of complex problems is that it provides an ordered, systematic, and quantified framework that yields reproducible results. One of its disadvantages is that it requires considerable time, effort, and planning. It is also highly specific to the value set of the decision maker for whom it is developed. The latter drawback can be mitigated, however, if sufficient time is available to develop decision analyses reflective of more than one stakeholder group. The Futures Group (1975: 135–149) effectively used decision analysis in the form of a systems-dynamics model (see Section 9.4.5). They were able to show that geothermal energy development would be more sensitive to availability of resources than to other decision factors and costs.

14.4.3 Policy Capture

The desirability of employing alternative value sets in the evaluation process was noted earlier (Section 14.2.1). If stakeholders are not incorporated into the process, some means must be employed to anticipate their judgment of the impacts. The Midwest Research Institute (1975), for example, used a direct survey of interested parties in their TA of integrated hog farming. The Futures Group (1975) employed formal interviews based on a variation of the Delphi process in their TA of geothermal energy.

A DECISION ANALYSIS UNDER RISK

A hypothetical public service company (HPSC) operates a 650-megawatt (MW) power generation plant in a county that has been designated by the U.S. Environmental Protection Agency (EPA) as a noncompliance area for sulfur-oxide emissions. This *designation*, as is frequently the case, is made on the basis of computer modeling studies. Because of the noncompliance designation, the HPSC will be required to install flue-gas scrubbing equipment at a capital cost of \$40 per installed kilowatt (kW) capacity (total capital cost = \$40/kW × 1000 kW/MW × 650 MW = \$26 million). The HPSC could choose to exert political pressure by showing adverse economic impact on the county, invalidity of the model, and so on, in an attempt to have the county *redesignated* as a compliance area. It could also choose to construct a county pollution-monitoring system in an attempt to demonstrate compliance (estimated capital cost \$0.3 million). What decisions should the HPSC make?

The first step is to construct the decision tree. The first decision facing the HPSC is whether to build the monitoring system and *sample* to determine the actual state of the air in the county (Figure 14.3). Suppose they choose not to sample; then they must decide whether to exert political pressure. Following this decision, is a chance fork at which EPA may or may not choose to redesignate the county. If the HPSC chooses to sample, the same flow of decision and chance occurs, except that an additional chance fork (county either in violation or not) is inserted (Figure 14.3).

Next, utilities must be assigned to each outcome at the tips of the tree. On the upper ("do not sample") portion of the tree, two payoffs exist: −\$26 million if no redesignation (and hence the HPSC has to install the scrubbers) and \$0 if the county is redesignated as a compliance area. For the tips of the lower ("sample") branch, the same two outcomes correspond to −\$26.3 million and −\$0.3 million, reflecting the additional cost of the monitoring system. These payoffs are entered in Figure 14.3.

Probabilities must now be assigned to all chance forks. Studies by HPSC of similar situations in comparable counties have determined the following probabilities: probability of redesignation with political pressure, $P(1) = .25$; probability of redesignation without political pressure, $P(2) = .10$; probability that sampling will show the county to be in violation of regulations, $P(3) = .4$; conditional probability of redesignation, with political pressure, if sampling shows the county in violation, $P(1|3) = .05$; and $P(2|3) = .02$. All other possibilities can be determined from these by using Bayes' rule,[5] and the definition of the probability of nonoccurrence.[6]

The final step is to compute expected utilities at the various decision forks. At the uppermost chance fork, the expected utility is

$$\begin{aligned}\text{Utility}_1 &= \$0 \times P(1) + (-\$26 \times P(1)) \\ &= \$0 \times .25 + (-\$26 \times .75) = -\$19.5 \text{ million.}\end{aligned}$$

The expected utility at the second chance fork is

$$\text{Utility}_2 = \$0 \times .10 + (-\$26 \times .90) = -\$23.4 \text{ million.}$$

Assuming that HPSC decision makers will seek to minimize their expected loss, they will decide to exert political pressure. Thus, the "no-political" pressure branch will not be followed, a decision that is noted on Figure 14.3 by a // symbol on that branch. Proceeding down the tree to the "sample/do not sample" decision node, the expected utility of the "do not sample" decision is −\$19.5 million. Branches emanating from the "sample" decision are analyzed in an analogous fashion to produce an expected utility for that decision of −\$19.8 million. All utilities computed for the various forks of the tree are displayed in Figure 14.3.

In this example, the HPSC decision makers should choose not to build the pollution-monitoring system and to exert whatever political pressure they can muster. This course of action will produce the smallest potential loss (i.e., largest utility). The results of the "sample" and "do not sample" branches are so close in this example, however, that it would be well to closely examine the sensitivity of the result to the probabilities estimated in the second step of the process. For example, if $P(3) = .35$ rather than .40, the "sample" branch becomes the more attractive of the two. This example is necessarily simplified and neglects any proclivity of the decision makers for risk-prone action, tax write-offs, or other advantages associated with the scrubbers or monitoring system, and so on. It is nonetheless illustrative of the decision-analysis process.

[5]Bayes' rule states that $P(i|j) = [P(i|j)]\, P(i)$.

[6]By definition, the sum of the probabilities of occurrence and nonoccurrence is unity. Therefore, $P(i) = 1 - P(i)$.

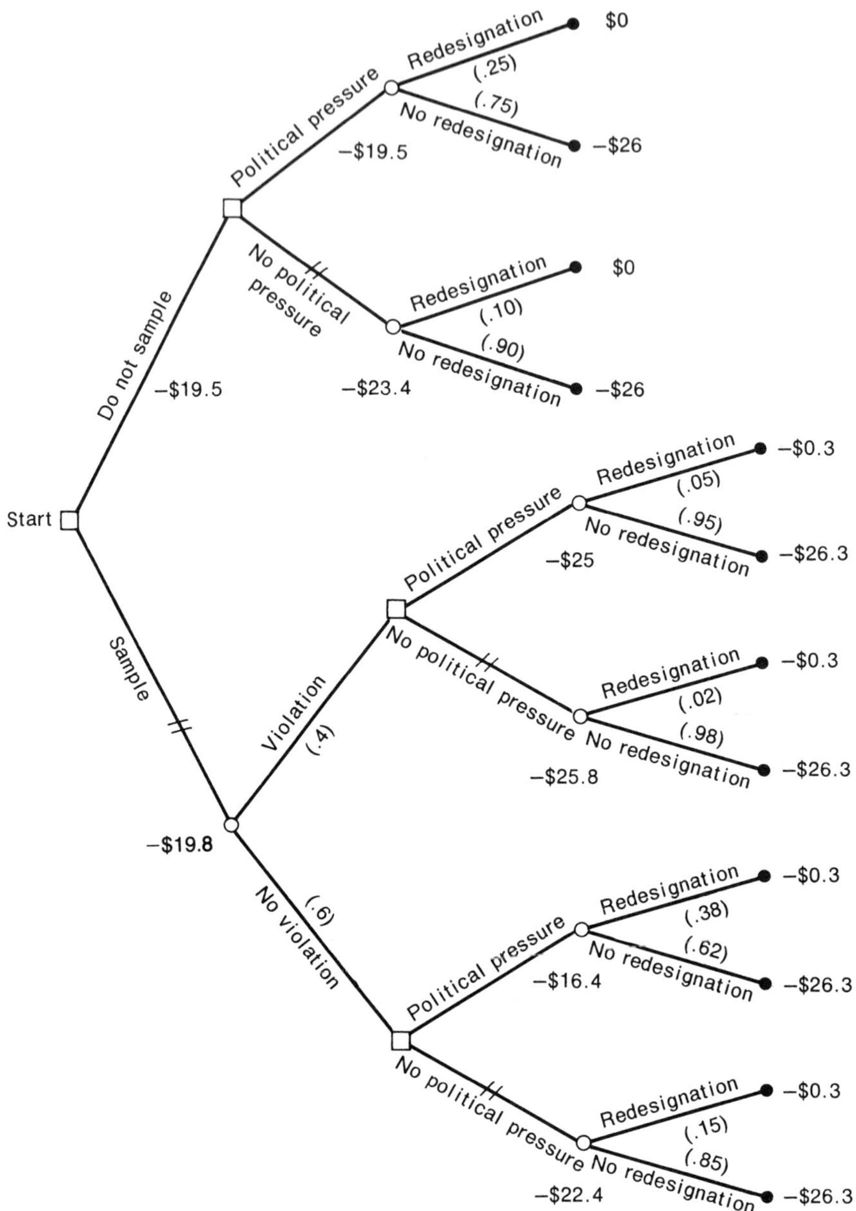

FIGURE 14.3. Decision analysis example for the HPSC Company. □, a decision fork; ○, a chance fork; ●, a payoff point; ( ), a probability; //, less favorable branch. Note: all monetary values are in millions of dollars ($=\times 10^6$).

Rather than seeking judgments of specific impacts, the evaluator may elect to model the judgment process. Such a model transcends impact specifics and can thus be employed to evaluate situations not included in the original assessment. "Policy capture," developed by Kenneth Hammond (see Hammond and Adelman 1976), provides a means to identify and quantify value judgments of various stakeholders. It can thus be used to evaluate the acceptability of trade-offs among alternatives (Crews and Johnson 1975). Hammond and Adelman (1976) describe such an application, in which the technique was used to select a bullet acceptable to both the Denver Police Department and the American Civil Liberties Union.

Policy capture attempts to weight the factors considered in reaching a decision. If the completed model is accurately constructed, it can serve as the basis for understanding present judgments, assessing their consistency, and predicting future judgments.

Simply put, policy capture constructs a mathematical model to parallel the actual decision process. This is accomplished by preparing a number of scenarios in which the factors assumed important to the judgment are varied systematically. One or more representatives of each stakeholder group judge the relative acceptability of each scenario. Multiple regression analysis (see Section 11.6.4) is then used to determine the weights implicitly placed by the stakeholder on each factor in reaching a judgment of acceptability. This process is illustrated in the cameo entitled "Policy Capture Computation."

Mitchell et al. (1975) note several general pitfalls associated with the policy-capture procedure. First, it is difficult to be sure that all factors pertinent to the decision process have been included. Also, the form in which the factors are presented can be critical. Second, a good deal of time is required of participants in the process. Frequently the process must be repeated several times to assure that a legitimate fit has been obtained. It is thus difficult to get high-level personnel to devote sufficient time to complete the policy-capture process. Finally, the linear regression normally used may not provide the most appropriate fit; indeed, Crews and Johnson (1975) present a nonlinear regression approach.

14.5 CONCLUDING OBSERVATIONS

Impacts, once identified and analyzed, must be evaluated in the light of societal values. Evaluation is intended to assess the costs and benefits proceeding from a technology or its alternatives and provide the foundation for policy formulation. It is important to realize that the

POLICY CAPTURE COMPUTATION

This cameo draws heavily on an example of policy capture presented by Mitchell et al. (1975), which was in turn based on Crews and Johnson (1975).

In this simple example, the problem is to determine an acceptable mix between two conflicting factors: national economic development (NED) and environmental quality (EQ). The assessor has selected two decision makers, A and B, as representative of stakeholder groups.

The factors identified for analysis are typically aggregate concepts, such as economics and environment, and are subject to trade-offs (i.e., it is possible to increase NED at the expense of EQ, and vice versa). To apply policy capture, numerical measures must be determined for each of the factors. For this example, the per capita gross national product in dollars will be used to measure NED ($\$1000 \leq \text{NED} \leq \$20{,}000$). Environmental quality will be measured by a dimensionless index, reflecting perceived air, water, and land quality. This index ranges from 0.1 (very polluted) to 0.9 (pristine). Current U.S. conditions are NED of about $10,000 and EQ of 0.7.

A number of scenarios must be developed for statistical inference. In this case, 20 alternative future scenarios are developed by combining various possible values of NED and EQ (Table 14.2).

Decision makers A and B are given the 20 scenarios and asked to rate them in order of their preference. In this example, A and B were asked to first score scenario 5 and then score all others relative to it. The preference scores, which could range from 1 to 1000 in this example (from least to most preferred), are summarized in Table 14.2. (The range is arbitrary but the values must be positive. The scores are standardized during analysis; i.e., the mean is subtracted from each scale, and the difference is divided by the standard deviation.)

The objective of the analysis is to determine how much each of the two factors NED and EQ contribute to the judgment of preference. The decision maker's policy is "captured" by determining the relative weight given to each factor in reaching a decision. A multiple regression equation is used to compute coefficients that determine these weights. The results of the analysis show that decision makers A and B use the following weights in reaching judgments of preference:

A. Preference = 0.8 (NED) + 0.5 (EQ)
B. Preference = 0.5 (NED) + 0.7 (EQ),

where NED and EQ are standardized values.

These preference equations are statistical (not deterministic), so they predict preferences imperfectly. For instance, applying the equations to scenario 5 yields:

$$\text{A. Calculated preference} = 0.8\left(\frac{10{,}000 - 11{,}300}{5948}\right) + 0.5\left(\frac{0.7 - 0.54}{0.23}\right)$$
$$= -0.175 \qquad + 0.348$$
$$= 0.173$$

$$\text{B. Calculated preference} = 0.5\left(\frac{10{,}000 - 11{,}300}{5948}\right) + 0.7\left(\frac{0.7 - 0.54}{0.23}\right)$$
$$= 0.378$$

These compare with standardized expressed preferences by A of [(600 − 527)/283] = 0.258 and by B of [(300 − 231)/252] = 0.274. The relative prediction error for A for scenario 5 is thus [(0.258 − 0.173)/0.258] = 0.33 and for B, [(0.274 − 0.378)/0.274] = 0.38. The errors vary for each scenario. These may reflect a certain inconsistency in A and B's individual judgments, with an acceptable margin of error in the equation, or an inappropriate model formulation. Careful review of the goodness of fit is needed, and, perhaps, further judgment data collection would be required (i.e., request A and B to rethink their rankings in response to the calculated preference rankings).

Whereas the confidence level that can be placed on the prediction of preference for any particular scenario is not high, the equation does indicate the general policy preferences of A and B. For example, A and B exhibit quite different policies with respect to economic development and environmental quality. Thus, they could be expected to evaluate the impacts of an innovation quite differently. If further iterations and discussions were held, however, it might be possible to get both A and B to agree on some acceptable combination of NED and EQ.

TABLE 14.2. **Scenarios with Planner Preferences**

	Factors		Preferences	
Scenario	NED ($)	EQ	Planner A	Planner B
1	20,000	0.9	1,000	900
2	17,500	0.9	950	800
3	15,000	0.7	900	500
4	12,000	0.7	825	400
5	10,000	0.7	600	300
6	20,000	0.6	850	300
7	7,000	0.8	500	300
8	13,000	0.6	600	250
9	10,000	0.5	500	180
10	7,000	0.5	400	100
11	12,000	0.4	500	100
12	5,000	0.4	150	90
13	20,000	0.3	700	80
14	15,000	0.3	600	70
15	10,000	0.3	500	60
16	17,500	0.1	325	20
17	7,000	0.2	95	50
18	4,000	0.5	200	30
19	2,500	0.5	150	40
20	1,000	0.9	200	50
Mean	11,300	0.54	527	231
Standard deviation	5,948	0.23	283	252

numerous decisions made in the conduct of the study, from the definition of scope to the selection of impacts for analysis, constitute de facto evaluation. Individually and collectively these decisions shape the study and its results as surely as does formal evaluation.

Impact evaluation is affected by external and internal value factors. Conflicting value sets and the degree of emphasis on current values and technology are issues arising in the societal context for the TA/EIA. Selection of evaluation criteria, definition of study scope and methods, disciplinary leanings, and compatibility between the values of the assessment team and society highlight internal value concerns.

Strategy for impact evaluation ought to begin with a commitment to full disclosure to elucidate team values and data manipulations. Impact evaluation can involve

the assessment team alone;

stakeholder representation (via team role playing, opportunity to comment, conferences, interviews with stakeholders, or direct participation); or

advocacy impact assessment, wherein a stakeholder group assesses the development from its own perspective.

The pros and cons of each are discussed.

Utility provides a basic criterion for impact evaluation. It is also important to consider equity and nonmaterialistic values seriously in performing evaluation.

Impacts are likely to require some form of integration for sensible evaluation. Alternative approaches are noted. The combination of quantitative and qualitative impacts is a difficult task.

There are numerous techniques available to execute the actual evaluation process. Dimensionless scaling, decision analysis, and policy capture have been discussed in this chapter. Whatever technique is used, the assessor should assure that it develops results in a logical and consistent manner that accounts for differences in the quantitative and qualitative nature of the impacts involved. Finally, the results of the evaluation should be presented to portray both the end result and the procedures by which it was reached.

RECOMMENDED READINGS

Arnstein, S. (1975) "A Working Model for Public Participation," *Public Administration Review* 35 (January/February), 70–73.

Ways and means for participatory assessment.

Baier, K. (1969) "What is Value? An Analysis of the Concept," in *Values and the Future* (Baier, K., and Rescher, N., eds.). New York: The Free Press.

The title speaks for itself.

Green, H. P. (1970) "The Adversary Process in Technology Assessment," *Technology and Society*, 5 163–167.

Discussion of TA as a tool for advocacy.

Sage, A. P. (1977) *Methodology for Large Scale Systems.* New York: McGraw-Hill.

Contains a clear and concise development of the mathematical foundations of decision analysis.

EXERCISES

14-1. As appropriate, divide into two groups. One group is to defend a "utilitarian" approach to impact evaluation; the other is to argue against it.

14-2. The hypothetical public service company (HPSC) plans to build a

650-MW power station in your county. This is a small plant and will employ approximately 180 people, half of whom will be hired within the county. The annual payroll will be approximately $2.5 million. The plant is to be coal fired and will use electrostatic precipitators to limit particulate emissions. The means for controlling sulfur-dioxide emissions have not been settled. Suppose, however, there are four options: no control, use low-sulfur coal, install flue-gas scrubbers, and a combination of the last two. For simplicity, assume 100% of the sulfur in the coal is emitted as sulfur dioxide and that scrubber systems will remove 90% of the sulfur dioxide from the stack gas. Characteristics of the four possible solutions are tabulated as follows:

Solution	*Sulfur content of coal (%)*	*Scrubber?*	*Daily coal (tons)*	*Delivered cost of coal ($/ton)*	*Cost per kW-hr for fuel ($)*
1	3.5	No	5000	18	0.00824
2	3.5	Yes	5000	18	0.00824
3	1.0	No	5500[a]	36	0.01810
4	1.0	Yes	5500[a]	36	0.01810

Solution	*Capital cost SO_2 removal per kW-hr ($)*	*Operating cost SO_2 removal per kW-hr ($)*	*Total cost per kW-hr ($)*	*Ash and sludge produced (yd^3/year)*
1	—	—	0.0824	202,800 (dry ash)
2	0.00103	0.0011	0.01037	746,200 (sludge and ash)[b]
3	—	—	0.0181	233,100 (dry ash)
4	0.00103	0.0011	0.02024	553,177 (sludge and ash)[b]

[a] Low-sulfur coal is generally approximately 10% lower in BTU/lb content.

[b] Sludge and ash mixture is 50% water. It must be stored in a landfill indefinitely.

1. As a group quickly sketch some of the impacts of a power station designed under the proposed solutions. You may wish to use some of the methods developed in earlier chapters, but don't attempt an elaborate analysis.

2. Subdivide into three groups. Each group is to evaluate the impacts determined in (1) from one of the following stakeholder group perspectives:

   a. HPSC decision makers;
   b. A county resident who will buy power from the HPSC; and

c. A decision maker employed by a nationally based environmental group.

Use a simple evaluation scheme, such as dimensionless scaling, and compare your results.

3. Structure a decision analysis tree for this issue.

4. How might one suitably address equity and nonmaterialistic concerns for this case?

14-3. Discuss the strong points and deficiencies you see in the schemes presented in Section 14.3.1 for participation in impact evaluation.

15

Policy Analysis

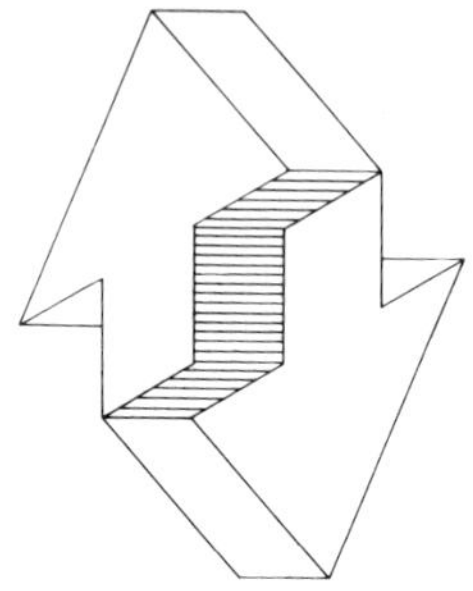

This chapter introduces the reader to the study of policy and to different models of policy making. Policy analysis is then specialized to the case of TA/EIA. Policy analysis objectives, characteristic parameters, and a strategy for performance in TA/EIA are considered in turn. Finally, the experience of assessment practitioners is discussed. This chapter provides the reader with an introduction to the policy making process and illustrates both how TA/EIA contributes to that process and how policy analysis may be performed in a TA/EIA.

15.1 INTRODUCTION

The primary purpose of assessments is to provide knowledge to aid in making decisions affecting technological developments. To this point we have developed approaches for understanding the present state and future prospects of the assessment subject and its broad environment, as well as the impacts resulting from the subject–environment interaction. We now turn to the problem of dealing with those impacts. Some impacts may prove desirable, and we may wish to enhance them. On the other hand, other consequences may be perceived as unwanted or even dangerous. In this case steps need to be taken to mitigate or eliminate these outcomes. "Policy" is involved in managing both the technological development and its impacts. This chapter treats both the policy process, to which TA/EIA contributes, and the development and analysis of policy options within a particular assessment. It begins with general policy considerations and then focuses on TA/EIA. Thus, the larger perspective of the policy process precedes its discussion in TA/EIA. This allows the reader to review a variety of approaches in understanding policy processes. We draw upon these rather eclectically, emphasizing the systems approach, in deriving a strategy for policy analysis in TA/EIA.

Policy analysis deserves special emphasis in TA/EIA. Not only is it a critical component in providing useful assessment information, but it has often been treated casually in practice. Time and budget constraints have often taken their greatest toll on policy analysis (too often saved for the last in the sequence of assessment activities). Consequently, we recommend that policy analysis commence early in the assessment activities and continue iteratively as the study develops. In this way the assessment team remains aware of the range of policy actions worthy of consideration, the relevant decision makers, and the stakeholders.

This chapter also touches on the issue of whether an assessment should produce explicit policy recommendations. This anticipates the issue of communication of results and the role of a TA/EIS in the larger social and political context. An assessment is usually judged on its outputs. Typically, the analysis of policy options and the possible formulation of recommendations is the keystone of those outputs.

15.2 POLICY AND POLICY STUDIES

15.2.1 A Description of Policy

The Bay of Pigs incident and the financial instabilities of the Social Security Program, the collapse of federal support for SST development but continuation of the educational Head Start program, alterations in the matching share of the federal aid highway program and Congressional committee reorganization—what do these have in common? They all reflected major federal governmental policy decisions. Some marked abrupt changes; others, more gradual evolutions of long-term governmental programs. Some pertained to executive branch affairs; others, to legislative matters; and still others, to both. Some were quite open; others were secretive. Some dealt with physical technologies; others were mainly social in nature. All of them reflected weighty public-sector decisions dependent to some extent on policy analysis for guidance as to likely economic, social, and political effects of alternative choices. Likewise, the equivalent of policy exists in nongovernmental institutions under such names as "corporate strategy" and is subject to the same scrutiny.

Wherever it exists, policy is difficult to define in any explicit and restrictive fashion. Lowi (1970: 317) concludes that major scholars perceive policy as "any output of any decision-maker, whether it be an individual or a collectivity." Public policy, the concern behind most TA/EIAs, is "what governments do" (Dolbeare 1973). Joseph Coates (1977b) notes that policy is not action in itself; rather, it is given meaning and set into action through legislation, programs and projects, regulations, taxes, and other operations of the instruments of government.

Policy making can be distinguished from decision making primarily by the importance of the choice involved. Policies are of a more fundamental nature with wider-ranging, less-reversible implications (i.e., "guiding principles"). Further distinctions among types of policies can be made. Dror (1971b) separates "metapolicies" (policies on making policy) from "megapolicies" (main postures and guidelines), in turn to be followed by specific "policies." Dolbeare's (1973) "fundamental policies" and "output policies" are akin to Dror's megapolicy and policy.

Lowi (1964) identifies policies by their characteristics as distributive, regulatory, or redistributive (taking from some to give to others). Categorizing policies in one way or another may begin to offer some insights as to relevant forms of policy studies. For instance, in discussing Congressional use of policy analysis, Schick (1976: 217) makes a telling point: When Congressmen are deciding issues that involve distributive policies, passing out benefits to particular groups, they "cannot be expected to renounce their political aims for the sake of efficiency; nor can they be expected to support analysis that conflicts with their assessment of political reality."

Policy covers a wide range of phenomena that can be studied by diverse approaches. We now turn to consider the study of policy.

15.2.2 Policy Studies

The term "policy analysis" posed a dilemma for us because it is conventionally used in two conflicting senses. According to the first, more encompassing usage, TA/EIA is a subset of policy analysis—a particular form that focuses on analysis of potential technological developments. According to the second, narrower usage, policy analysis is a component activity within a TA/EIA—wherein one considers policy options generated in the study. To resolve this conflict, *we will use policy analysis in the second sense, as a component step in TA/EIA*. To enable us to address certain general considerations that strongly bear upon TA/EIA, *we substitute the term "policy studies" for policy analysis in the first, encompassing sense*. However, the reader should expect to encounter the term used either way in other sources.

Policy studies address varied aspects of policy formulation, implementation, effects, and methodology. Examples stretch from "quick and dirty" pragmatic advice on hot issues to grand and scholarly discourses on long-standing concerns, and even on policy making itself (Dror 1971a). Naturally, both the form and the content of such diverse efforts vary greatly.

Many students of policy have increasingly turned their attention toward "scientific" methodologies. This has led to employment of quantitative research skills and attempts to establish "policy sciences," which has been defined as "a new supradiscipline, oriented toward the improvement of policymaking" (Dror 1971b; see also Lasswell 1971). However, other analysts continue to rely heavily on informal forms of study—insightful case analysis, tacit (implicit) sources of knowledge, and common-sense conclusions based on understanding of the political forces at play, the parties at interest on a given issue, the legal and institutional constraints, and the general nature of the consequences of alternative policies.

Policy studies are not the sole prerogative of one scholarly discipline. They may be closely identified with political science and public administration, somewhat less with economics, and to some extent as well with management science and various subject areas (e.g., transportation policy analysis). Policy studies are approached from a wide range of perspectives, depending on the disciplinary outlooks and interests of the analysts. There are also a number of significantly differing conceptual frameworks in which policy is viewed. We now consider several of these.

15.3 MODELS OF THE POLICY PROCESS

Six distinctive perspectives on the policy process contribute useful and complementary insights: namely, the (1) *rational*, (2) *institutional*, (3) *equilibrium*, (4) *elite*, (5) *incremental*, and (6) *systems* models.

The *rational* model involves a firm commitment to a particular form of utilitarian reasoning. Policy is seen as efficient goal achievement. "A policy is rational if the ratio between the values it achieves and those it sacrifices is positive and higher than other alternatives" (Dye 1975: 27). One assumes that social values are known and weighted, policy alternatives and their consequences are fully known, and achievement–sacrifice ratios are calculable. This model offers a logical basis on which to judge policy alternatives. There are difficulties with this model when there are major disagreements about societal values. It deals inadequately with values that are difficult to quantify as, for example, in issues regarding ecological balance, urban aesthetics, or community cohesion.

The *institutional model* holds that one comes to understand public policy by seeing what institutions actually do, how they legitimate policy, how they make policy routine through administrative action, and how, in Lowi's formulation (1970: 314–324), they coerce the body politic. This perspective introduces important and realistic policy considerations. This model suffers from a possible narrowness of focus.

The issue of power is explicitly addressed in *group equilibrium models*. Public policy is seen as a resultant of power balances between competing interest groups within the society. Policy analysis becomes the identification of the often overlapping parties at interest and the description of the complex group struggles in which they engage. It becomes necessary to identify coalitions and to track responses to group pressures and clashes, as seen in negotiation and compromise.

Another theory of policy, the *elite model*, often associated with the writings of C. Wright Mills, holds that policy reflects the preferences of elites. Elites are powerful minorities within the body politic who possess the wealth, connections, and shared values to decisively influence government officials and administrators. Public policy thus mirrors the demands of these elites while the vast majority of citizens remain powerless. The stability of a system is maintained in part because the elites are often "public regarding." Some mobility between elites and nonelites is recognized so that stagnation of elites does not occur. The institutional, group equilibrium, and elite models emphasize different participants in the policy process. Depending on the situation at hand, one or the other of these three facets may be dominant.

The ponderous and routine character of the daily workings of the government bureaucracy dictates that change, when it occurs, does so very slowly. The *incremental model* of policy recognizes this fact. Policy analysis thus amounts to a "muddling through" (Lindblom 1959: 78–88). The inherent conservatism of government action in the face of limited time, money, and knowledge means that past policies and standard operating procedures are bound to be preferred in most cases. Past commitments of time and money lead to more of the same. "How can we abandon the project after having already spent x million dollars on it?" Additionally, because incrementalism is likely to create the least political hassle—"fine tuning" existing policies isn't very controversial—it tends to be preferred by policy makers. Incrementalism seems to be hardest pressed when a major crisis is encountered. The incrementalist model avoids pretense about "ideal" policy tools and concentrates on the way policy implementation is believed to occur.

The *systems model* (Easton 1953, 1965) complements the incrementalist view. The sphere of political action is described by a systems model with inputs, categorized as demands and supports, and outputs in the form of policy decisions and actions. The system consists of the political arena situated within the broader social environment. The function of the political system is to transform the *demands*, generated by perceived needs of individuals or groups in society, and *supports*, reflecting acceptance of prior policy decisions, into *outputs*. This produces what Easton calls "the authoritative allocation of values." Actions taken by the political system are judged effective only if they enable the system to survive. The system ensures its survival by "(1) producing reasonable, satisfying outputs, (2) relying on deep rooted attachments to the system itself, and (3) rising, or threatening to use force" (Dye 1972: 19).

The systems model usefully focuses attention on the feedback character of the political process, addresses the effects of the broader environment on the political process, and notes that the system will generate policy to preserve itself. One weakness lies in the difficulty in operationalizing key concepts. As developed in the "policy sciences" by Dror (cf. 1971a,b), the systems view also suffers from a tendency to ignore the "small realities," so well captured in the incrementalist view.

None of the models we have presented is totally correct, and none is absolutely wrong. Each offers insight into the policy process that can be exploited in policy analysis. The institutional, group equilibrium, and elite models introduce some of the major potential actors in the policy process and identify their roles. The rational model underscores the fact that assessment can incorporate objective policy criteria. Incrementalism, whether we like it or not, usefully characterizes much bureaucratic and political behavior. The notion of assessment as a program of ongoing partial studies, rather than a "one-shot," complete systems analysis (Section 4.4), tries to take account of incremental considerations. Overall, however, the model most compatible with our purposes is the systems model. It deals best with the complexity of the technological delivery systems (TDSs) being studied. However, we suggest that the treatment of policy in any particular situation is best achieved by using the models most appropriate to the case at hand.

15.4 POLICY ANALYSIS IN TA/EIA

15.4.1 Objectives of Policy Analysis

In a sentence, policy analysis is the consideration of what is likely to happen under alternative courses of action. It requires a sound sense of the policy actions and a tacit understanding of the political context. In TA /EIA, policy analysis relates to the management of the technology in question in any of the following ways:

1. identifying areas requiring additional research and development prior to implementation of the technology;
2. determining actions needed to implement the technology;
3. proposing additional institutional controls that should accompany the implementation;
4. pinpointing specific ways to modify the technology in light of problems uncovered in the analysis;
5. proposing ways to better monitor the effects of the technology;

TABLE 15.1. Policy Analysis Parameters

Parameter	Range	
Content	1. Physical technology	2. Social technology
Locus	1. Public	2. Private
Political sensitivity	1. Low	2. High
Time dimension	1. Retrospective	2. Near-term prospective
	3. Long-term prospective	
Focus	1. Policy-making process	2. Policy-implementation process
	3. Policy outputs	4. Policy impacts
Analytical stance	1. Explanatory	2. Neutral, action oriented
	3. Advocacy, action oriented	
Information type	1. Qualitative	2. Quantitative

6. encouraging the development of complementary technologies to the ones in question; and
7. developing ways to reduce, retard, or even stop inappropriate technology.

15.4.2 Policy-Analysis Parameters

We now turn to the description of important properties of policy analysis that have implications for its conduct. Table 15.1 contains a set of parameters useful in describing policy analysis with emphasis on TA/EIA (recall also Table 5.1, p. 74, on the content and process dimensions of a TA/EIA). The first three parameters (content, locus, and political sensitivity) concern the subject matter and the next four (time dimension, focus, analytical stance, and information type), the study orientation.

Explicit methods, that is, analytical techniques, are not included in the typology; these will be considered as a function of the combination of parameters in each study. The following discussions are intended to suggest the gist of these dimensions. Interactions among the parameters in specific situations are intriguing in their implications for the conduct of policy analyses.

Content. Analysis of certain technological policies may demand substantive technical expertise and awareness of the characteristics of the particular TDS involved. One might surmise that analysis of "high-technology" issues would be more dependent on technical expertise and less easily opened to public participation than analysis of "lower technologies" or social technology issues.

Locus. Obvious differences between public and private policy include the policy objectives (i.e., political vs economic—there is usually little market mechanism at work in the public sector) and the complexity of objective functions (public policy objectives are often multiple in nature, difficult to specify, and conflicting). The overlap between public and private policy deserves attention. Corporate policy is formed in the shadow of governmental incentives, regulations, taxes, and even competition; and governmental policy with respect to a technology may need to address incentives for innovation and diffusion, and regulatory mechanisms to influence private sector actions.

Political Sensitivity. It is obvious that policy analysis must be sensitive to political considerations in formulating policy options and drawing conclusions. The character of the prime users for a TA/EIA must be taken into account. A Congressional committee may be unlikely to welcome overly explicit conclusions on delicate issues that threaten its sovereignty and room for developing workable compromise. On the other hand, a governmental agency or corporation may welcome more specific findings that ease its task in justifying a course of action.

Time Dimension. Retrospective analysis is instructive in understanding the effects of previous policy actions. It is by no means a simple task (Tarr 1977). However, a prospective orientation is necessary for providing insight into preferable policy alternatives. Unfortunately, the methodology of "futures" studies does not currently possess a strong theoretical base (see Chapters 6 and 7). The distinction between short-term assessments (e.g., 1 year) and long-term assessments (e.g., 20 years) is considerable, as our treatment of forecasting has indicated. The former are likely to engender more interest from stakeholders and decision makers, to be subject to relatively precise forecasts, to be performed under time duress, and to have clear policy relevance. Some implications for policy analysis are that short-term assessments are more liable to generate interest in stakeholder participation, to generate controversy, and to produce immediate policy actions.

Focus. Analysis may focus on various aspects of the policy process. Studies of policy-making and policy-implementation processes typically encounter difficulties in obtaining adequate inside information to describe what "really" matters (e.g., in institutional analysis; Chapter 13). "Policy outputs" refer to the direct and intended effects of policy actions. Obviously, it is important to perceive these correctly if one is interested in pursuing the less direct impacts. For instance, if one miscalculates the implications of a particular governmental regulation on the number of automobiles produced, one will misjudge the air-pollution implications as well.

Analytical Stance. Most TA/EIAs are directed to producing action options, rather than to explaining policy implications as such. The distinction between neutral and advocacy orientations is treated in Section 15.4.4.

Information Type. Two key distinctions can be drawn: (1) that between tacit, "inside" information gathered in informal ways and qualitative information collected according to sound social science standards and (2) that between qualitative and quantitative data. Applicable techniques of policy analysis depend on the type of information used. Meaningful analysis cannot go beyond the data available, but the temptation to employ particular analytical techniques, without concern for the data scaling and error characteristics, is great (Goldberger and Duncan 1973).

This typology should be taken as a set of dimensions hypothesized to be important to the conduct of policy analysis. There is support for the usefulness of these dimensions in at least four respects.

First, in considering the requirements and the aims of a TA/EIA, potential sponsors or performers might specify the sort of study they need (e.g., what time horizon, neutral stance or not, situation-specific or general, and how large an effort). Also they might determine what they want to know about policy making for the TDS, implementation strategies, and policy outputs.

Second, consideration of alternative parameter states may open up possibilities for enrichment of policy analysis. For instance, what scholarly groups share interests in complex sociotechnical systems? How have various disciplines dealt with poor quality data; for instance, could work on measurement error in social forecasting assist technological forecasting? Are counterpart insights on policy-formulation processes or ways to measure policy effects transferable between public- and private-sector analyses?

Third, strong support for an eclectic approach follows from the classification exercise. Particularly, there appears to be great advantage in combining multiple analytical styles for the study of an issue. "Macro"- and "micro"-scale foci may yield complementary information. For instance, a qualitative case study on a particular siting may usefully complement a quantitative assessment of nuclear power on a national basis. Deliberate inclusion of alternative models, possibly dealing with feedback in different ways, or drawing upon different types of information, may put each in better perspective.

Fourth, consideration of the importance of technological content appears timely. What difference does the technological content of a policy matter make in assessment? Should one analyze a social technology (e.g., no-fault automobile insurance) differently from a physical one (e.g., plutonium recycling)?

Discussion of this typology was intended as a starting point for categorization of the rich variety of activities that can take place under the label of policy analysis. We now further focus our attention to recommend a strategy for policy analysis within TA/EIA.

15.4.3 A Policy Analysis Model for TA/EIA

In developing a strategy for policy analysis in TA /EIA, we begin with the components of an assessment (Chapter 4) and the policy study parameters just discussed. Now we introduce the notion of the *policy system* (Kash 1977: 148–151). The system contains two components, the *policy sector* and the *policy community*. The policy sector is defined and limited by the set of substantive actions potentially available regarding the subject of the assessment. The policy community consists of those actors who either define the boundaries of the policy sector or who have a stake in the outcome of the policy choices. In any study the delineation of the policy system begins, implicitly, at the start of the study and proceeds iteratively as the study develops. In terms of the assessment components in this book, the treatment of the policy system begins explicitly with the identification of impacts and policy considerations (Chapter 8). The development of knowledge about the impacts and the policy system proceeds iteratively, as shown in Figure 15.1.

The *assessors' perspective* (Figure 15.1) consists of the goals, objectives, values, and assumptions about themselves and the subject of the assessment that the assessors bring to the treatment of impacts and policies. This perspective is the basis from which impact and policy-system identification proceeds. The assessors should make their perspective explicit for themselves and for users of the assessment.

Dror (1971b) emphasizes the desirability of *innovativeness* in identifying the *policy system*. The fewer current assumptions that are taken for granted, the greater the likelihood of a real contribution to policy development. All too often, assessments complacently accept the status quo of social aspirations, technologies, and, particularly, institutional arrangements—yielding a strong conservative bias. Furthermore, as J. Coates (1977c) has noted, in the adversarial policy process in the United States, there is no constituency for the future.

Given a sound description and understanding of the policy system, one draws upon intuition, experience of any available sources, and brainstorming to *identify policy options*. The list of options should begin as comprehensively as possible, incorporating a broad set of novel possibilities. It could include policies largely institutional in nature (creating a new federal agency), regulatory policies (setting environmental standards), fiscal policies (offering tax incentives), and policies dealing with technological aspects (or alternatives to technological development), social changes, or legal obstacles. Clearly,

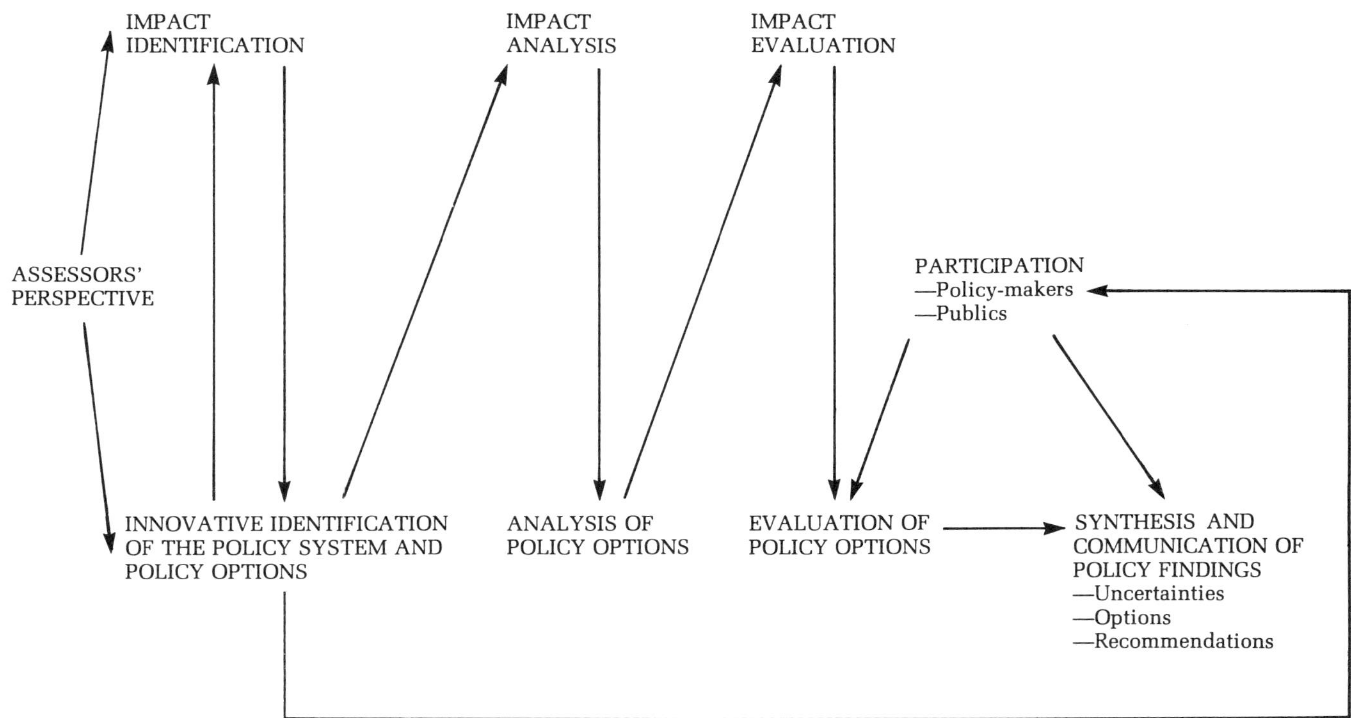

FIGURE 15.1. A strategy for policy analysis in TA/EIA.

the range of possible policy options is far richer than the range of technological alternatives confronting the assessors.

Having opened up a range of policy alternatives, one must begin to foreclose them to achieve a viable analysis. Selection criteria include a quick assessment of the *feasibility* of a given policy option and its *consistency* with assumed goals and values. One must consider both the policy implications for the development in question and for its impacts. *Interactions* among multiple policy actions cannot be neglected—one can compensate for the shortcomings of another or vitiate its effectiveness. Indeed, the federal patent system may grant the bases for new monopolies, to be followed by enforcement of the federal antitrust laws to break them up. If the richness of potential policies is too great to handle in the TA/EIA, one might elect to focus on several *extreme* options to highlight the range of possible actions and their effects.

Most critically, the delineation of policy options must be realistic with regard to the mandate for the TA/EIA. An assessment may be constrained by guidelines in its focus and scope. This does not prevent innovativeness in identifying options, but it may pose a limitation in pursuing them. Furthermore, the eventual policy output of the study should relate to actual policy processes. It must recognize (J. Coates 1977b: 8)

> the limited boundaries of limited authority of any potential user of a study, input, or diagnosis, or analysis. Against that background policy advice that is global in scale, decoupled, remote, or unrelated to the forces on public policy decisionmaking in the United States, is beside the point in the short run.

The assessor must decide on the appropriate mix of novelty and practicality, sweeping scope and practical detail.

Before moving beyond policy option identification, we reemphasize that this should begin early in the TA/EIA (hence our inclusion of this important step with impact identification in Chapter 8). The outputs of the *impact identification, analysis, and evaluation* efforts (hopefully formulated with due regard for policy considerations) contribute to policy analysis (Figure 15.1). These assessment components should be tightly coupled. Unfortunately, as Armstrong and Harman (1977) have observed, the coupling of policy analysis to the impact assessment phase, and to earlier phases, such as technology description and forecast, is often weak. Indeed, half the TAs they studied used the early-phase results only intuitively in the development of policy considerations.

Analysis of the *policy options* incorporates insights gained throughout the TA/EIA to address the likely effects of various policies and combinations of policies. Depending upon the specifics of the TA/EIA, this may draw upon various of the techniques discussed previously in this book (see Table 5.3, p. 82). In particular, quantitative decision analysis methods (discussed in Chapter 14) may prove useful. We prefer more subjective, qualitative approaches for the policy analysis step, but infusion of "objective" information can be very valuable. Indeed, multiple analytical approaches are desirable. Blending quantitative with qualitative findings, and expert with stakeholder review, can strengthen the policy analysis. As Dror (1971b) notes, it is also appropriate to recognize nonrational elements that seriously affect policy processes (e.g., ideologies, mass social phenomena), along with rational considerations (e.g., tallies of costs and benefits).

White et al. (1976, 1977) propose five criteria—effectiveness, efficiency, equity, flexibility, and implementability—for selection among alternative policies. These are shown in Table 15.2 together with general questions that make them operational. Both quantitative and qualitative measures may be useful in answering these questions.

The issue of *participation* in TA/EIA bears on *policy analysis* and, especially, on *evaluation*. It is worth considering the inclusion of policy makers in the assessment process to facilitate mutual understanding, study relevance, and study receptivity. Likewise, involvement of stakeholders in the assessment can be used to broaden its perspective, to enhance its credibility, and to generate workable policy alternatives.

There are widely differing views on how to handle the issue of participation in TA/EIA. For instance:

- the Jet Propulsion Lab's (1975) automobile power-plant assessment in which the sponsor (Ford Motor Co.) and decision makers were excluded from even review of draft documents to protect study credibility;
- EIS open hearing and review requirements on draft documents;
- Kash's (1977) emphasis on stakeholder involvement early on and continuing throughout the assessment process;
- the A. D. Little, Inc. solar energy assessment's (1975) establishment of a citizen-participation panel.

Should the assessors decide to exclude direct participation, it is still valuable for the assessment team to anticipate stakeholder responses to impacts and policies. Role playing may be useful in this regard to generate a stronger sense of stakeholder perspectives.

EIA, and also TA, can often contribute to the generation of better

TABLE 15.2. Evaluation Criteria[a]

Criterion	What the Criterion Reflects
Effectiveness	Achievement of objective Avoid or mitigate the problem or issue? Short- or long-term resolution or solution? Dependency on state-of-society assumptions?
Efficiency	Costs, risks, and benefits Economic costs, risks, benefits? Social costs, risks, benefits? Environmental costs and risks? Reversible/irreversible, short- or long-term?
Equity	Distribution of costs, risks, and benefits Who will benefit? experience costs? assume risks? Geographically? Sectorially? Socially?
Flexibility	Applicability/adaptability Are local and regional differences accommodated? Are social and sectorial differences taken into account? How difficult will it be to administer? How difficult will it be to change?
Implementability	Adoptability/acceptability Can it be implemented within existing laws, regulations, and programs? Can it be implemented by a single agency or level of government? Is it compatible with existing societal values? Is it likely to generate significant opposition?

[a] After White et al. (1976, 1977).

policies through an "active" stance. Rather than merely identifying impacts and policy options, it can be fruitful to modify policies (and thereby impacts) by taking into account participant responses. For instance, rather than cataloging a community's displeasure at the impacts of a highway, the assessors may be able to help resolve the problems. By exploring policy options with participating stakeholders, the assessment may generate more satisfactory action possibilities (e.g., covering the highway and providing public playgrounds on that surface).

As we consider the evaluation of policy options, it is useful to puncture the dual illusions that participation is *the* thing to do and that value considerations (no assessment can be value neutral) preclude evaluations of options by assessors. J. Coates (1977b: 35) notes that participation is subject to abuse:

> The pressures of public involvement resulting from legal challenge and legislative mandate have led many agencies to convert public participation into a kind of ceremonial unguent which once applied will facilitate whatever they want to do in the first place. The usually bureaucratic effort at participation is pro-forma, stilted, uninforming, non-responsive, quasi-legal, not structured to result in or influence action and not structured to inform decisionmaking.

Rein (1971: 309) puts the values issue in a clear perspective:

> The close relationship between values and modes of analysis does not invalidate policy analysis, but it does imply that there will never be one "true analysis." Every analysis must be judged good or bad within the framework of its value assumptions. The study of policy can be most insightful when it examines afresh the critical assumption on which action proceeds. One such assumption is the context within which the analysis is framed, including definitions of and choices between constraints and options, which are typically governed by belief or opportunity or both.

Evaluation of policy options should be done in the TA/EIA, but it should be done openly and explicitly, documenting the reasons and supporting evidence. Selectivity is implicit in the identification, analysis, and evaluation of options. All lists of options exclude as well as include; they would be useless if they did not. However, selection should not be biased. If we distinguish between *policy advisors* and *stakeholders*, it is clear that balanced policy is unlikely to come from a stakeholder. Consequently, unless an adversary assessment is involved, it is usually desirable for the TA/EIA team to be perceived by the stakeholders as unbiased, with no vested interest in the recommended policy outcome. This distinction has been clearly articulated by Kash and White (see Section 15.5). After having demonstrated their lack of bias, they nevertheless were not timid about going beyond a listing of policy options to give policy advice.

The final element of our policy analysis model is *synthesis and communication of policy findings* (Figure 15.1). TA/EIAs deal with prospective decisions under uncertainty. There is little chance that information will develop so clearly in the course of the assessment that previously cloudy choices will become crystal clear. Nonetheless, reduction of the uncertainty associated with a technology, its potential impacts, and pertinent policy options is probably the main means whereby TA/EIA informs the policy process.

The OECD (1975: 32) recommends that the findings of the TA/EIA be thoroughly integrated into complete action options. We agree that synthesis of impact and policy analyses is a worthy objective. A decision maker or a stakeholder can more readily relate to a coherent program alternative than to a series of disjoint observations on particular impacts and policy possibilities.

We thus come to the issue of whether or not to make policy recommendations. The step of policy advocacy in a technology assessment is controversial. It may be misconstrued as reflecting self-interests of the assessors. It may encourage the assessors to go much further than the analysis would justify. It may crystallize the assessors in a particular stance and thereby compromise their future credibility. Finally, it may be politically naive, thereby undermining a sound assessment (Enzer 1975: 108–109). Most technology assessments have avoided making specific policy recommendations. On the other hand, Enzer, in advocating recommendations, points out that not driving toward these may reduce an impact assessment to triviality. One guideline for the assessor, who will probably have arrived at a preferred policy option, is to realize that "the facts often speak for themselves." Discussion of the pros and cons of each of several alternatives is likely to convey implicitly the assessors' preferences. A danger in making several specific recommendations is to undermine the importance of the policy maker. This also risks alienating portions of the desired policy community so as to make the assessment a matter of contention and thereby less credible. On the other hand, the dangers in not taking this last step to recommendations lie in people ignoring the TA/EIA due to blandness, or misinterpreting it to meet their own purposes. The choice of which path to follow may largely depend on the character of the sponsor, the primary intended users, and the assessors themselves.

15.5 SOME LESSONS FROM ACTUAL EXPERIENCE

Probably the best way to convey the essence of policy analysis within an impact assessment is to describe how assessors reflect on their experiences. One group well respected for its policy sense, The Science and Public Policy Program of the University of Oklahoma, has performed several large-scale impact assessments. The following points are abstracted from a discussion by Kash (1977) of what they consider to be important principles for effective policy analysis.

1. The objective of providing policy advice is the starting point.
2. Credibility in the role of policy analyst depends on having knowledge of the substance of the activity for which policy is

being made, and on having policy makers aware that you have that knowledge.

3. The policy realm is most concerned with those substantive (e.g., technological) activities on which there is disagreement, rather than on those beyond the state of the art (no useful policy could be set anyway) or those on which there is general agreement and present implementation; focus on this middle domain where uncertainty exists but practical actions are feasible.
4. Both vested interest and expertise play a major role in resolving activities on which there is disagreement; the most useful policy research reflects understanding of both the substantive and the political issues, but reflects no obvious vested interest.
5. Interdisciplinary research is useful in identifying the substantive issues of disagreement, the relevant policy sector, and the policy community (those who participate either because of expertise or a stake in the policy outcomes)—it is basic to the identification of policy options.
6. Interaction with the policy community during the study helps to identify policy options by gaining a thorough understanding of both the policy sector and the policy community. It also alerts potential users of the study. This interaction can take several forms:

   a. dissemination of early draft papers and the succeeding iterations reflecting an embarrassing initial level of ignorance—demonstrating that the assessors do not start with a predetermined position;
   b. an oversight committee, with members representing each major element of the policy community; and
   c. use of paid consultants from every sector of the policy community, plus interviews with people across the whole range of the policy community.

7. Broad dissemination of the findings should be pursued through multiple approaches, including provision of advice based on the study, at the convenience of the interested user.
8. Specific, concise recommendations that could be implemented are most useful; generalizations that must be translated into specifics by the users appear to get lost.
9. Results need to be timely.

Arnstein and Christakis (1975: 118–119) have pieced together recommendations from the Oklahoma group, Hittman Associates [who performed a technology assessment of advanced automotive propulsion systems (Harvey and Menchen 1974)], and other technology assessors, regarding policy analysis. They note that:

1. Many painful iterations of policy considerations with the parties at interest are needed.
2. Cost overruns frequently constrict the resources available to perform the policy analysis task (which typically occurs relatively late in the project).
3. There is a tendency for both analysts and sponsoring agencies to temper their politically sensitive conclusions; to counter this, it may be necessary for the assessment community to develop its own group identity, standards, and internal policing to buttress itself against this censorship.

We concur with this summary statement from Arnstein and Christakis (1975: 115):

> Policy analysis represents the most critical challenge to technology assessment (TA), because the most comprehensive and scientific data gathering and data manipulation is of little value to the policy-maker unless it has relevant and cogent policy implications. Despite its criticality—or perhaps because of it—policy-related research in general and concluding policy analysis chapters of TA studies in particular have left much to be desired.
>
> Among the analytic "defects" cited by various workshop participants are verbosity, general "mushiness," inconclusiveness, high level of platitudes and abstraction, elitist bias, lack of perspective, lack of relevance to the policy-maker, lack of political feasibility, and propensity for including trivial alternatives.
>
> Among the lessons learned by the assessors is that good policy advice cannot be offered without a thorough understanding of the legal/institutional context of a technology. In this respect, there was general agreement that it would be cost-effective for the public policy analyst(s) on the team to be involved at the outset of the study in the problem definition in the first iteration, rather than being used as an add-on after the impact analysis has been completed.
>
> Workshop participants also agreed that high-quality policy analysis requires explicit identification of those who have the authority to set policy in relation to the technology being assessed.

In conclusion, we note that TA/EIAs are not the sole form of information available to address policy issues. Various other forms of policy studies are possible. These may complement TA/EIAs. Such possibilities, which have relative advantages and disadvantages, include pilot experiments and monitoring of impacts of implemented policies immediately after the fact. These deserve consideration as alternatives or complements to impact studies in the formulation of policy.

15.6 CONCLUDING OBSERVATIONS

The first part of this chapter (Sections 15.2 and 15.3) presents an introduction to policy, policy studies, and models of the policy process. The last part (Sections 15.4 and 15.5) discusses an approach to policy analysis in TA/EIA, including pertinent experience of practitioners. Critical points discussed include the following:

Policies are guiding principles in both the public and private sectors.

Policy studies address policy formulation, implementation, effects, and methodology.

Models of the policy process include the rational model, the institutional model, the group equilibrium model, the elite model, the incremental model, and the systems model. Each has something to contribute to policy consideration in TA/EIA; the systems model appears particularly useful.

In TA/EIA, policy analysis consists of a thoughtful consideration of what is likely to happen under alternative courses of action.

Figure 15.1 sketches a strategy for policy analyses in TA/EIA. Policy analysis should commence at the beginning of the assessment and be deepened through iteration. It rests on a knowledge of possible impacts, and so can be developed as these become better known.

Policy analysis should usually include participation by policy makers and impacted publics—the policy community.

Making explicit recommendations is the option of the assessors and should depend on study context. Explicit recommendations are usually desirable if the assessment team can convince its audience that it is unbiased.

Policy analysis is a most important part of any assessment. It requires sufficient time and resources to be done well.

RECOMMENDED READINGS

Braybrooke, D. and Lindblom, C. E. (1963) *A Strategy of Decision*. New York: The Free Press.

The classic presentation of the incremental view of policy.

Dror, Y. (1971) *Design for Policy Sciences*. New York: American Elsevier.

A well-developed framework to relate the systems approach to policy.

Dye, T. R. (1975) *Understanding Public Policy*, 2nd ed. Englewood Cliffs, NJ: Prentice-Hall.

A good discussion of models of the policy process.

Kash, D. (1977) "Observations on Interdisciplinary Studies and Government Roles," in *Conference Proceedings: Adapting Science to Social Needs*. Washington, DC: American Association for the Advancement of Science, AAAS Report No. 76-R-8, pp. 147–178.

Excellent insights into the practical aspects of policy analysis in TA/EIA.

Lowi, T. J. (1976) *Poliscide*. New York: Macmillan.

Sobering demonstration of the possibility of coercion in the policy process.

Tribe, L. H. (1972) "Policy Science: Analysis or Ideology," *Philosophy and Public Affairs 2* (1), 66–110.

Excellent demonstration that ideological biases frequently underlie so-called value-free analysis.

EXERCISES

15-1. Should the assessment team adopt a rule concerning team membership stipulating that any scientific or technical expert with a vested interest in one or more of the policy options identified in the policy analysis be disqualified from team membership? Give reasonable arguments pro and con.

15-2. Discuss the claim made by some TA/EIA practitioners that participation by citizen groups or other special interest representatives in the actual process of producing a TA/EIA should be discouraged since these individuals are likely to be biased.

15-3. Identify who comprises the interest sectors: (1) scientists and technologists, (2) decision makers, and (3) stakeholders—in the following policy issues:

Issue—Are societal goals best reached by (a) more automation or (b) less automation?

Issue—Is social injustice better remedied by (a) the private sector or (b) the public sector?

Issue—Can a solution to urban problems be better attained by (a) technological fix or (b) social programs?

Issue—Is the energy shortage better remedied by (a) more energy production or (b) less energy usage?

15-4. Policy analysis critique.

1. Obtain a particular TA or EIA for a number of persons to review. Ideally, some persons might review a TA while others look at an EIA.
2. Briefly describe the policy analysis done in the TA or EIA.

3. Discuss the adequacy of the policy analysis. Consider, in particular, its:
   a. comprehensiveness in addressing the full range of policy possibilities (the policy sector);
   b. innovativeness in considering novel possibilities (policy options);
   c. analysis of policy alternatives (how thoroughly and validly the effects of various policies were determined);
   d. evaluation of policy alternatives; and
   e. credibility to various users of the study.

15-5. (Project-suitable) Select a technology or project. Identify the relevant policy sector and policy community. Analyze and evaluate the substantive actions available to deal with the subject of the assessment.

16

Communication of Results

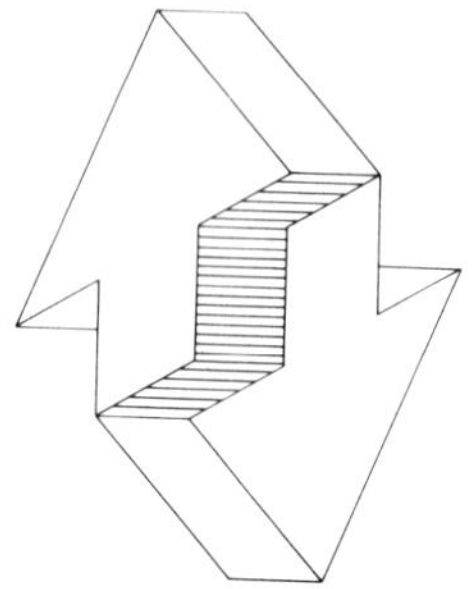

"Communication of Results" deals with the question of how to make the results of a TA/EIA useful. Achieving a useful assessment requires identification of the intended audiences and planning to meet their needs throughout the assessment process (not only after the report is written). This chapter characterizes the potential assessment users and their relationships with TA/EIA producers. It suggests effective actions to be taken at various stages of the assessment process to enhance the prospects of its utilization.

16.1 INTRODUCTION

In Chapter 4 two prime objectives were advanced for TA/EIAs—validity and utility. As we move to the 10th component of the TA/EIA process, communication of results, the emphasis shifts from concerns about validity to utility.

Effective communication of results cannot await preparation of the final assessment report. At the very start, assessors and their sponsors ought to begin to consider their intended results and prospective users of those results. Communication by the assessment team with users should proceed during the course of the TA/EIA. (The assessment team's internal communication processes are discussed in Chapter 17.)

The chapter first introduces some basic concepts, then discusses the uses of TA/EIA. It then reverts to the very beginning of the assessment process. We next consider the research product itself, and how its characteristics affect the uses made of it. Issues surrounding efforts at dissemination of results are then treated. Finally, some novel approaches to communication are introduced, followed by concluding observations.

16.2 TWO COMMUNITIES

Research used in policy making necessarily involves two communities: the producers of the research and its potential users. The *producers* of a study (i.e., the assessors) are relatively obvious—the principal investigator, the study team, and supporting personnel. The set of *potential users* is more diffuse and quite difficult to identify. Bureaucrats, lawmakers and their staffs, private and public interest groups, other researchers, and affected parties may be candidate users of any given policy study. Utilization of a given study depends on the establishment of an effective flow of information between these two communities. It is useful to address communication of results in this light (Figure 16.1), with special focus on the barriers that may impede this flow.

Caplan et al. (1975) have classified theories about utilizing social-science knowledge in public-policy deliberations into three types: (1) knowledge-specific theories, (2) policy-maker-constraint theories, and (3) two-communities theories. In knowledge-specific theories, research producers create the major barriers to utilization—such as disciplinary narrowness, ideological bias, overly quantitative orientation, lack of policy orientation, and inadequate or inappropriate theory, data, and

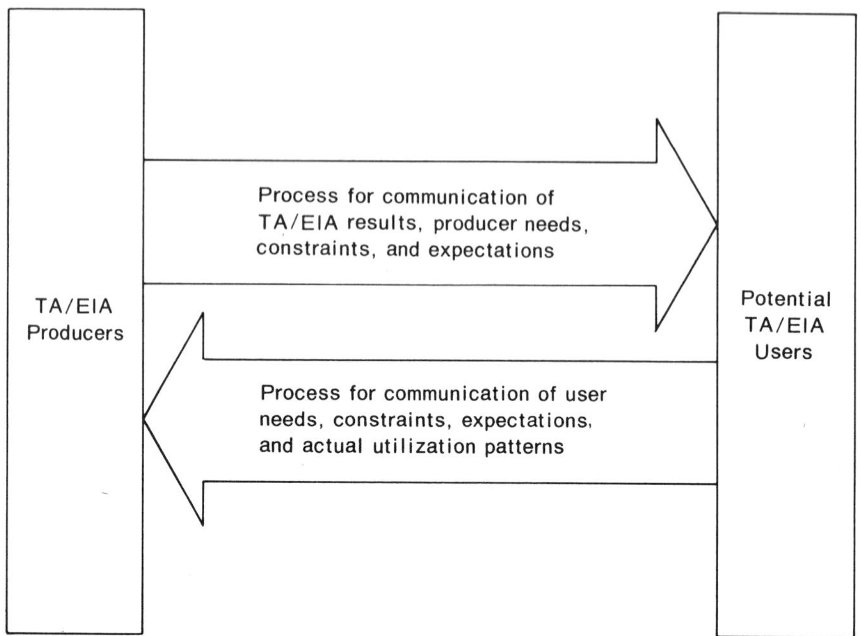

FIGURE 16.1. The TA /EIA utilization system.
Source: Michael et al. (1976).

methodology. In policy-maker-constraint theories the main barriers to utilization lie in the demand for oversimplified information, the time limitations of policy actions, and organizational and political factors that can override systematic analyses.

The two-communities perspective emphasizes the mismatches between the producers and potential users. These include differences in language, values, goals, methods, and standards of quality and significance. In this view, the utilization process is improved by eliminating mismatches through mutual efforts. We agree with Berg et al. (1978)[1] that communication barriers are the easiest to overcome. We discuss opening the communication flow channels from producers to users, and from users to producers (Figure 16.1), after a discussion of the uses of TA /EIA.

[1]This chapter relies heavily on their empirical study of the utilization of TAs.

16.3 USES OF TA/EIA

The two major categories of assessment use (Berg et al. 1978) are (1) to influence thinking (conceptual consequences) and (2) to have identifiable impact on policy decisions (instrumental consequences). Potential conceptual consequences include efforts to:

- increase understanding of the issues, particularly in a broader context;
- raise new questions, and offer new perspectives;
- make a broad range of information available in one concise, organized source;
- provide information for use in secondary documents such as memoranda, professional articles, handbooks, and research proposals;
- identify problem areas and business opportunities for corporations;
- broaden the thinking style of persons brought into contact with the assessment process;

Instrumental consequences can include:

- stimulate legislative activity, such as hearings and review of pending legislative provisions;
- alter the planning processes of organizations required to perform assessments;
- contribute arguments for advocates, as in lobbying with a legislature, executive agency, or regulatory body;
- generate research in associated areas.

In their survey of 280 potential TA users, Berg et al. (1978) inquired about the uses of particular studies. Results indicated that TA information was most used for background information and least, for decision making. In a parallel set of measures, TAs most influenced organizations by bringing issues to the attention of decision makers and least, in selecting policy options. The results indicate that TAs are no more than moderately influential; certainly, EISs share a similar record. Impact assessments serve to raise issues and provide information, but only rarely do they directly lead to policy actions. This finding should put TA/EIA efforts in a realistic perspective. Assessors should not have excessively high expectations of exerting policy influence. However, we believe that improved assessments, attentive particularly to the issues raised in this chapter, can improve the track record.

16.4 THE COMMUNICATION PROCESS

Neither the EIS procedures nor their less formal TA counterparts assure that any use will be made of assessment studies. The burden for successful utilization thus falls upon the producers and potential users—with a special place for study sponsors. Except when a specific target user is involved, the producers (and sponsors) must usually take the initiative if a study is to be highly useful. This initiative should begin before the study, continue throughout, and end some time after study completion.

16.4.1 Before the Study

Responsibility rests upon both producer and sponsor to orient the assessment to potential users. Very often the sponsor will be the primary user, as in TAs for corporations or executive operating agencies. In EIA the main user is the agency responsible for the development and preparation of the required EIS.

The cameo entitled "Some Potential Users of a TA/EIA" provides an initial checklist for identifying other users. Willeke (1974) has described processes for identifying various "publics" that can be extended to other potential users. Identification entails locating the publics, determining their interests and capabilities, and learning how to communicate with them. This can be accomplished through:

self-identification (e.g., through petition, public hearing, protest, publicity);

staff identification (e.g., through general lists, analysis of associations, historical and demographic analysis, field interview, and analysis of affected parties); and

third-party identification (e.g., through asking some person or group to identify others who should be involved; "snowball" interviews are a special form in which persons identified are, in turn, contacted to seek more persons).

After identifying the potential study users, the intended study users should be chosen. By and large, governmental studies must go public. However, they may be *targeted* to a particular user, or they may be oriented to a wide range of users. Privately sponsored studies may find

SOME POTENTIAL USERS OF A TA/EIA

The sponsoring organization

Other executive agencies (e.g., EPA, Department of Transportation)
- Various divisions or bureaus (e.g., Federal Highway Administration Office of Socio-Economic Analyses)

Other executive branch units (e.g., Federal Communications Commission, the Office of Management and Budget)

Legislative bodies
- Particular committees or subcommittees (e.g., Senate Committee on Commerce, Science, and Transportation)
- Particular congressmen
- Interested staff members

Private interests
- Trade associations (e.g., The American Petroleum Institute)
- Particular firms
- Labor unions

Public interests
- Established interest groups (e.g., The Sierra Club)
- Issue-oriented groups (e.g., NORML—National Organization for Reform of Marijuana Laws)
- Parties directly or indirectly affected by the technology in question
- Influential individuals (e.g., Ralph Nader)

The research community
- Those involved in basic or applied, scientific and technological matters related to the subject of the TA/EIA (e.g., nuclear engineers for the GESMO study)
- Those addressing directly or indirectly related policy matters (e.g., solar-energy analysts for the GESMO study)

it in the sponsor's interest to make results public to contribute to policy deliberations (and public relations too). In any event, one must consider the needs and capabilities of the intended users. Some audiences are highly organized, with strong technical staffs; others lack both funds and technical sophistication. Some groups bring a keen awareness of an issue's importance; others must be alerted to its salience. Some groups may want background information (for use in lobbying);

others may want policy options evaluated. The general public is unorganized and diffuse. The TA/EIA should be matched to its intended users' needs, including that of broadening their perspectives.

For example, consider Congress as a target user. Congress is not interested in long lists of impacts, numerically weighted or not, with elaborate cross-impact analyses (Christopher T. Hill 1977, private communication). Assessments must concisely relate to the real policy issues whose nature is highly variable, for instance:

"go/no go" decisions [e.g., the Anti-Ballistic Missile (ABM) program].

broad programmatic dialogue instead of single decisions (e.g., energy assessments contribute mostly to the ongoing broad energy policy deliberations).

control of impacts attendant to a "go" decision (e.g., Alaska pipeline design modifications to control some impacts brought up in the environmental impact analyses)—sometimes this reduces to concern over one or a very few specific impacts and what to do about them.

a new policy option worthy of consideration (usually Congressional interests dictate the options with the assessment "policy-oriented" to help choose among these).

Policy analyses must take account of the political realities that constrain choices. For instance, realizing that one Senator has the major jurisdiction over energy legislation, while another controls regulatory options, may have bearing on what constitutes a viable policy option. Useful impact assessments must mesh with the operating mode of the intended users; Congressional users operate under the traditions of politics and law, not those of science.

In essence, then, it is highly desirable to consider the intended users at the outset of a TA/EIA. Consideration of their needs should lead to communication throughout the study.

16.4.2 During the Study

What is the appropriate degree and form of interaction between the producers and the users during the TA/EIA? The answer arises from the needs and capabilities of the intended users. In some cases broad interaction may be inappropriate. A proprietary corporate study might want to restrict access (at the risk of diminished information input). The Jet Propulsion Laboratory (JPL) (1975) TA of automotive engine possibilities was sponsored by Ford Motor Company, which agreed to a "hands-off" posture. Ford officials might have preferred more of a say in problem definition, but they endorsed the performers' opinion that noninteraction boosted public credibility. Despite such situations, we believe that a highly interactive assessment process is generally advantageous.

Table 16.1 [from Berg et al. (1978)] shows how five forms of producer–user interaction serve four valuable functions. It was also found that interaction in early phases and greater overall interaction contributed to utilization.

Active participation in the study is the most extreme form of interac-

TABLE 16.1. Forms of Producer–User Interaction and the Functions These Can Serve[a]

	Valuable Functions			
Form of interaction by the potential users	Input information and opinion to the study	Increase awareness of, and familiarity with, the study	Provide opportunity to evaluate the study and its producers	Increase the eventual utilization of the study's results
Serve as active participants in the study	+	+	+	+
Act as members of advisory or oversight committees	+	+	+	−
Review drafts of the TA/EIAs	+	+	+	+
Interact with the producers in conferences and briefings	√	+	+	√
Give inputs such as technical data and beliefs and attitudes about impacts	+	+	√	√
Obtain information from the TA/EIA through individual contacts with the producers	−	+	+	+

[a] This table is reconstructed with Berg's help and interpretation from Berg et al.'s (1978) findings. Some inference by the authors was necessary. The sign (+) indicates the form of interaction is likely to serve the function; (−), that it is not; and (√) is intermediate.

tion. Sherry Arnstein (1975) led an interesting experiment that involved six public interest representatives in a solar energy TA. The so-called PIGAP (public-interest-group advisory panel) interacted sharply with the assessment team, illuminating advantages and difficulties of active participation. In this case the values and immediate concerns of public interest groups contrasted with the research-oriented attitudes of contract research professionals. Arnstein (1973) has also described citizen participation in EIS preparation (Figure 16.2).

One of the surprising findings by Berg et al. (1978) is that advisory committee members used the TA study less than any user group, including those who do not interact directly with the assessors. The explanation appears to be that the TA does not increase their knowledge, or they become sufficiently familiar with the issues while serving as advisors.

The review of draft materials mandated by the federal EIA process can also provide a strong means of interaction in TAs. It has proved the most effective of the currently institutionalized mechanisms for interaction.

To comment briefly on the other interaction forms:

Conferences need not await report completion. A. D. Little, Inc. (1975) used a midproject conference effectively to make potential users aware of the implications of electronic funds transfer for themselves, to obtain their inputs, and to broaden the base of eventual study users.

There are many ways to obtain factual or opinion inputs including face-to-face, phone, or mail contact with the range of interested parties. "The Oklahoma Strategy" (see cameo, p. 413) incorporates inputs particularly effectively.

Direct transmittal of information to users also need not await report completion. It is the most effective (but not necessarily cost-effective) way to foster utilization of study results (Berg et al. 1978).

Berg et al. (1978) found that TA use is unlikely unless users view the methods employed in the study as credible. Table 16.2 indicates how their surveyed users rated 21 techniques, many of which have been introduced in this book.

The content of impact assessments is of potential interest to a wide user audience. The five top-rated components indicate the users' desire

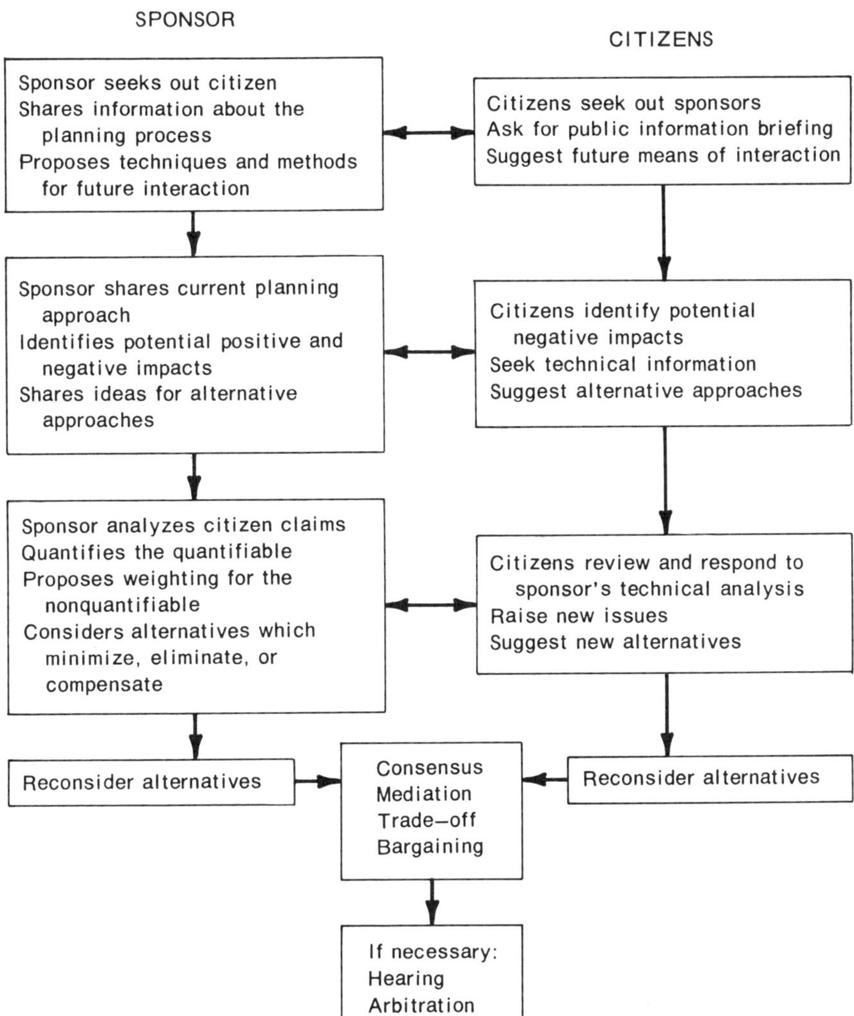

FIGURE 16.2. Participation in the EIS process.
Source: Arnstein (1973).

to have sufficient information to assess the completeness and underlying assumptions of the assessment:

identification of key areas of uncertainty (97% of 189 users said this was very useful or essential);

identification of key assumptions in analysis (97%);

TABLE 16.2. User and Producer Ratings of the Credibility of the Information and Analyses Generated by 21 Methodologies [a]

Ranking		Percent rating credibility as moderate or greater		Difference	Percent not familiar with method		
User (N = 181)	Prod. (N = 30)	User	Prod.		User	Prod.	Method
1	1	84	97	−13	1	0	Statistical analysis
2	2	76	90	−14	3	0	Case study
3	3	68	83	−15	3	0	Cost–benefit analysis
4	5	68	79	−11	5	0	Survey research
5	6	64	73	− 9	12	10	Market analysis
6	4	58	80	−22	6	3	Environmental impact analysis
7	12	57	62	− 5	8	10	Trend analysis
8	8	54	70	−16	21	10	Relevance or probability trees
9	13	53	60	− 7	28	13	Decision analysis
10	7	53	73	−20	19	7	Input–output analysis
11	21	50	34	16	40	45	Evaluation research
12	9	49	69	−20[b]	1	0	Public opinion polling
13	14	48	55	− 7	2	3	Economic forecasting
14	11	46	65	−19[b]	3	0	Games and simulation modeling
15	17	44	52	− 8	10	3	Scenario generation
16	10	44	69	−25[b]	9	10	Technological forecasting
17	19	40	43	− 3	32	20	Systems-dynamics models
18	16	40	53	−13	30	0	Delphi techniques
19	18	39	48	− 9	1	3	Congressional or executive agency hearings
20	15	38	53	−15	45	23	Cross-impact analysis
21	20	36	38	− 2	2	0	Brainstorming

[a] From Berg et al. (1978: 120).

[b] Statistically significant differences between responses of potential users and producers at a level of 0.05 or better based on Mann–Whitney U Test.

identification and analysis of technological alternatives (90%);

identification and analysis of direct impacts (87%); and

identification of external factors or omitted variables that could affect results (85%).

We summarize other considerations for producer–user interaction (Berg et al., 1978: 138–141):

Appropriate type of interaction depends on the resources available, the nature and timeliness of the study focus, and the information needs and characteristics of its potential users (e.g., in "early warning"—as opposed to timely, policy-urgent—assessments, interaction must stimulate interest in the topic).

Interaction with potential users does incur costs as well as risks to the maintenance of assessment integrity.

Experienced upper-level decision makers may be a most skeptical audience, but most responsive to personal interaction. Interaction may be most important for policy-oriented assessments.

Improved procedures to provide user feedback are needed, especially for producers who are not institutionally tied to their primary audience [e.g., TAs for the National Science Foundation (NSF)].

Early and continuing involvement between producers and users appears to be a useful strategy for increasing familiarity, confidence, and utilization.

16.4.3 The Report

The following points reflect our impressions of the important properties of written reports affecting the communicability of a TA/EIA.

Length (level of technical detail) should match with intended users—researchers require thorough documentation; the policy community appreciates brevity. Preparing different reports for different audiences merits consideration.

Summary is always valuable.

Integration of the content matter, beyond good organization, is desirable.

Writing style: comprehensibility and motivation to read are certainly influenced by style. Careful, concise prose with a minimum of jargon is a goal; a good technical writer may help, particularly if a participant during the study.

Graphics do help.

There is room for creativity in the preparation of written documents. Pat Johnson, as TA program manager at NSF, has explored user workshops in which the producers, helped by a representative group of users, boil down the report document to produce a lively 30-page document more digestible by busy users (cf. Farhar et al. 1977). Also, the presentation of findings of a Caltech (California Institute of Technology, Pasadena) nuclear power plant siting assessment, as draft legislation for California, proved an effective way to communicate with legislators (Weingart 1977).

16.4.4 After the Study

All too often, especially in the case of final EISs, there is no further effort at communication after report preparation. (NEPA puts no requirement on the active dissemination of EISs.) Evidence indicates that reward for further effort in dissemination of TAs can be great. "The Oklahoma Strategy" (see cameo of that title) exemplifies the importance of timeliness, good user contacts, and poststudy interaction with users.

In considering poststudy communication measures, it may be helpful to match user sources of information with producer channels of dissemination. The most important user sources, according to Berg's study, are *informal communication,* inside and outside of organizations; *professional* or trade journals and newsletters; *formal organizational* channels; and *newspapers*. Contact with other researchers, hearings and reports, and professional meetings are noted less frequently. This suggests that development of informal contacts, reviews in professional publications, and newspaper coverage may be good means to generate interest in TA/EIA findings.

On the other side, assessors have used a variety of dissemination mechanisms. Among the more common ones are:

direct mailing of the project report, or its publication as a commercially published book or as a technical report disseminated through the National Technical Information Service or Government Printing Office (these last two mechanisms have not proven very useful).

dissemination of an executive summary or nontechnical report, interim reports, press releases, and publication of derived articles.

informal oral communications, formal briefings, testimony at hearings, and participation in conferences. (It has been said that information becomes salient to the bureaucracy only when it is written; to the Congress, only when spoken.)

We suggest that both traditional and novel dissemination methods be considered. The output information should be matched to the users' information preferences. This requires *resources.* The following recommendations to NSF, made at a workshop of TA producers (Arnstein and Christakis 1975: 140), seem to apply to most sponsors:

[Sponsors] should allow extra funds to teams for post-assessment dissemination/utilization activities such as preparation of

THE OKLAHOMA STRATEGY

[Kash (1977: 152–155) describes the communication measures used by the Science and Public Policy Group of the University of Oklahoma in preparing the assessment of outer continental shelf (OCS) oil and gas technologies (Kash et al. 1973)]

Continuous Contact with the Policy Community

To assure that our findings would be given consideration in this policy community the minimum requirement was for us to communicate that we had substantive knowledge and no vested interest in the outcome. We did this by maintaining continuous contact with all of the above components of the policy community [congressional committees and OTA, executive agencies, petroleum industry groups, environmental interest groups]. I would note here that our lack of a stake in policy outcomes was regularly questioned. Being from Oklahoma, the environmentalists were generally suspicious that we were puppets of the industry. Being from a university the industry generally suspected that we had an anti-industry–environmentalist bias.

Our interaction with the policy community during the study had five components:

1. We appointed an oversight committee with members representing each of the four major elements of the policy community plus two professors. Individual members of this group repeatedly reviewed the papers we wrote. Additionally, we had the group together for three two-day review sessions covering overall drafts of the study at an early, middle, and final stage of the study.
2. We made extensive use of consultants from every sector of the policy community.
3. We used repeated interviews with people from the whole range of the OCS oil and gas policy community.
4. We sent out papers to anyone we could get to read them.
5. We held a one-week conference at a boys' school in the middle of Maine. About 80 people attended from all points in the policy community.

The above activities had two purposes. One was to get all the help we could in identifying policy options by gaining a thorough understanding both of the policy sector and of the policy community. The other was to alert and inform potential users of our study.

(continued)

THE OKLAHOMA STRATEGY *(continued)*

Members of this policy community find few things more distasteful than advice by ambush. Stated differently, policy communities work by evolving consensus and they like nothing less than surprises.

Our pattern of continuous contact had a major payoff we had not foreseen. To appreciate this point, I should note that no one on our research team knew anything about offshore oil operations in advance of the study. Some of our initial papers reflected a level of ignorance and shoddy quality that was embarrassing in the extreme. Most of us on the team were certain that airing such papers would ruin our reputations forever. As it turned out, showing our early work, warts and all, was a major factor in establishing our credibility. At one time or another we misunderstood nearly everything and were on every side of every issue.

The effect of this evolution was to demonstrate that we did not start out with a predetermined position and build a case to support it. Further, the quality, the understanding, and the hard data improved with each draft, and we began to build a reputation for having the capability to learn.

The final draft of the study was submitted to twenty-five people for review. We picked ten reviewers and the National Science Foundation, which funded the study, selected fifteen. Reviewers ranged from people with clear vested interests to academic experts in specialized areas. The reviews were distinctly favorable. Of the group, only one was overwhelmingly negative and it was by a political scientist.

This favorable set of reviews set in motion a kind of cumulative pattern of noise throughout the policy community. The rumor system was now building this up. One experienced bureaucrat told me, at this point, "You've got it made. It is now a part of the conventional wisdom of people involved in this area that yours is an important well-done policy study. Even those people who don't like it will not be able to write it off." He went on to note that most people will never read it in total, rather they will only use those elements they have an interest in.

Review and Distribution of Results

All of the preceding might have happened and the study would still have been filed and forgotten had it not been for two other factors. Most important was the fact that two days after the study was delivered to NSF on April 14, the President went on television and proposed that resource leasing on the OCS would be tripled. At the same time, he directed a one-year study of OCS oil and gas operations by the Council on Environmental Quality (CEQ).

Another major factor was that a strategically placed civil servant had read the study and decided it warranted major attention as a starting point for the presidentially directed study which was to be carried out by CEQ. The first step in that study was a series of public hearings at six different locations in the U.S. The announcement of the hearings designated our study as one of three background documents.

In preparation for those hearings we were requested to prepare a one-day briefing to be held at the auditorium of the National Academy of Sciences. NSF agreed to purchase 1500 copies of the book which resulted from the study for distribution to interested parties.

In the study, we had made 39 specific recommendations and these were the basis around which the Washington briefing was organized. Part of the briefing format involved critiques by representatives of government, industry, and environmental interest groups who had no previous advisory role in the study. Fifteen hundred people were sent letters of invitation. Some two hundred showed up.

Every component of the briefing made use of slides and we made every effort to make the briefing concise. A terribly difficult task for academics, I might note.

In conjunction with the briefing, we held a press conference which attracted the major wire services as well as several major publications.

Following the briefing, NSF provided us with support which allowed us to offer advice to anyone who was interested in the work or in using us as advisors. We followed a policy of offering our assistance to anyone who was interested at no cost and on short notice.

Given the development of the Arab oil boycott and the associated emphasis on OCS oil we were nearly overwhelmed with requests. Part of the explanation for our role as policy advisors goes back to an early point in this paper. We were perceived as being knowledgeable about OCS oil and gas operations, but more importantly we were seen as the only group with that kind of knowledge who did not have a stake in what happened. The central point is that there was almost no substantive expertise outside the industry and the Department of Interior.

(continued)

THE OKLAHOMA STRATEGY *(continued)*

The other element that made us attractive was that our recommendations were specific. That is, they proposed discrete actions that could be implemented. Both the executive agencies and the congressional committees found specific policy options that could be adopted immediately, that is, the recommendations provided courses of action that could be used to respond to the growing although very unspecific pressures for some kind of action. Portions of the recommendations appeared in six Senate bills and we were asked to comment on various other legislatively proposed options.

magazine articles, participation in agency and Congressional briefing sessions, sponsorship of public forums, and stimulation of television and newspaper coverage. . . .

[Sponsors] should allow extra funds, on the order of 10% of the grant or contract, to enable core members of a TA team to provide two annual updates of a study. This device would keep the commitment down to a modest level and, at the same time, obligate the core team to update its information.

16.5 INNOVATIVE APPROACHES

A number of unconventional means to facilitate communication with users have been mentioned earlier. This section adds some others worthy of consideration.

Berg et al. (1978) identify several techniques that have been used in TA presentations. Radio and television programs, videotapes, and slide presentations reflect nonprint media. Reports are often made available through information-retrieval systems. Most notably, a handful of producers have hand-delivered copies of their report to important potential users. These assessors were enthusiastic about the effort (mostly focused on the Washington, D.C. area) and reported favorable feedback.

Others have suggested potential value in gaming simulations and computer graphics. These could aid in cutting through complexity to convey the essence of any policy implications. Such approaches appear particularly suited to TA/EIAs that make use of formal computer models.

The Jet Propulsion Lab is participating in an interesting follow-up effort to their automotive-power systems-evaluation study (JPL 1975) that implements several of the suggestions of the previous section. The Energy Research and Development Administration supported JPL to perform further research. Considering the TA as a basis, they are to:

carry on a *continuing assessment*,

prepare a series of *annual reports*, and

as an early task, compile and *publish critiques* of the original TA as a basis for constructive dialogue in industry and government circles; about 50 critiques are included in this volume, including detailed technical reviews by the original TA sponsor (Ford Motor Co.), General Motors, and others (Williams 1977).

An interesting "during the study" approach has been used at the Georgia Institute of Technology, Atlanta, in a preliminary screening TA for NSF on the subject of the "built environment" (Martin and Willeke 1978). The assessment team employed a *communication specialist* as an auxiliary team member throughout the study. Reporters, technical writers, or other specialists could serve in various ways to improve communicability of results, thus facilitating team dialogue to enhance integration, reducing jargon, offering insights into the communication process and abstracting interesting findings for communication purposes.

16.6 CONCLUDING OBSERVATIONS

Effective communication of TA/EIA findings requires consideration of assessors and study users, and ways to facilitate information exchange in both directions between these two communities. Three factors that greatly affect the communication to the prospective users are (1) the level of knowledge about the assessment held by the potential user beforehand, (2) the amount of new knowledge the assessment provided, and (3) whether the assessment was compatible with the interests of the user's organization.

TA/EIAs can have a number of uses, both in influencing thinking and inducing policy actions. Four characteristics of user organizations that are conducive to their use of TA information for policy making are that they (1) regard the TA subject matter as important, (2) do long-range planning, (3) have made or plan to make decisions relevant to the issues addressed, and (4) are receptive to externally produced information.

Utility can be enhanced by working toward communication with users at all stages of the assessment process.

Before the study: The intended study audience should be identified and its information needs understood.

During the study: Carefully considered interaction between potential users and study producers can take several forms with different purposes. Study methods, completeness, and assumptions should be weighed against user perceptions of what makes for a credible assessment.

The report itself: This should match user needs, with attention to length, summary preparation, integration, writing style, and graphics.

After the study: A variety of written, oral, and "nontraditional" dissemination means should be selected to best match the intended users' interests.

RECOMMENDED READINGS

Berg, M. R., Brudney, J. L., Fuller, T. D., Michael, D. N., and Roth, B. K. (1978) *Factors Affecting Utilization of Technology Assessment Studies in Policy-Making*. Report to the National Science Foundation; Center for Research on Utilization of Scientific Knowledge, Institute for Social Research, The University of Michigan, Ann Arbor, MI.

The definitive study of the utilization of TA results.

EXERCISES

16-1. The "two communities" notion that underlies this chapter was initially formulated to describe the interaction between social scientist producers and users of social science information. TA and EIA producers and users deal with technological and environmental information in addition to social information. In what ways, if any, might relationships between producers and users of TAs and EISs differ from those between producers and users of social studies?

16-2. "The Oklahoma Strategy" reflects a rather radically interactive approach to producer–user communication in a TA. Compare and contrast their process with that of a typical federal EIA. (You might want to review EIS preparation guidelines; see Chapter 10.)

16-3. (Project-suitable) Consider a TA or program EIS with which you have become familiar in previous assignments (or consider the GESMO study, Chapter 5 Appendix). Propose specific ways in which to further disseminate findings of that study. Consider both traditional and innovative measures.

Now do the same, supposing that you are a group different from the one that prepared the study. As a group exercise, small groups could perform this for different organizations, such as an environmental group, a consortium of utility companies, or an ad hoc body formed to deal with the specific issue under assessment.

16-4. Consider a potential assessment of ways to reduce consumer demand on the U.S. health-care delivery system. The present system exhibits spiraling costs as supply does not seem to match demand. Furthermore, serious questions could be raised as to how much current medical practice contributes to "health." The proposed assessment is to evaluate policy options to reduce demand via three different approaches: (1) traditional medical system practice, (2) a "behavioral" approach that emphasizes the consumer's responsibility for his (her) own health, and (3) a "far-out" approach that emphasizes alternatives to traditional medicine (e.g., acupuncture, spiritual aids, and environmental improvements). Identify the potential *users* of such a TA. Identify the potential *uses* of the TA findings. Then construct a matrix to indicate in what uses five "target" users (of your choice) would be most interested.

16-5. For the prospective TA of Exercise 16-4, describe how you would involve five "target" users in each step of the TA. Assume that you are a private contract research firm located in Washington, D.C. and that you have about 18 months and support for four person-years of effort to perform the TA (of which 10% is set aside specifically to foster good communication with the users).

17
Project Management

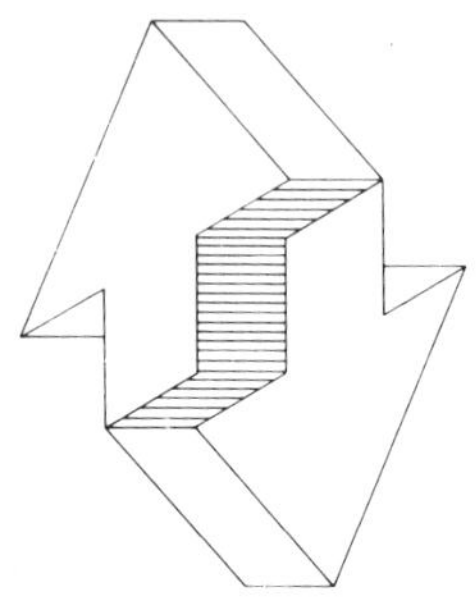

Project management addresses the actual conduct of a TA/EIA—how to get people to work together to produce a quality assessment. Management concerns overlay the whole assessment process. Only through successful project direction can the study be successfully performed. This chapter indicates how a number of structural and procedural factors influence the performance of an assessment. Specific suggestions, based heavily on project experience, are made on likely areas of conflict, suitable leadership methods, and appropriate communication patterns. Common scheduling techniques are presented to aid in planning the study. Four ways to integrate component analyses into an effective TA/EIA are discussed.

17.1 INTRODUCTION

The successful execution of a TA/EIA is not easy. Interviews with participants in 24 TAs for a TA methodology study (Rossini et al. 1978) uncovered some chilling tales. A number of participants and their employers parted company over the failure to produce a satisfactory product. In one study, no report ever emerged. We repeatedly heard stories of long days and short weekends. Such pressures take their toll—in the course of one TA, five team members were divorced.

This chapter draws together suggestions for management of TA/EIA that have worked in practice and identifies pitfalls to be avoided. As a brief primer on managing a TA/EIA, it covers both the sponsor's and the performers' perspectives.

The elements of this chapter relate to the *conduct* of the assessment, in contrast to Chapters 4–16 that dealt with the *content*. In the actual course of a TA/EIA, management overlays the whole process. It is treated at this point in the book as we step back from the particulars of the assessment to address more general concerns.

17.2 MANAGEMENT FROM THE SPONSOR'S PERSPECTIVE

The sponsor of an assessment may represent the same or a different organization than the performer. The sponsor is largely responsible for setting the objectives and providing the resources to conduct the assessment, subject to external constraints such as EIS regulations. Three areas of sponsor responsibility are considered: (1) *topic selection* (problem definition), (2) *selection of the performer* of the assessment, and (3) *assurance of satisfactory performance.*

The sponsor may have wide discretion in selecting assessment topics [as exemplified by the National Science Foundation's (NSF's) TA program], some latitude [typical for federal agencies such as the Environmental Protection Agency (EPA) assessing regional energy development], or little or no say (as in the preparation of required EISs). Issues in selection of an assessment topic, when leeway exists, include perceived magnitude of impact of the technology (importance of assessment), potential policy relevance (likelihood of providing valuable and timely information), and feasibility of assessment (sufficient information, sufficient study resources). In particular, the appropriateness of a TA/EIA to the interests and policy responsibilities of the sponsor is important.

Selection of the performer of the TA/EIA may likewise present wide options to the sponsor, or none at all. In many cases, the performer is restricted to be a particular unit in the agency. More interesting is the situation where some choice between "in-house" and "out-house" assessors exists [e.g., for the Office of Technology Assessment (OTA)]. In-house assessment is typically favored for simpler study procurement and tighter managerial control over the effort, as well as assessors' familiarity with the issue in question, access to internal information, and credibility within the agency. Contracted assessments may have advantages in terms of staffing flexibility, fresh perspective, and greater credibility with parties concerned by the issue under assessment.

The primary mechanism for selection of external performers is the request for proposal (RFP). The RFP specifies the task to be accomplished, sometimes closely constraining study design [e.g., U.S. EPA's Appalachian energy TA solicitation (1976)], sometimes leaving much latitude to the potential performer (e.g., U.S. NSF's 1974 RFP for assessments in selected areas). The sponsor reviews the proposals received, sometimes urges the best candidates to make certain revisions, and then selects the performer. Other selection mechanisms are also possible. By use of a "statement of capabilities," a small number of potentially suitable performers are identified and then invited to prepare a study plan or formal proposal. Some sponsors will consider unsolicited proposals or invite suggestions within defined program areas [e.g., U.S. NSF TA program announcement (1977)]. The cameo entitled "Nine Criteria for the Selection of TA/EIA Performers" notes characteristics that a sponsor might seek in evaluating prospective assessors. Many of these are discussed in the following sections.

Those preparing proposals for TA/EIAs would also do well to consider these criteria. The proposal represents an effort to sell one's assessment capabilities to a sponsor. As such, it indicates what is to be analyzed, by whom, how, and why the chosen approach is best. The proposal should usually indicate the users of the assessment and how the study will meet their needs. It should suggest a plausible schedule of tasks and the budget to accomplish them. Although experience indicates that such schedules should be loosely interpreted, the proposal does begin to set the structural features of the assessment.

After selection of a performer, the sponsor's principal means for assuring quality performance is monitoring the progress of the TA/EIA. Usually a representative of the sponsor, the project monitor, stays in close contact with the actual performers. By keeping abreast of the study's progress, the monitor can help overcome difficulties, secure needed information, and allow for appropriate revision of the study plan. The aim is to assure successful performance, on time and within budget.

NINE CRITERIA FOR THE SELECTION OF TA/EIA PERFORMERS

1. Contractor organization structure conducive to TA /EIA performance (possessing flexible personnel and allowing easy interchange among organizational units).
2. Successful interdisciplinary research experience of the organization and the project team.
3. Leadership qualities of the project manager (principal investigator).
4. Inclusion in the core-team[1] of an appropriate range of expertise (as well as evidence that the proposed performers reflect the actual core team).
5. Core-team experience in TA /EIA and in working with each other.
6. Evidence that the core team can accomplish tasks critical to the assessment (such as social-impact assessment or water-quality modeling).
7. Indications that the performers will secure a credible balance of assessment inputs to obtain the necessary knowledge of the topic.
8. A management plan indicative of effective scheduling, iteration, and communication.
9. A budget realistic to accomplish the full assessment effort needed.

17.3 STRUCTURAL FEATURES OF THE ASSESSMENT

We now focus in on the performer and the features that influence the successful accomplishment of the TA /EIA.[2] In this section, the concern is with those features more or less "given" with respect to conduct of the study. This includes boundary conditions (such as study requirements and organizational attributes) and project features not subject to easy revision (characteristics of the project team).

[1]The *core team* is the small group who play central roles throughout the research, as opposed to those who contribute a specific piece of research for others to integrate into the assessment.

[2]Results in this and the following section are based on a study of the research process used in 24 NSF-sponsored TAs (Rossini et al. 1978).

17.3.1 Boundary Conditions

The *character of the topic* involved affects the conduct of the study. For instance, highly technical topics may be more difficult to integrate into a coherent assessment than topics that are technologically unsophisticated. The nature of the study also dictates much about its conduct. The focus, duration, and available resources vary from mini-TAs to EISs to macro-TAs.

Budgetary resources are a dominant influence on the conduct of a TA/EIA. Each task of the assessment will involve trade-offs between the benefits of better information and analysis, and the costs in money, time, and personnel. Whether to construct a formidable computer model of air quality in an impacted basin, or merely to rely on current estimates, must be weighed against other uses of available resources. Low budgets have seriously hampered some TA/EIAs, particularly in the crucial areas of iteration and policy analysis. Too much time and money were invested in earlier phases of the assessment, leaving these critical phases to suffer. Unfortunately, the proposer is caught in a double bind. The lower the proposed budget, the more attractive the proposal is to the sponsor, while at the same time the less likely it is to produce a satisfactory assessment.

Organizational context is another important boundary condition since impact assessments do not fit comfortably in all research institutions. Organizations aid TA/EIA work by making researchers from various disciplines available, rewarding its successful performance, and being flexible with personnel time and resources as the project unfolds. Small, relatively unstructured organizations (private firms or university institutes) tend to be most supportive of the interdisciplinary work involved. In contrast, status and rewards in academic departments traditionally go to disciplinary research, and divisional barriers and rigid time accounting reduce operating flexibility in some large-contract research organizations. This is not to say that good TA/EIA work cannot be done in such environments, but only that there may be inhibiting factors.

Organizational separations within the core team, and between the core team and subcontractors and consultants, are relatively common in TA/EIA. Limited interaction and resultant lack of mutual understanding are potential problems. Physical separation and different perspectives can cause subcontractors' work to end up poorly integrated into the assessment. Consultants, too, pose potential problems. Well-known, senior individuals are attractive consultants, but such people rarely produce major substantive contributions. Senior consultants most effectively provide personal insights and telling criticisms of the work of others.

17.3.2 Project Team Characteristics

We now consider attributes that are more easily changed than the boundary conditions, but that are still relatively fixed for a given study. These include characteristics of project leaders and research teams, their styles of seeking knowledge, and their inherent values.

An important factor to consider is the style of the leader. Hill (1970: 11) notes three commonly identified *leadership styles:*

1. nondirective, permissive, *laissez-faire* style where the leader relinquishes any influence in setting goals to the group;
2. *democratic,* a participatory, group-centered, human-relations-oriented style where the supervisor encourages a mutual relationship with subordinates;
3. autocratic, *authoritarian,* boss-centered, task-centered, close and punitive style where the supervisor allows his subordinates little or no influence in the setting up of work procedures.

Rossini et al. (1978) found that the democratic (facilitating) style is most effective. Here the leader acts as a first among equals. Open discussion is encouraged and expertise within the team explicitly acknowledged, but the leader takes responsibility for effecting closure. Stature within the organization and respect of the project team members make the leader's job easier.

The size of the *core research team* may be an important factor. Review of TA experiences indicates that a core team size of three to five professionals is desirable for a well-integrated study of around 6 person-years of effort. Such a team is large enough for effective division of labor and small enough for good communication. Team size and organization would require modification for larger or smaller studies.

The selection of study participants involves trade-offs among personnel availability, disciplinary expertise, and interpersonal skills. If needed expertise is unavailable within the organization, a subcontractor or consultant can help. The ability to work together is very important. Prior experience in interdisciplinary research and working in group research is usually to be preferred to disciplinary expertise. The team can always secure specific pieces of information. Team stability is also a consideration. Turnovers in project leaders and core team members are costly in lost time and continuity. To paraphrase Samuel Estep, of the University of Michigan Law School (private communication from this lawyer experienced in TA/EIA, 1977), a participant should be

intelligent in the culture of academic life and contract research;

blessed with common sense and a stable ego (resistant to attacks on one's ideas without interpreting them as personal attacks);

open minded with intellectual curiosity and a willingness to question one's own value assumptions;

self-disciplined with a hard-nosed result orientation;

not afraid to ask "stupid" questions, nor easily "snowed"; and

able to get along with others, and willing to accept responsibility for mistakes.

The rewards for doing TA/EIA are basically similar to those for doing any other form of research. The *motivations* of individual team members will be colored by their particular circumstances. The following factors were most often noted as major motivations for individuals participating in the 24 NSF-sponsored TAs:

interest in technology assessment,

interest in the subject matter,

the learning experience and doing professional work,

job rewards, and

new contracts and grants (project leaders only).

Ideally, the rewards available will be compatible with the desires of the participants and the demands of the assessment. For instance, publication of findings as books and scholarly articles may be an attractive consideration to the participants (particularly academics) and to the organization, as well as an effective means for dissemination.

Differences in research approach among members of an interdisciplinary team may impede effective communication. Four primary areas of *epistemological*[3] difference within the studies worthy of managerial attention surfaced:

1. *Data vs speculation.* Some researchers find it difficult to project future states, preferring to have data in hand to support any claims. This can cause serious problems, particularly for studying long-range impacts and issues of high uncertainty.

[3]Epistemological differences relate to different cognitive frameworks for structuring the world, determining what in it is problematic, and what is important.

2. *Social-impact assessment (SIA).* While agreeing on the need for SIA, serious differences of opinion arise concerning how to do it. Some physical scientists and engineers see SIA as "soft," and hence suitably handled at the level of common sense. They sometimes clash with data-oriented social scientists whose disciplines indicate rigorous approaches toward SIA.
3. *Use of assessment techniques.* The majority of performers ignore these techniques, perceiving them as recasting existing information with a false impression of precision. Others feel that techniques, such as cross-impact matrices, aid in focusing analyses and generating new insights—although the outputs of the exercises are never more than "guesstimates." A small minority of practitioners value quantitative results of any sort as adding precision and credibility to the study.
4. *Economists.* Economists stand out as difficult colleagues because of their specialized jargon, methodological preoccupations, unrealistic data requirements, and disregard for the contributions of other disciplines. However, economics is essential to TA. It is important to have economists attuned to the broader institutional and behavioral aspects of economics, as well as policy implications. Sometimes surrogate economists, such as systems analysts and management scientists, are used.

In addition to these four areas, a TA/EIA manager ought to give thought to the basic values, thinking styles, and modes of inquiry of the participants. Members of the same discipline may have significantly different viewpoints (e.g., humanistic psychologists see things differently from behavioral psychologists). Research professionals can clash dramatically with lay participants (e.g., parties at interest) due to their thinking styles, often amplifying differences of opinion on the issue under assessment. Citizen participation is thus a volatile matter (Arnstein 1975), but one that deserves continuing attention because of its importance. Value assumptions (e.g., "growth is good") of the team members ought to be explored, resolved (if possible), and made explicit for the team and for study users.

17.4 PROCESS FEATURES OF THE ASSESSMENT

There are considerations in the conduct of the TA/EIA more subject to direct managerial control than the structural features just discussed. Three such process features are emphasized in this section: (1) scheduling, (2) team communication, and (3) integration.

17.4.1 Project Scheduling

Articulating the study plan begins in the proposal stage. Whereas the sponsor, or legal requirements, largely specify the topic of assessment, the performers usually have the main say in setting the study plan. This plan details the tasks to be accomplished, the personnel to perform them, and the general flow of activities.

Bounding the study (i.e., setting the limits and form of the inquiry) affects the scope of the tasks to be performed in the TA/EIA. As discussed in Chapter 5, bounding is an ongoing process that takes place over the course of the study. Armstrong and Harman (1977) suggest about 20% of the time be devoted to information gathering before the main features of the study are set. Even then, the TA/EIA should retain flexibility to respond to surprises (e.g., a new policy initiative, foreclosure of an alternative, or discovery of an unexpected major impact area) by modifying the study tasks.

The outcome of an ongoing and flexible bounding process is a well-targeted TA/EIA. The assessors are caught between the need to produce the study on time and within budget, and the conflicting pressure to retain study flexibility and iterate. Project scheduling is the vehicle to reconcile these aims.

There are two main elements involved in project scheduling—events and activities. An event is an occurrence identified with a point in time, such as the beginning of a task. An activity is the process of progressing from one event to another. Planning the flow of project events and activities helps to utilize resources effectively, to determine whether progress is adequate, and to take corrective steps to assure a quality product on time and within budget.

Several techniques are useful in project scheduling. Among these are Gantt charts, milestone charts, and network scheduling.

The *Gantt chart* is a bar graph indicating the ordering and duration of project activities. The horizontal axis (abscissa) represents time, and the vertical axis (ordinate) exhibits the tasks to be performed (Silverman 1976). Figure 17.1 illustrates a Gantt chart for a TA/EIA. Notice that the Gantt chart does not mention outlay of resources, nor does it indicate any relationships among the activities. It is poorly suited to deal with uncertainties in the timing of activities and, most crucially, with the iteration of activities.

The *milestone chart* indicates events, rather than activities, over the course of the project. In contrast to Figure 17.1, a milestone chart might show events such as "completion of the impact analysis" and "draft

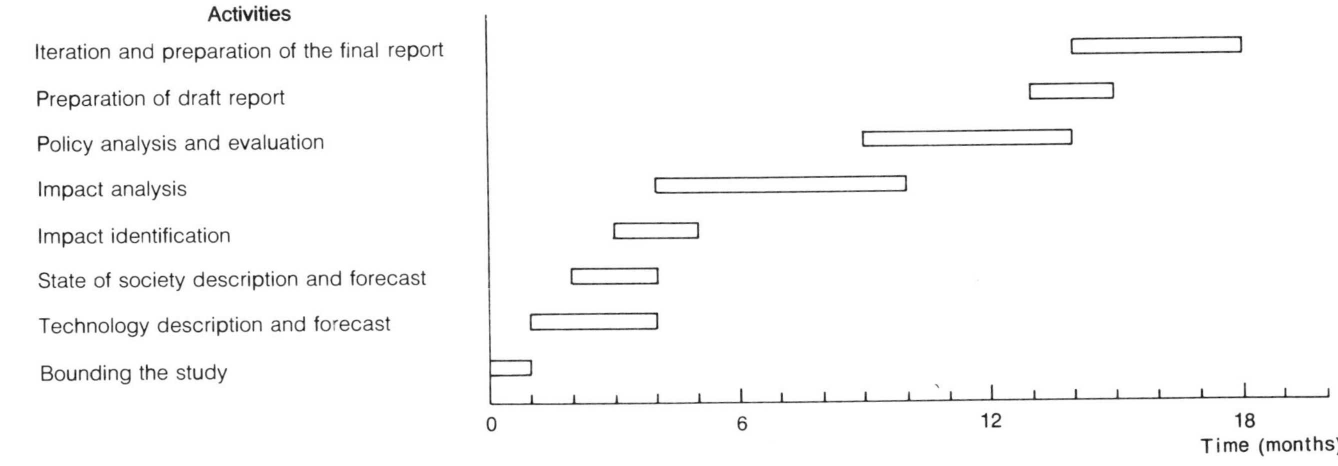

FIGURE 17.1. A simple Gantt chart.

NETWORK-SCHEDULING STEPS

1. List the activities that must be performed.
2. List the events that terminate each activity (and possibly those that begin it as well).
3. Determine the relationships among these activities and construct the network showing these dependencies (use forward and /or backward planning).
4. Estimate the expected time needed to complete each activity (possibly use the three-time estimation technique).
5. Make the PERT chart.
6. Identify the critical path (i.e., the longest temporal path through the network); note allowable slack times for events not on the critical path.

report." Milestone events provide checkpoints for the project team and monitor to evaluate progress. They are often combined with Gantt charts to indicate both activities and events. Milestones suffer limitations similar to activity charts in their linearity and failure to indicate interrelationships.

Network techniques were developed to show the interrelationships among project tasks better. The program evaluation and review technique (PERT) emphasizes timing of events, and the critical path method (CPM) arrow diagrams emphasize timing of activities. Cost and manpower considerations can also be interworked with network scheduling to provide integrated management control information (Archibald and Villoria 1967). The PERT approach has proven more useful in managing research projects, so we consider it briefly.

Figure 17.2 presents a PERT chart. The cameo entitled "Network-Scheduling Steps" indicates the steps involved in preparation of such a schedule. The time needed for each activity is often estimated by the *three-time estimation technique*:

the *optimistic* estimate—the shortest feasible time to complete an activity;

the *pessimistic* estimate—the longest time to complete an activity assuming all of the possible delays and setbacks occur; and

the *most likely* estimate—assumes that normal conditions will be encountered and that the normal amount of things will go wrong.

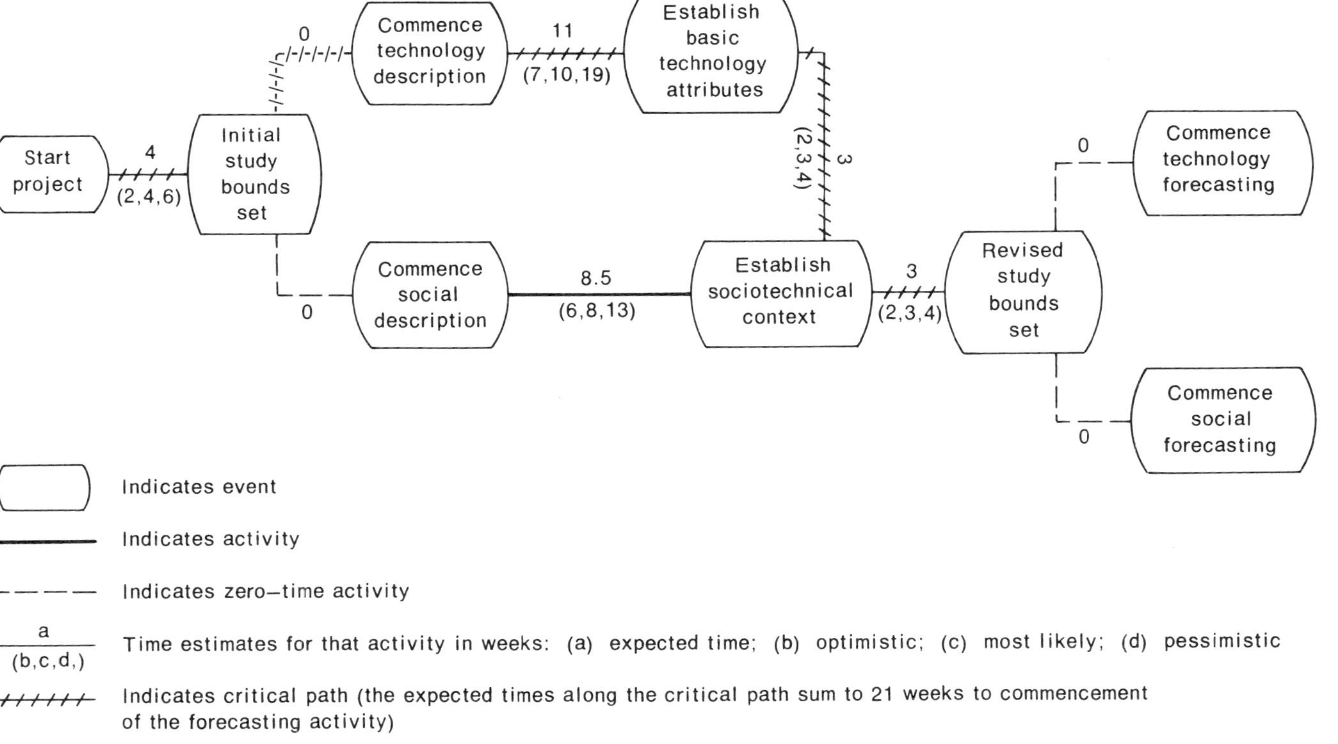

FIGURE 17.2. A partial PERT chart. Note: An alternative network scheduling format describes the activities along the arrows and only shows events as numbered entities, also not separating "commence activity" and "complete activity" into two events.

The *expected* activity duration time is (Archibald and Villoria 1967: 81):

$$\text{Expected} = \frac{(\text{Optimistic}) + 4(\text{Most likely}) + (\text{Pessimistic})}{6}$$

Making the PERT chart can proceed via forward planning ("What comes next?") and/or backward planning ("What must be completed before the object event can occur?"). A network can include parallel paths to shorten the time required to do a series of tasks (e.g., technological description and social description in Figure 17.2). Note that times in the schedule are chronological; activities can be expedited by having more people work together on a task. Constraints that require completion of one task before another can begin are indicated as zero-time activities (e.g., between initial study bounds and commencement of technological description in Figure 17.2). In some cases only a part of an activity may need to be complete for another to begin. This useful scheduling insight can be shown in a PERT diagram by separating the activity into two parts (e.g., technology description and technology forecasting in Figure 17.2). The PERT networks should not contain feedback loops. Instead, iteration of activities should be shown by repetition further down the time path (e.g., initial and revised study bounding in Figure 17.2).

The CPM calculation finds the most time-consuming path. This determines the time to complete the entire project, with other activities having some slack (see Figure 17.2).

Complex schedules may involve the integration of separate networks and require computer assistance in the computation of the critical path. Such procedures are not covered here because they are not warranted in TA/EIA. In a flexible project such as a TA/EIA, a schedule is a guide rather than a fixed procedure. Project scheduling is most helpful in building in time and resources to iterate the assessment and to respond to unplanned study developments.

One specific worthy of emphasis is the scheduling of iteration. The folk wisdom of TA advocates writing the draft final report at the very beginning of the project and then redoing it twice as the study advances. This is a potentially effective way to bound the project and to achieve eventual closure. Certainly, great changes ought to take place between the first draft and the final report, but the early draft can provide a starting point.

17.4.2 Communication Patterns Among the Project Team

TA/EIAs are group efforts. The various parts of the study, done by individual team members, relate closely to one another. Therefore, it is

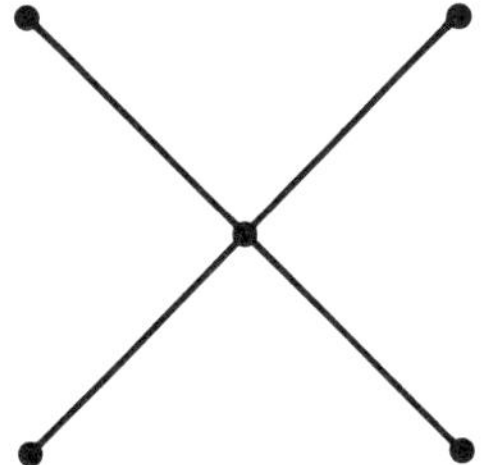
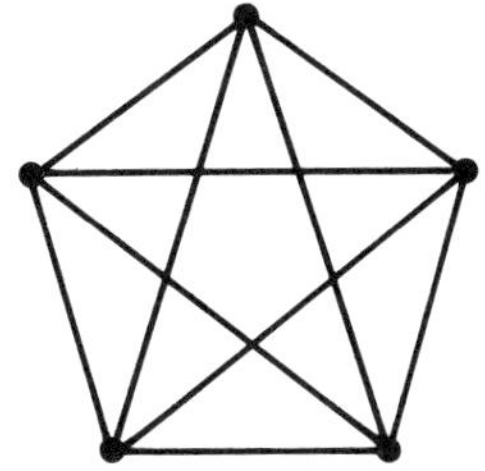
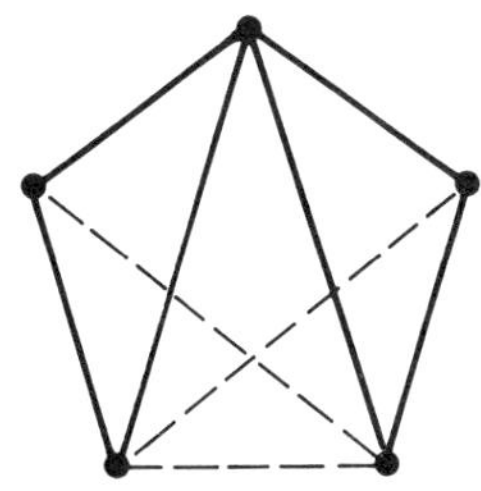

(a) Hub and spokes (b) All channel (c) An intermediate form

FIGURE 17.3. Communication patterns for a five-member group. Strength of linkage: ——— strong, – – – weak.

important that good intraproject communication exist among the participants, both to exchange information, and to bring together the various perspectives involved in the study.

Communication patterns vary from phase to phase as the character of project activities changes. Interaction patterns vary according to the nature of the assessment, the form and content of inputs (e.g., solicitation of expert opinion) and outputs (e.g., presentations by the project team), and the number and backgrounds of the team members. Actual communication patterns are never simple; however, it is instructive to consider two polar opposites—the "hub and spokes" and the "all channel" (illustrated in Figure 17.3, along with a more typical intermediate form).

Bavelas (1950) and Guetzkow and Simon (1955) have found the hub and spokes to be the most effective arrangement for communicating simple factual information and performing simple tasks. As the complexity of the task increases, the all-channel configuration becomes more appropriate. The additional complexity of the all-channel pattern over the hub-and-spokes scheme increases quickly as the number of participants increases.[4] The maintenance of these additional links requires time and effort, which detract from individual research efforts.

The establishment of a core team of three to five members for a project of around 6 person-years of effort allows the full communication essential to achieve a well-integrated, interdisciplinary analysis. On larger projects whose complexity and technical sophistication require a greater range of expertise, it may be more effective to create a hierarchical team structure to augment the core team's capabilities than

[4]For N participants, the number of links in the hub-and-spokes pattern is $(N - 1)$. In the all-channel configuration it is $\frac{1}{2}N(N - 1)$.

to have a large core team. Figure 17.4 suggests two of many possible communication arrangements. In Figure 17.4a, the project leader coordinates the acquisition of most of the auxiliary information needed by the core team; in Figure 17.4b, each core team member takes responsibility for a given topical area, using whatever communication structures are appropriate. [Alternatively, Hiltz and Turoff (1977) have proposed computer conferencing for up to 40 participants to pool their insights in a TA.]

Second, despite the relatively large amounts of time and personnel required for all-channel communication, that mode seems preferable to the hub-and-spokes pattern in the conduct of TA. Open channels of communication lead to fruitful interchange of ideas and encourage an integrated assessment. Despite its organizational neatness, the hub-and-spokes arrangement seems to cruelly overburden the leader. It does not foster interaction with colleagues working on other parts of the assessment. This interaction is essential to enrich the component analyses, gain maximum assessment insight, and yield a cohesive product.

FIGURE 17.4. Possible hierarchical communication patterns for a large TA/EIA.

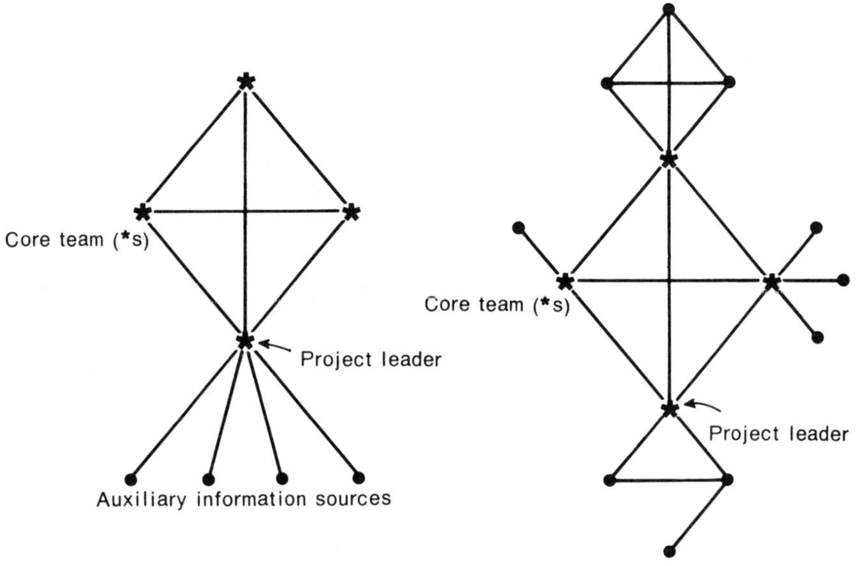

Our recommendation is that the project leader assure that all channels are available, but not insist that all be used. This might be called an "*any-channel*" pattern. Development of "esprit de corps," physical proximity of core participants, and long-standing interpersonal relationships are conducive to its establishment.

17.4.3 Integration

The full value of TA/EIA requires integration of the component analyses: technological and social forecasts; environmental, economic, and social impact analyses; and policy analysis. However, the inherent complexity of such assessments invites fragmentation. The disciplinary training of researchers does not match well with the flow of consequences of a technology that interlinks and cuts across disciplines. Understanding and expressing the interrelationships between various impacts and policies to better understand the issue as a whole is referred to as *integration*.

In Chapter 4 we argued that validity and utility are prime criteria on which to judge a TA/EIA. Both are served by integration. A valid assessment must capture the significant interactions, such as those between the economy and the environment. Separated analyses are unlikely to accomplish this; it is necessary to consider the mutual effects. A well-integrated assessment is also likely to be more credible and useful, as it presents a coherent portrayal of the consequences of a development and its alternatives.

Different degrees of integration can be specified. The most basic corresponds to well-conducted *multidisciplinary* assessment. This yields a report with properly collated component analyses and an explanation of their relationships to each other, written in a consistent vocabulary and style, and linked by means of a summary. Going beyond that, one moves to an *interdisciplinary* integration in which one finds a common view underlying the component analyses—specifically, a shared framework providing systematic linkage among the parts of the assessment. Interdisciplinary integration is required for EISs by the Council on Environmental Quality (CEQ) NEPA regulations (U.S. CEQ 1978: 55995).

Four generic approaches (Figure 17.5) to integrating a TA/EIA are identifiable. Each has merits and demerits, and, in practice, one is likely to witness combinations.

As practiced by Don Kash, Jack White, and their co-workers at the University of Olahoma [see Kash (1977)], *common group learning* (Figure 17.5a) emphasizes the project team. Specific expertise is less critical than group functioning; experts may be brought in to educate

(a) Common group learning

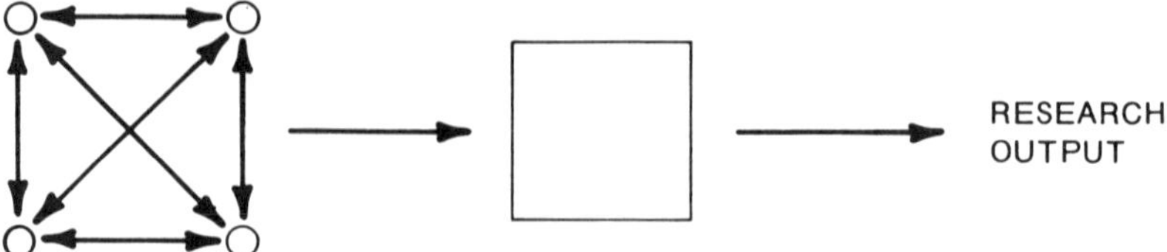

1. Intensive group interaction 2. Common group knowledge

(b) Modeling

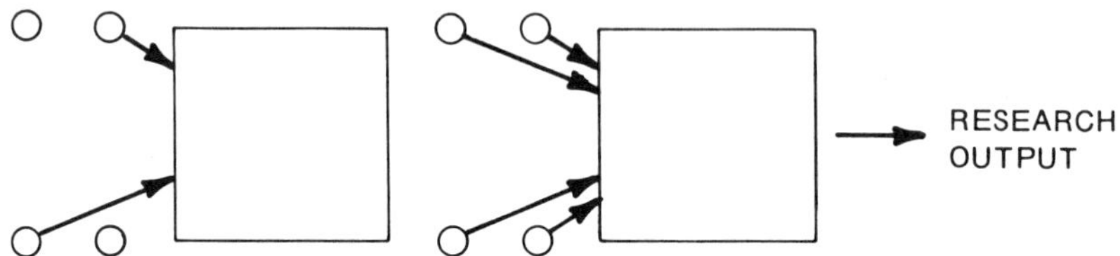

1. Model created by certain individuals 2. Individuals contribute information to the model and use it in establishing findings

(c) Negotiation among experts

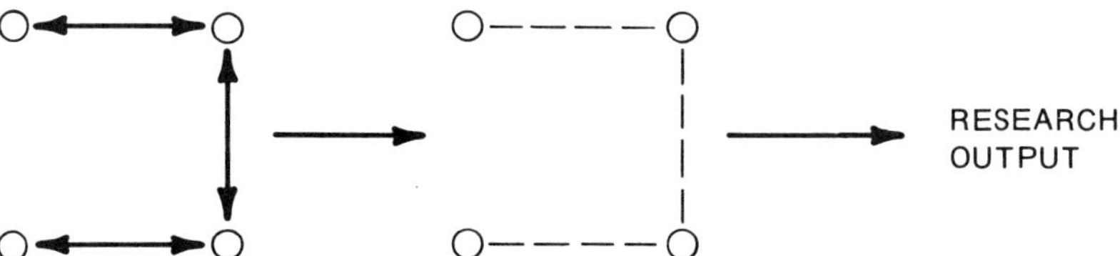

1. Pairwise interaction at boundaries between component experts 2. Better informed and interrelated analysis

(d) Integration by leader

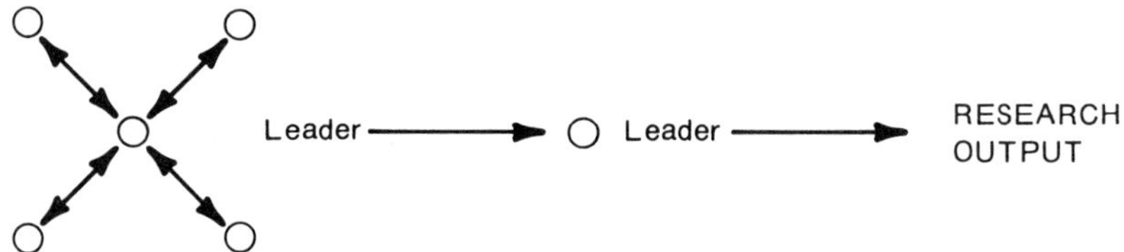

1. Pairwise interaction only between the leader and other individuals 2. Leader acquires composite knowledge and synthesizes findings

FIGURE 17.5. Four approaches to integration within TA/EIA. ○, individuals who possess particular expertise; □, repositories of knowledge other than individuals.

the group. The group learning process is extremely demanding and progresses slowly at first; the familiarization process can be described as "wallowing" in the issues. Research tasks are first assigned to team members with expertise or interest in that particular area, leading to working papers. These are critiqued by the entire group in team meetings as well as by outside reviewers and are then rewritten as many times as needed, often by someone with a different background. Closure is effected by the project leader, but the report is the product of the team. The key to this strategy is development of a common core of knowledge by the team itself.

A second approach (Figure 17.5b) relies on a *model*, often computerized, to provide a basis for integrating information. In TAs, empirical economic models were most frequently used to compare the requirements, costs, and payoffs of alternative developments [cf. Harvey and Menchen (1974) on automotive propulsion systems and Enzer (1974) on no-fault automobile insurance]. As integrative frameworks, models tend to narrow the focus to a relatively few factors of interest, often emphasizing what is most quantifiable. Component analyses feed information into the model, which then yields projections, evaluations, and/or policy option comparisons. A model should not be allowed to dominate the assessment and predetermine conclusions; rather it should be recognized as one source of useful information. That information, in turn, needs to be integrated with insights that transcend the model.

Negotiation among experts involves the division of project tasks according to individual areas of expertise. Negotiation takes place at the boundaries, where one area substantively affects another's conclusions (Figure 17.5c). For instance, in one TA a lawyer reviewed an economic analysis and pointed out its failure to consider institutional realities adequately. The economist then redid the analysis, incorporating the insights of the lawyer. In this mode of integration, disciplinary depth of knowledge is preserved; there is no question of nonexperts redoing an analysis.

Integration by the project leader (Figure 17.5d) is the fourth type identified. In its extreme form, the leader interacts individually with each team member in the hub-and-spokes communication pattern to understand and assimilate that member's contribution. This places a great burden on the leader. Consequently, assessments prepared with this strategy are likely to be more multidisciplinary than interdisciplinary. In the 24 TAs, the participants interviewed indicated dissatisfaction with the results obtained by this approach. Integration by a leader appears better suited to simpler problems than to TAs or major EIAs.

17.5 CONCLUDING OBSERVATIONS

This chapter considers the conduct of TA/EIA from a managerial perspective. It briefly addresses the role of the sponsor in selecting topics to be assessed and performers to do the work.

Shifting to project management per se, factors of concern are divided into structural features (relatively set) and process features (relatively controllable). Structurally, the nature of the assessment and the budget affect its performance, as does the extent to which the performing organization supports such interdisciplinary research. Subcontractors and consultants working on a project require special management. As to the process itself, we recommend that the project leader use a democratic/facilitating style. Ideally, the core project team should be motivated by interest in assessment and the topic under assessment, and should be suited for policy-related group research.

Four potential epistemological problem areas deserve managerial attention:

- the willingness of professionals to speculate on the future rather than remaining tied to data in hand;
- whether to perform social impact assessment following the procedures of social scientists or using common-sense brainstorming;
- whether to use formal assessment techniques; and
- how to relate the economic analysis to the rest of the assessment.

Participants may also differ sharply on value-laden issues, and if they don't, this is a sign of narrow constitution of the project team and a need for broader inputs.

Three techniques—Gantt charts, milestone charts, and network scheduling (PERT)—are described as aids in project scheduling. We recommend their flexible use to build in time and resources for effective iteration and response to unplanned issues.

In full-scale assessments, a core team of three to five members, assisted by auxiliary researchers and consultants as necessary, can achieve satisfactory communication and integration. "Any-channel" communication patterns are preferred.

We recommend integration of the component analyses to yield a more valid and usable study. Four integration strategies offer complementary strengths, to be used in combination in most TA/EIAs:

- Common-group learning effectively builds shared group knowledge and deemphasizes expertise. It is particularly effective for policy considerations.

Modeling tends to be more effective for technical and economic matters and less so for policy and qualitative considerations.

Very sophisticated analyses, typically of impacts, demand undiluted expertise on the team. They are suitable for negotiation among experts.

Integration by leader is most useful for small-scale assessments.

RECOMMENDED READINGS

Arnstein, S. R., and Christakis, A. (1975) *Perspectives on Technology Assessment.* Jerusalem: Science and Technology Publishers.

Cumulates insights of theorists and practitioners on the practice of TA.

Rossini, F. A., Porter, A. L., Kelly, P., and Chubin, D. E. (1978) *Frameworks and Factors Affecting Integration within Technology Assessments,* Report to the National Science Foundation; Atlanta: Georgia Institute of Technology.

An in-depth exploration of the process of performing TAs.

Winder, J. S., Jr. and Allen, R. H. (1975) *The Environmental Impact Assessment Project: A Critical Appraisal.* Washington, D.C.: The Institute of Ecology. [Current address, The Institute of Ecology, Butler University, Indianapolis, IN].

Consideration of EIS preparation, primarily from the perspective of how to improve sponsor guidelines.

EXERCISES

17-1. Discuss the human value concerns likely to influence the preparation of an EIS on the siting of a new nuclear power plant. Briefly, how would you manage the EIA to cope with these?

17-2. Divide into two groups to consider the epistemological problem areas:
Group A is to prepare arguments in favor of: reasonable speculation on the future, common-sense social-impact assessment (SIA), use of a noneconomist (e.g., someone with an M.B.A. degree) to perform the economic analyses, and reliance on formal assessment techniques (e.g., those listed in Table 5.3, p. 82).
Group B is to counter with arguments for sticking close to defensible data bases, reliance on social scientists to conduct SIAs and economists to conduct economic analyses, and use of traditional disciplinary analytical techniques instead of formal assessment techniques.

17-3. Suppose you are a vice-president of a large-contract research organization. You are requested to step in to help salvage a TA that has used up its budget and only has a collection of draft working papers to show for it (the leader has been fired). The sponsor is furious, and your firm's reputation is on the line. What would you do? (This is based on a composite of actual experiences.)

17-4. Select a TA/EIA topic. Compare the likely merits of an academic and a large-contract research organization as an assessor. (Consider the nine criteria for selecting a performer.)

17-5. Select a topic for a TA/EIA. Prepare a PERT chart to show how you would schedule the 10 basic steps of the TA/EIA (Chapter 4).

17-6. (Project-suitable)[5]

1. Select an assessment topic and assume a certain budget and time period available to conduct an assessment. Preferably working in groups of about five members, sketch out the *design* of the assessment:
   a. reach a workable definition of the *subject;*
   b. identify the potential *users* and their prime concerns;
   c. indicate what *impact areas* will be emphasized;
   d. suggest which *analytical techniques* to use;
   e. consider what *data* are needed and how to obtain them; and
   f. propose a strategy for *policy analysis.*

2. As a group, reflect on how one might *manage* the assessment study you have sketched out. Prepare a *research plan* covering the following points:
   a. determine *core team personnel* needed to perform that assessment;
   b. list in roughly sequential order the *activities,* to be performed and *milestone events* in the TA/EIA (optionally, prepare a PERT chart indicating how these tasks fit together);
   c. indicate how you would *divide* the tasks among the team members;
   d. estimate the person-months of work/activity and in total; and
   e. discuss your *management strategy* to accomplish all this, indicating leadership style, communication arrangements, work settings, and approaches to integration.

3. If more than one group has prepared a design and research plan, compare these. Discuss whether any assessment components have been left out, which plans are likely to yield the most valid findings, and how the research plans might be improved.

17-7.[6] 1. Use an actual "group experience," such as that gained in Exercise 17-6, to focus on *group behavior.* Then *individually:*
   a. Write down each person's first and last names and what you know of their background.

[5]Exercises 17-6 and 17-7 were originally developed by Alan Porter for the Industrial Management Center short course on Technology Assessment, Hilton Head, South Carolina, December, 1975.

[6]See footnote 5.

b. Select a key term used in the group's discussions and, individually, write down its definition.
c. Describe how the group selected a leader and how that person led the group.
d. Describe the roles performed by each of the other group participants (e.g., did someone act as secretary, "devil's advocate," "expert," "clock-watcher," etc.).

2. Now focus on the group's *communication patterns*. Then *individually:*
   a. Indicate the major information flows in a sketch like Figure 17.3c.
   b. Note what, if any, subgroups are formed.
   c. Describe who dominated the discussion and why.
   d. Describe who said relatively little, and speculate why.
   e. Suggest what, if anything, should be done to foster more equal participation.
   f. Judge to what extent the group processes affected the outcome of the group's discussion (i.e., the TA/EIA design and research plan produced).

3. Discuss your individual observations as a group.

4. Extrapolate from the actual group process just studied to consider concerns about *group behavior in a TA/EIA* (e.g., for an assessment in which your group will participate or for the TA/EIA planned in Exercise 17-6):
   a. Discuss the likelihood of eliciting cooperation from participants as a function of leadership style and rewards for participation.
   b. Set forth criteria for assigning people to tasks (experience, technical skills, personalities, etc.).
   c. What tasks should be individual, subgroup, or whole group?
   d. When should whole group meetings be held, and how should they be conducted?
   e. How would you structure any outside advice to be most effective (advisory group meetings, individual consultations, review of documents)?

5. If more than one group has performed the exercise, contrast their strategies. Discuss how well the core team and others would interact, how integration of component analyses might proceed, and how the final report would be compiled.

18

Evaluation of Technology Assessments and Environmental Impact Statements

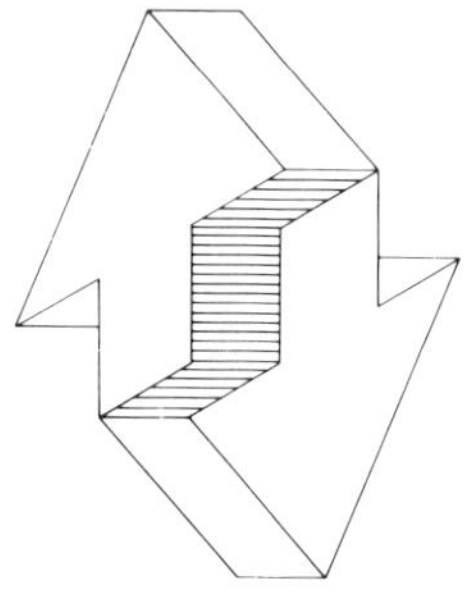

This chapter takes a hard look at the TA/EIA process. It addresses the question of what makes a good assessment and provides explicit criteria against which to judge technology assessments and environmental impact statements. It directs attention to the major concerns in the evaluation of TA/EISs, namely, what should be considered, by whom, and when. This guidance is directed, first of all, to prospective evaluators of TA/EISs. The points raised should also be of concern to performers and users of impact assessments.

18.1 INTRODUCTION

Chapters 4−17 presented guidance on how to perform TA/EIA. In contrast, Chapters 18−20 take a critical look at TA/EIA. The present chapter concerns the evaluation of specific TA/EISs, whereas Chapter 19 critiques TA/EIA more generally. Finally, Chapter 20 anticipates the future for TA/EIA.

Why should one wish to evaluate an impact statement? Certainly, those preparing the statements do not typically encourage their evaluation by others; people usually prefer not to be critically scrutinized. Evaluation consumes resources that could be spent for more assessment activities. Furthermore, evaluations are difficult to perform. However, there are two potential payoffs that make evaluation desirable: (1) the "assessment of assessments," to help potential users judge the worth of particular TA/EIAs and (2) the improvement of assessments in general, through feedback as to strengths and weaknesses. Such evaluation information can benefit sponsors, users, and performers of impact assessments. As Figure 18.1 indicates, review of TA/EISs can encourage a higher quality of effort.

On a more serious note, Schindler (1976), writing as an environmental scientist, has driven home the dangers of uncritical acceptance of impact assessments. The lack of hard scrutiny by scientific peers contributes to a literature of unassessed reports. The absence of viable feedback permits the use of outdated techniques and results in weak studies. Without ongoing evaluation efforts, TA/EIAs may lack both the validity and credibility to contribute to policy formulation.

Evaluation has developed into a field of its own. In contrast to TA/EIA, which anticipates future effects, "evaluation research" weighs the

FIGURE 18.1 The necessity for evaluation.

CAPITOL GAMES

By JAMES STEVENSON

degree to which current programs, often social, accomplish their objectives. This has proven a very difficult and costly task, since social actions are inherently complex. To deal with this complexity, social scientists have evolved methodology to balance evaluation design validity against the pragmatic constraints of dealing with operating programs.[1]

In discussing the evaluation of TA /EIAs, we first consider the objectives of TA /EIA, followed by discussion of the means to evaluate how well particular TA /EISs satisfy those objectives. We attempt to apply the techniques of evaluation research to study the effectiveness of impact assessment by determining parallels between social programs and TA /EIA. Where these parallels do not exist (as in the case of "validity"), we develop other criteria consistent with TA /EIA objectives.

18.2 CRITERIA AND PURPOSES FOR THE EVALUATION OF TA/EIAS

As discussed in Chapter 4, there are two major objectives for TA /EIA:

Validity—is the TA /EIA well grounded on the principles of evidence, is it able to withstand criticism, and does it correspond to real happenings over time?

Utility—is the TA /EIA useful in setting policies?

A third objective may be present in some cases—the *advancement of TA/EIA methodology.*

To borrow from Weiss' (1972) treatment of the evaluation of social programs, fruitful evaluation of the utility of a TA /EIA requires one to

1. learn the assessment's objectives;
2. describe the essential features of the assessment activity; and
3. investigate the assessment's outcomes, including how the various assessment components contributed to the study's final outcomes.

On the other hand, evaluation of validity (and methodology), where the analogy with social program evaluation is weak, requires careful scrutiny of the TA /EIAs for internal consistency, agreement with reality, and, possibly, improvement over past studies.

It may be helpful to suggest basic questions that could be raised concerning a given TA /EIA evaluation. These should vary considerably

[1]Scriven (1967) conceptualizes key evaluation issues. Weiss (1972) overviews the evaluation process clearly and concisely. Riecken and Boruch (1974) advance realistic considerations in evaluation efforts. Cook and Campbell (1979) lay out the methodological design possibilities for evaluation.

TYPICAL EVALUATION QUESTIONS

1. Pertaining to internal assessment validity:
   a. Are the appropriate available data used?
   b. Are analyses conducted according to accepted principles, and to sufficient depth?
   c. Is the study comprehensive in its breadth of coverage so as to incorporate most all relevant considerations? Does it integrate these considerations to draw reasonable conclusions?
   d. Are inferences sound and supported?
   e. Do projections conform with actual developments? Over what time frame?

2. Pertaining to utility of the TA/EIA:
   a. Is the study pertinent to issues at hand?
   b. Were the potential study users identified?
   c. Is it readable and understandable to those interested in the subject in question?
   d. Is it credible to various potential users?
   e. Does it present a balanced point of view, or does it favor a particular viewpoint? In what ways does this affect conclusions drawn?
   f. Do the parties at interest use the study?
   g. Is the study influential on user attitudes and behaviors? Does it affect decision making and, if so, in what regard?
   h. Over what time period does the study retain its usefulness? Is it quickly outdated?
   i. Is the study worth its cost in resources and in time?

3. Pertaining to methodological improvement:
   a. Does the study document its techniques sufficiently to be of methodological use?
   b. Are there novel elements to the study? Do these have implications for other applications or extensions?
   c. Does it introduce new techniques?

from case to case. The specific concerns of any given evaluation are a complex function of the reasons for doing the evaluation, the content-specific nature of the TA/EIA being evaluated, and the resources available to the evaluators. Evaluative issues include those noted as "Typical Evaluation Questions" (see cameo).

Evaluation of TA/EIAs addresses the questions posed in the cameo concerning validity, utility, and methodological improvement. The nature of a given evaluation will depend on which of these is of primary concern.

An advisory committee commonly evaluates the *validity* of a study in progress. Internal study attributes, such as scientific propriety and thoroughness of coverage, are matters of concern. This review depends on the judgments of a few selected experts. In contrast, it may be possible to objectively weigh the correctness of TA/EIA conclusions, after the fact. Predictions can be compared with real occurrences, although this must be done with care.

Other evaluations address the extent of *utilization* of TA/EIAs. Utility must be judged on the basis of the *attitudes* of potential users. Rarely is it possible to demonstrate that a TA/EIA caused a behavioral response (a policy action) on its own. There has been concern with the effectiveness of TA/EIAs, and, consequently, some attempts at evaluation. For instance, in a study funded by the National Science Foundation (NSF), the Research Triangle Institute (Burger 1975) found that the TA of offshore oil operations (Kash et al. 1973) was highly useful in a variety of decision-making contexts.

Some TAs have actually been designed for the primary purpose of furthering methodological development. The MITRE Corporation (Jones 1971) studies involved the performance of five mini-TAs, followed by a methodological overview based on the lessons learned from doing them.

Evaluations also differ in their relative emphasis on *substance* or *procedure*. The distinction is strongest for EISs. The Institute of Ecology's (TIE) reviews of EISs were substantive in character, critiquing the validity of conclusions reached (Fletcher and Baldwin 1973). Some evaluations reflect a highly specific interest. A utility company's review of a draft EIS considers its stake in the subject. Conversely, an environmental group's review of the same EIS is likely to focus on different factors due to its different concerns. Most challenges to EISs have been based on inadequate attention to required issues. Court rulings of "inadequacy" have usually been on these procedural grounds, rather than on the validity of the EIS findings.

It is a further step to ask whether a development has proceeded, despite an EIS's identification of negative impacts. This is essentially an issue of utility, rather than of validity. [It also goes beyond the National Environmental Policy Act's (NEPA's) requirements.]

18.3 DESIGN OF EVALUATIONS

In developing an evaluation, one wants to determine the effectiveness of a TA/EIA at reasonable cost. Trade-offs among convenience, feasibility, and strength of inference dominate the design of evaluations. The strongest inferences (i.e., the most clear-cut answers to the questions posed) arise where the evaluator can control all other factors that might affect the outcome. For instance, if one wanted to know the usefulness of a particular TA/EIS, it would be best to begin with two comparable groups of parties at interest, equivalent in all significant respects (probably accomplished through random assignment to a group). Only one group would be provided with the study; all contaminating communication with the other group would be prevented, and the resultant actions and opinions of the two groups would be compared. This is obviously an impossible ideal. However, there are realistic means to rule out many plausible alternative explanations for observed differences so that the evaluator can estimate the effects of the study in question.

Table 18.1 lists four important dimensions for the evaluation of TA/EIAs. The evaluator faces choices on each to develop a viable evaluation. Combinations can vary greatly. For example, the overall TIE evaluation of the EIA process weighed the immediate merits of 10 completed (draft or final) EISs. This evaluation was (1) summative, (2) post hoc, (3) of multiple studies, and (4) immediate. In contrast, the advisory committee for a TA, such as that on alternative automotive power plants (Harvey and Menchen 1974), evaluated progress during the course of the study to assure balance and credibility. This evaluation was (1) formative, (2) a priori, (3) of a single study, and (4) immediate. Each of these four dimensions is now discussed.

The distinction between formative and summative evaluation (Scriven 1967) is basic. *Formative* evaluation provides feedback to the study while it is in progress. It addresses validity in terms of currently available knowledge and method, as well as internal consistency and completeness. It also considers the potential utility of the study by identifying the prospective users and matching anticipated study outputs to their needs. Oversight committees are the logical performers of formative evaluations, but their limited commitments mean that such evaluations are not elaborate. An alternative evaluator is the project monitor,

TABLE 18.1. Evaluation Dimensions

Formative	vs	Summative
A priori design	vs	Post hoc design
Single TA/EIA evaluated	vs	Multiple TA/EIAs evaluated
Immediate evaluation	vs	Extended-term evaluation

where one exists, but here, too, resources are limited and emotional ties may be strong. The project team itself may conduct evaluations for its own use (a "devil's advocate" can sharpen investigation).

Summative evaluations weigh the merits of a study after its completion. They can be used to judge the effectiveness of assessors, particular types of studies, or particular methods. In an ongoing program, results of evaluations of earlier studies can guide the development of later studies.

The second dimension refers to the time at which the evaluation is designed. *A priori* evaluations are developed as an integral part of the TA/EIA; *post hoc* ones are constructed after the TA/EIA has been designed (usually completed). A priori designs facilitate documentation of methods, procedures, and evidence of use. Furthermore, specific methodologies can be built into the TA/EIA for evaluation of its effectiveness. Many more options for strong comparisons exist when one designs the evaluation beforehand. Post hoc evaluation, in contrast, is easily manageable; indeed, it reflects normal review processes. Such hindsight is more convenient and cheaper than a priori design, but it is likely to suffer from a lack of study records prepared with evaluation in mind, and it is likely to yield weaker inferences.

Single TA/EIAs can be evaluated or comparisons can be drawn among multiple TA/EIAs. *Single-study* evaluations are most appropriate to help interested parties judge the validity of a particular study. As a rule, final EISs follow draft EISs, enabling comparisons of substantive changes that take place. *Multiple-study* evaluations offer a wealth of possibilities for advancing the state of the art of impact assessment. Comparing alternative study strategies, methods, performers, utilization mechanisms, and so on can reveal what works and when. Many variations are possible. For instance, the NSF supported two simultaneous TAs of the same topic, earthquake prediction, in order to compare a mini-TA (Jones and Jones 1975) with a full-scale one (Weisbecker et al. 1977). The Office of Technology Assessment (OTA) has supported a set of five studies on various aspects of potential shortages in natural materials.

The time of evaluation is also variable. *Immediate* evaluations provide information to potential users on the validity of a study in terms of scientific propriety, absence of bias, and so on. This can aid users in interpreting the TA/EIA's findings. *Extended term* evaluations can take place long after completion of a study to assess the realism of its projections or its usefulness over time. Measurements of user opinions or attitudes toward the TA/EIA can also take place at different points in time. Prior measurements can establish baseline attitudes toward the assessment subject. Measurements while the TA/EIA is underway can

TWO EVALUATION DESIGNS[2]

1. (An a priori, summative evaluation of multiple TA/EIAs, addressed to *validity*)

   A set of TA/EIAs of the same topic is designed with comparison in mind. The TA/EIA work plans are reviewed beforehand by a set of professionals to judge their internal validity and feasibility. Review is repeated immediately after the studies are completed. It is followed by a series of reviews of how well the studies' findings match with real developments over a series of, say, 15 years.

   Such an evaluation would provide insight into (a) the ability of professionals to judge what makes for a valid study, (b) the relative effectiveness of the studies being compared, and (c) the degree to which the studies were able to anticipate actual developments and key policy issues. The set of studies could be designed to compare: (a) different TA/EIA formulations (technology-oriented vs policy-oriented), (b) different types of assessors (in-house vs out-of-house), (c) different study strategies (highly quantitative vs very qualitative, or tightly integrated vs multidisciplinary), and/or (d) different levels of effort (macro- vs micro-TA/EIAs). The design could be amplified to compare assessment programs, that is, sequences of assessments (see Section 4.4).

2. (An a priori, summative evaluation of two TA/EIAs addressed to *utility*)

   The responses of potential users to two TA/EIAs of the same topic are compared. The users are randomly assigned to three groups. All users are surveyed as to their opinions on the subject under assessment before seeing either of the assessments. One group is provided with one of the TA/EIAs; a second group, with the other one; and the third, with none. All users are again surveyed as to their opinions on the subject. All groups are then provided with the TA/EIAs they had not yet seen. Again, their opinions are surveyed.

   This evaluation design compares the extent to which each of the TA/EIAs affects potential users. It could be augmented by asking the users what it was about the studies they found persuasive. The design could be modified to compare different modes of presentation (e.g., a series of briefings vs reading a final report). The studies being compared could be varied along any number of characteristics [see (1) above].

[2]Porter and Rossini (1977) enumerate a range of alternative evaluation designs for TAs and technology forecasts.

help include current policy concerns in the study. Shortly after completion, users can be surveyed to ascertain the effects, if any, of the TA/EIA's findings. Long after, observations can attest to the lasting value of the study's findings. Explicit predictions can be verified. Scenarios can be compared with actual occurrences in terms of their plausibility, and the degree to which they spanned actual events and trends. Finally, professionals can judge the contribution of particular TA/EIAs to methodological advances.

It should be clear that an evaluation design involves a combination of the four dimensions, plus an understanding of the intended evaluation purposes. To provide a sense of the possibilities for imaginative and powerful evaluations, the cameo entitled "Two Evaluation Designs" is offered. The range of possible evaluation strategies is limited by available resources and by the human factor (Section 18.4).

18.4 PRACTICAL CONSIDERATIONS: THE HUMAN ELEMENT IN EVALUATION

Evaluation carries with it the implication of possible censure. Evaluations thus often lack the cooperation of involved parties and can even be subject to sabotage. It is thus important to clearly justify the evaluation (what can it contribute to which decisions?) and make it palatable. For instance, an agency sponsoring a TA might be quite interested in learning if that study is valid—provided the information is kept confidential. A negative evaluation could reflect on both the sponsor's managerial abilities and the assessor's performance (Connolly 1976). In the situation where the evaluation cannot be kept confidential, or is being performed for a party with power over the agency (e.g., Congressional oversight), the TA/EIA's producers are likely to have mixed emotions. If they anticipate that the evaluation will be favorable, they will support it; if not, they will try to undermine it.

Such human considerations raise the issue of whether an evaluation should be conducted "in-house" or "out-of-house." *In-house* evaluations tend to be less costly, faster, and more likely to elicit cooperation. *Out-of-house* (contracted) evaluations can draw on professional evaluators, may be less biased, and are likely to be more credible. The choice depends on the intended uses of the evaluation. For instance, an in-house formative evaluation by trusted colleagues may be very effective feedback in the course of a TA/EIA. However, an agency's summative evaluation of an EIS it has sponsored may not carry much weight with outsiders.

18.5 CURRENT EVALUATION PRACTICE

Various formal and informal evaluation efforts for TA have been noted in this chapter. Most common has been the use of oversight advisory committees for formative evaluation. Some individual TAs have been critiqued in detail [e.g., Oettinger and Shapiro's (1974) review of Dickson's *The Video Telephone* (1974)].

Most attempts to assess multiple TAs have been descriptive studies. V. Coates (1972; to be updated shortly), Jones (1973), Jones and Jones (1977), and Breslow et al. (1972) cataloged TA-related activities and compared certain properties of these studies. A recent series of studies sponsored by NSF carried evaluation further. Armstrong and Harman (1977) recommended an overall "study strategy" based on study of a number of NSF-sponsored TAs. Rossini et al. (1978) studied the interdisciplinary process involved in preparing a TA, and Berg et al. (1978) evaluated the utilization of TAs as viewed both from the producers' and users' perspectives. The common objective of these evaluations has been to assess the adequacy of certain aspects of TAs to date, and to suggest ways to improve the assessment process.

Considerable attention has been directed to the review and comment procedures in the EIS process. Statutes provide the opportunity for critique of draft EISs. Substantive comment is provided by concerned government agencies and private parties. In addition, procedural evaluation has been effectively employed to assure study depth and breadth. Witness the court intervention in the Trans-Alaska Pipeline EIS, which determined the Department of the Interior's first EIS to be inadequate. As of June 30, 1975, 332 court cases related to NEPA's EIS provisions had been completed with another 322 pending (U.S. CEQ 1976a: 123–124). A majority of the cases concerned allegations that an EIS was required. Twenty-three injunctions were issued for inadequate EISs; another 32 allegations of inadequate EIS were dismissed.

In an evaluation program aimed at both critique of individual EISs and improvement of agency EIA processes, the Environmental Protection Agency (EPA) has systematically commented on federal EISs. The EPA provides summary feedback in terms of both the perceived environmental impact and the adequacy of the impact statement. The comments serve as input to individual policy decisions. In addition they provide formative feedback to the agencies' EIA process.

Another EIS evaluation effort has focused on detailed analyses of selected EISs. TIE put together teams of volunteers to review 10 EISs for validity. For instance, the Garrison Diversion Unit EIS (Pearson et al. 1975) of the Bureau of Reclamation (summarized in Chapter 5 Appendix) was reviewed. The TIE evaluation found serious deficiencies in the EIS in overstating economic justifications, incompleteness of analysis of water-quality impacts, and inadequate discussion of social effects, among others. They called for a moratorium on the project pending preparation of a new EIS on the total Garrison project. Based on such evaluations of individual EISs, TIE has prepared suggestions for revised CEQ guidelines (Andrews et al. 1977). Thus, these evaluations have implications both for policy in particular cases and general EIS procedure.

18.6 CONCLUDING OBSERVATIONS

Formal evaluation of every TA/EIS is not justified. Some form of feedback (formative evaluation) can be helpful, but it need not be elaborate. Any policy-relevant TA/EIS will generate informal, if not formal, evaluation of its findings (summative evaluation). A mechanism to elicit greater peer review of TA/EIA studies would be desirable to assure scientific quality standards.

Rather than a mediocre effort to evaluate everything, we suggest carefully focused TA/EIA evaluation programs that address only selected TA/EIAs, possibly matched for comparative purposes. These programs should have definite objectives, and should be of sufficient duration to gather and take advantage of time-series information. Such programs definitely imply predesigned evaluation components for the selected assessments. They should be oriented to improving specific elements in producing valid and useful impact assessments, *not* to evaluating individuals and placing blame for shortcomings in particular TA/EIAs. These evaluations should involve maximum control of confounding factors even though this requires commitments extending over multiple TA/EIAs and considerable resources. Such evaluations may go far in understanding what factors make differences in the quality of TA/EIAs.

Some additional key points of this chapter are as follows:

Summative evaluations can help interested parties judge the findings of individual TA/EIAs.

Evaluation while a study is in progress (formative evaluation) can help improve it.

Evaluation research is a field in its own right with a considerable body of analytical expertise.

A series of evaluation questions dealing with TA/EIA validity, utility, and methodological improvement is introduced.

Four dimensions that differentiate evaluations are discussed: formative vs summative, a priori vs post hoc, single-study vs multiple studies, and immediate vs long-term.

Two sample designs illustrate the potential uses of evaluations of multiple TA/EIAs in determining validity and utility.

Human resistance can ruin evaluation efforts unless evaluations are carefully thought out.

Some useful evaluations of TAs and EIAs have already been performed.

RECOMMENDED READINGS

Cook, T. D., and Campbell, D. T. (1979) *The Design and Conduct of Quasi-Experiments and True Experiments in Field Settings*. New York: Rand-McNally.

An excellent discourse on studying the effects of real treatments, basic to any attempt at evaluation.

Porter, A. L., and Rossini, F. A. (1977) "Evaluation Designs for Technology Assessments and Forecasts," *Technological Forecasting and Social Change* 10, 369–380.

Discusses a number of specific designs for evaluating assessments.

Weiss, C. H. (1972) *Evaluation Research*. Englewood Cliffs, NJ: Prentice-Hall.

The basic primer on the practice of evaluation.

Winder, J. S., Jr., and Allen, R. H. (1975). *The Environmental Impact Assessment Project: A Critical Appraisal*. Washington, DC: The Institute of Ecology (Now at Butler University, Indianapolis, IN).

A report on an outstanding effort to evaluate EISs. [See also Andrews et al. (1977) *Substantive Guidance for Environmental Impact Assessment*. (Washington, DC: The Institute of Ecology)] (Now at Butler University, Indianapolis, IN).

EXERCISES

18-1. What is evaluation research?

18-2. What is the difference between formative and summative evaluation?

18-3. What are key differences between performing an evaluation of a TA/EIA aimed at determining its validity and an evaluation aimed at utility concerns?

18-4. Table 18.1 presents four dimensions on which evaluations can be importantly differentiated. Indicate three likely combinations and explain why these are suitable (e.g., formative evaluations will tend to be designed a priori, . . .).

18-5. Read an evaluation research literature selection, such as Carol Weiss' (1972) book, the book by Riecken and Boruch (1974), or review issues of *Evaluation Quarterly* or *Evaluation Magazine*. Discuss four operationally important similarities and four important differences between evaluation research and TA/EIA. (Alternatively, discuss similarities and differences between evaluation of social programs and evaluation of TA/EIAs.)

18-6. (Project-suitable) Advance a specific evaluation plan to determine what degree of quantitative analysis is most suitable in preparing EISs for the Corps of Engineers. (Alternatively, pose the same issue for another federal or state agency, or pose a similar issue for the preparation of TAs for the Congress.)

18-7. (Project-suitable) Select a particular final EIS or TA. Attempt to answer the "Typical Evaluation Questions" posed in Section 18.2. You will not be able to answer many of these with high certainty, but provide a good guess. Some will not be answerable at all. (Alternatively, address these questions to a TA/EIA project on which you are working.)

19

Critiques of TA/EIA

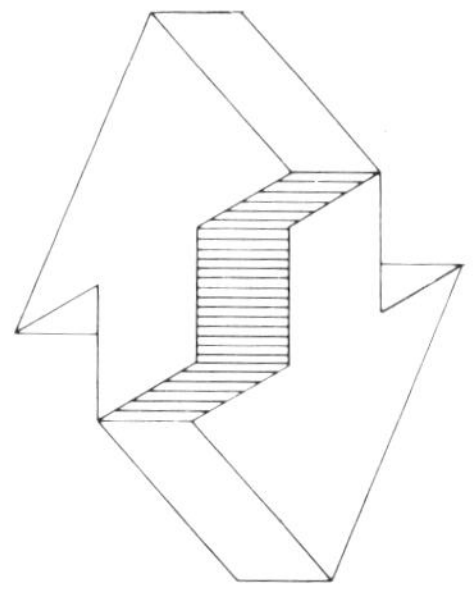

This chapter addresses generic criticisms of TA/EIA. It may be considered as an assessment of the assessment process itself. The chapter is divided into three sections. The first raises substantive criticisms of the entire TA/EIA approach and its implementation. The second poses questions about the execution of impact assessment—methodological criticisms. The third critiques the key institutional mechanisms for TA/EIA in the United States. Overall, this chapter raises concerns about TA/EIA that merit reflection by its practitioners and users.

19.1 INTRODUCTION

We can reflect on the course of development of TA/EIA with relative ease because of its brief duration. In Chapter 3 we outlined a number of critical events in the evolution of TA/EIA. In the present chapter we present a brief synopsis and analysis of critical reservations that have accompanied the development of the TA/EIA enterprise.

Some reservations about TA/EIA have been voiced at a fundamental conceptual level. Challenges have been raised to the basic philosophy underlying TA/EIA from both those concerned *for* the continued development of technology and those concerned *about* that continued development. At the level of actual performance of impact assessments, other objections have been raised. These run the gamut from worries about the intrusion of values into assessment to the techniques available to draw inferences about the future. In addition, there are concerns over the institutional arrangements, focusing on the Office of Technology Assessment (OTA) and the EIS requirements of the National Environmental Policy Act of 1969 (NEPA).

By and large, the present book is a positive statement in favor of TA/EIA. It attempts to indicate viable ways to perform impact assessment so as to produce useful results. This chapter can be viewed as somewhat of a counterbalance to the rest of the book. It considers the negative aspects, in order to sensitize the TA/EIA community to real concerns.

Despite our advocacy of TA/EIA, we have attempted to be sensitive to criticisms that have been leveled at it. Thus limitations, as well as strengths, of the approaches and techniques considered have been treated in conjunction with each other throughout the book. We hope that the reader finds that the treatment of TA/EIA meets most of the operational objections raised here. To the extent that this is so, many of the criticisms developed in this chapter will have been muted. Similarly, in the case of the institutional concerns, we believe that efforts to improve OTA and NEPA are progressing. These take into account many of the points to be discussed.

An overriding rationale for inclusion of this chapter is to encourage the reader to seek improved approaches. Whether within the rubric of TA/EIA or outside it, better ways to anticipate the future are vital to the effective social management of our technological society.

19.2 SUBSTANTIVE CRITIQUES OF TA/EIA

We first turn to the substantive criticisms of TA/EIA. These question the legitimacy of the approach itself. Impact assessment depends implicitly on the assessors' underlying notions of technological and social change (Chapter 2). The TA/EIA never takes place in a sociopolitical vacuum, but both shapes and is itself shaped by the values and visions of society. This section concerns itself with two general issues that bear upon the performance of TA/EIA in its sociotechnical context: (1) general attitudes toward technology and (2) the relation of TA/EIA to the political process.

19.2.1 Attitudes Toward Technology

Critiques of TA/EIA with respect to attitudes toward technology come from both those concerned for the continued development of technology and those concerned about such continued development.

The former have characterized TA/EIA by such phrases as "technology harassment" or "technology arrestment." The gist of the charge is that TA/EIA could be a tool of those radically opposed to all sorts of technological advance. The assessment process could inhibit timely innovation and stifle the technological progress so essential to the well-being of our society.

Evidence seems to weigh against this charge. TA has not been a tool used to derail deserving technological developments; instead, it is a sensible mechanism to avoid costly, unforeseen blunders that could undermine investments. It just makes good sense to "look before you leap," particularly when developments involve extensive commitments and potentially serious societal consequences.

EIA receives good marks from governmental reviewers of the process (U.S. CEQ 1977: 117). The process of EIS preparation has become more routine and better integrated with agency planning. Thus, delays are not as severe as they once were, although further efforts to speed up the preparation process are underway [cf. agency-wide regulations; U.S. CEQ (1978)]. The number of developments stopped by injunctions over EISs is relatively small (U.S. CEQ 1977: 121). Industry is certainly not enamored of the EIA process, but it is learning to live with it. As industrial planners foresee the need to prepare environmental assessments, plans are apt to be increasingly environmentally sound, and EIS difficulties correspondingly lessened.

Those concerned about technological development may have a stronger case in their suspicions about TA/EIA. The essence of these critiques is that assessors are too accepting of the "technological imperative"— that we are properly a highly technological society and will continue to be.

It has been claimed that TA/EIA is guilty of a narrowness of viewpoint and an intolerance of opposing views. How does this criticism apply? TA/EIA deliberately avoids generalized criticisms of the present technological society and does not attempt to pass judgment on the ways we have "got where we are." It is not concerned with finding the causal factors that have created present undesirable impacts of technology [although there has been some work at retrospective TA; see Tarr (1977)]. Its vision is directed toward the future. Such a limitation is not inherently bad, but unfortunately, assessors have generally been intolerant of those who do attempt a critique of contemporary technological society. Reading transcripts of the hearings prior to the establishment of the OTA, one encounters phrases such as "bleeding-heart humanists" or "technophobiacs" used to characterize those who call into question the adequacy of "technological fixes" for societal ills. Winner (1972: 39) notes that Mesthene's call for more and better expert analyses of technological change (see Chapter 2) is itself accompanied by a dismissal of the penetrating, though admittedly pessimistic, writings of Ellul, Mumford, Marcuse, and Arendt as "demonic visions."

Clearly, the styles of the two communities (assessors and critics of technology) are different. Assessors, as we use the term, restrict their focus to specific technological developments, for which they seek to make well-supported analyses of likely impacts (e.g., firm data, defensible models). Critics of technology in its generic sense typically intellectualize broadly, with little recourse to a technical data base. It is thus not surprising that the technological critics are disappointed with the restricted sense of "technology assessment" being practiced. One could hope for more constructive communication between these two communities.

An extension of this criticism of TA/EIA accuses it of actual ideological bias. Dickson (1974) argues that modern societies, including both nonsocialist and socialist powers, are captives to what he terms the "ideology of industrialization."[1] Among the features of the ideology of industrialization are the tendency to equate progress with industrial development; a faith in applied science as the driving force of desirable change; and tendencies toward increased capitalization, technological sophistication, and centralization. Also involved is the belief that the citizens of the nation should be prepared to bear numerous social costs such as major environmental changes, urban in-migration, technological unemployment, and a pace of living set by the rhythms of industry itself.

Beliefs such as those just mentioned bear directly on TA/EIA, since the assessor is most likely to regard them as definitive of its operational

[1]An ideology consists of a body of beliefs possessing philosophical presuppositions, a measure of internal coherence, and principles whereby it is set apart from opposing viewpoints.

context. Education and employment patterns of the majority of TA practitioners reinforce this approach. They are usually technically trained with good grounding in the physical sciences and to a lesser degree in the social sciences. McDermott (1969: 27) observes that it is not characteristic of those so trained to possess a high degree of ideological sophistication and its practical counterpart, a drive toward political power. What is likely to be found is a strong action orientation, a commitment to problem solving, and a general belief that ideological issues are too often the arena for idle speculation. Philosophers, however, never tire of telling their students, that ignorance of one's philosophical underpinnings makes one a captive to them. TA/EIA practitioners need to reflect on the beliefs that underlie their vision of society.

19.2.2 Relation of TA/EIA to the Political Process

As indicated at the close of the last section, ideological concerns lead to political concerns. In this regard as well, there have been several significant criticisms leveled against TA/EIA. These range from charges that TA/EIA is antidemocratic to rebuttals that it is just the opposite. Those with yet another perspective wonder what the fuss is all about, doubting that TA/EIA is very relevant to the policy process at all.

Some critics claim that TA/EIA possesses antidemocratic tendencies. It is claimed that TA/EIA excludes more of the citizenry from the process of sociopolitical decision making than are excluded at present. How might this argument be supported? Berg (1975b: 21–32) argues that TA acquires political power because it generates the knowledge necessary for implementing change. TA/EIA performers possessing such information thus assume a special place. TA/EIA practitioners are then in a position to influence policy decisions, either by specific recommendation or by the isolation of feasible policy options.

The TA/EIA expert can also unwittingly contribute to an "authoritative allocation of values" (Carroll 1971: 648). Consider some examples. Technical questions about waste disposal carry implicit value choices about acceptable environments, obligations to posterity, importance of aesthetics, and other issues. Energy-use projections, an important part of many TA/EIAs, depend on models of consumption patterns, importance of conservation strategies, and so on. Selection of feasible models reflects value choices and, to an extent, shapes the future itself. Technical questions of product design entail value choices concerning reliability levels, mean life cycle, and ease of repair (even who can do the repair, expert or consumer).

Who should decide such matters? Ideally, in the open, democratic society, citizen consensus represents value preferences. If those with power gained from the possession of technical skills and information co-opt value selection, then a less democratic policy process will result.

Many practitioners of TA/EIA are sensitive to such criticisms. This is not surprising when one recalls the tenor of the times when TA/EIA first took shape. The 1960s witnessed widespread unrest fostered in part by a disillusionment with expertise in general. It is clear from his early testimony (see Chapter 3) that Congressman Daddario intended that TA would both inform and reinvolve the citizen in the creation of the future.

Democratic use of TA/EIA can take place in several ways. First, there is the possibility of using impact assessment information to inform concerned stakeholders of the likely consequences of a development. They, in turn, may seek redress through the mechanisms of representative government or direct action (e.g., popular protests, court actions). Second, there is the opportunity for direct involvement in the assessment process by concerned parties. As discussed in some detail in Chapter 16 and elsewhere in the book, however, participatory assessment has been difficult to accomplish. When one considers the alternative of no impact assessment, TA/EIA certainly offers the potential of a more informed citizenry and, hence, a citizenry better disposed to protect its perceived interests.

A different concern is that TA and EIA do not bear significantly on the policy process. There are several reasons for this. For example, TA is not systematically performed. Indeed, observers of OTA's performance have commented on the gaping holes in its coverage of the significant and controversial issues before Congress (Casper 1978). Nor have the federal Executive Branch TA efforts been exhaustive. On the other hand, it is asserted that EIAs are often pro forma documents prepared to justify decisions, not to influence them. The requirement to prepare EISs on so many issues (e.g., some 1000 federal EISs per year), it is claimed, reduces them to routine status. The fact that the party responsible for preparation is the proponent also casts doubt on the role of EIA. This is, of course, a multifaceted issue. Proponent preparation counters the "technology harassment" notion and offers institutional process advantages in integrating EIA into agency planning. Chapter 16 discusses issues in the utilization of TA/EIA in the policy process.

19.3 METHODOLOGICAL CRITIQUES OF TA/EIA

We now turn to more pragmatic concerns about the performance of valid impact assessment. These begin with the operational implications of the value issues raised in the previous section. They continue through the methodology used to anticipate future effects of technological developments. This section does not discuss these difficulties technique by technique because such considerations appear throughout the book. Some attention, however, is devoted to cost–benefit analysis and the assignment of probabilities as examples of serious methodological concerns.

A special concern is the proper role of values in the assessment process. Most practitioners accept that societal values necessarily affect impact assessment. There is real disagreement, however, on how to inject societal preferences systematically into a TA/EIA (see Chapter 14). Some claim that the assessment team can adequately sensitize itself to the concerns and desires of the publics and act as their ombudsmen. Certainly, special interest groups are biased, often militant, and emotional, whereas the assessment should be impartial, rational, and detached. A common position is that public input is necessary, but that substantive public participation in the assessment itself will undermine study quality. On the other hand, some suggest that full participation is preferable to trusting the paternalism of a technical elite (the assessment team). We suggest a full disclosure policy. The practitioners of TA/EIA need be aware of their own intellectual frameworks and implicit values, and let their study users know where they stand.

Students of interdisciplinary research point out that to produce an integrated and useful TA/EIA, even in the best of circumstances, certain problems endemic to all interdisciplinary collaboration must be overcome. While the members of the project team may share a technocratic outlook, their disciplinary training has provided them with different "cognitive maps" (Petrie 1976: 11). Thus technical questions tend to get framed in the terms of one's own discipline. Each discipline provides a model or paradigm that serves to define basic concepts, modes of inquiry, and observational categories. These circumscribe a range of problems and demarcate the discipline from others. Standards of proof and explanation acquire a kind of certification by common usage within each discipline, but these standards differ among disciplines. For the TA team to work effectively, each team member must, to some extent, be able to step outside this particular disciplinary world and communicate with colleagues, on a substantive level. For example, "value" means different things to an economist and sociologist. Efforts to communicate across disciplinary boundaries can be painful and ego deflating.

Because of such difficulties, a study may lack *integration*. That is, the different disciplinary components may never be combined in any effective way. Sections of the TA may be combined in piecemeal fashion and thus be virtually useless for information or decision making. The user needs policy options based on a synthesis of economic, environmental, social, political, and legal factors. Chapter 17 presents our recommendations on how to integrate the component pieces of an impact assessment.

Another methodological problem area is the tendency for most TAs and EIAs to neglect social description, social forecasting, the analysis of social and political impacts, and policy analysis. This is true in absolute terms and relative to both technology description and forecasting, and analysis of impacts in areas where quantitative data exist (e.g., economics, technology, and some aspects of the environment). Consideration of the assessment summaries in the Chapter 5 Appendix illustrates this tendency. The omission of social factors and policy analysis in a significant fraction of TA/EIA studies presents a serious problem that can undermine both the validity and utility of these studies.

Another challenge to TA/EIA methodology has been issued by Lynn White, Jr. (1974). He criticizes it for lacking an historical perspective. An inadequate grounding in the history of technological development may lead to a lack of concern for the generative causes of present technological problems. Furthermore, because long-term assessments must generate a range of possible future scenarios, an inadequate appreciation of the radical differences in the past and present may lead to a lack of imagination about the future.

Analytical limitations hamper the performance of TA/EIA. Taking full account of history is often made difficult by the absence of suitable data. Data over time and across disciplines are typically hard to obtain. Further, differences in data quality are likely to make analysis and synthesis awkward. Much of the crucial insight into policy possibilities may come from tacit knowledge that is most difficult to treat in a replicable scientific manner. Moreover, there are no perfect techniques for anticipating the future, whatever the available information. In sum, limitations in data and methods severely restrict the possibility of strong inferences about the future.

As a rejoinder, the TA/EIA practitioner can rightfully point out that some information is better than none at all. Being aware of the limitations in futures methods, one can make useful projections and enumerate likely possibilities worthy of consideration in the formulation of policy.

Various weapons in the TA/EIA arsenal have come in for specific criticism. None perhaps more than cost–benefit analysis (Chapter 11). Although we make use of this powerful tool, one must be aware of its limitations. For example, British economist Streeten (1971) believes that cost–benefit variables are too often assigned arbitrary weightings by economists. How is one to assign the relative importance of two conflicting objectives such as industrial growth and environmental protection? The technical solution may mask an imposed weighting by a minority, when the question is more properly political, demanding the cumbersome dialogue of a democratic society.

Hazel Henderson (1974) offers additional criticisms (some of which were touched on in Chapters 11 and 14):

> Another serious flaw of cost/benefit analyses is that they average out costs and benefits per capita. This tells us nothing about who bears the costs and who gets the benefits; whose ox is gored, such as when neighborhoods are disrupted by highways or sports arenas, and who gets the contracts, the bond issue business and the jobs. It is seldom the people under the flight paths at airports that also fly in planes and profit from building them or operating them. . . .
>
> There are other more technical problems with cost/benefit analyses which I shall just mention briefly. They assume that adequate information is available to all parties and they accept the existing distribution of income as a given, although these two factors can disenfranchise many citizens, such as those without economic or political power, or adequate information on costs, health effects or long-range risks of a particular technology or development. In addition, cost and benefit ratios can be completely different depending on what rate of discount is used, i.e. the assumption of what interest rates will prevail over the lifetime of the project. Such arbitrary assumptions can overstate the costs and understate the benefits or vice versa, and are currently the subject of hot debate among economists.

Another technique where difficulties arise is the assignment of conditional probabilities (see Chapters 9 and 14). The process by which these probabilities are assigned often boils down to "best judgment" without any firm evidence. However, once these probabilities are expressed as numbers, they take on a life of their own and are manipulated according to formal rules. The results, expressed numerically, reflect only the quality of estimation of the original probabilities, and do not, thereby, become truly precise or more believable. More generally, the tension over the proper degree of quantification presents itself to each assessment team.

In the early days of the TA/EIA era, there was considerable grumbling over shoddy assessment practices. Contractors sometimes produced some poor-quality efforts in a vacuum of experience from which to gauge what was or was not acceptable work. Both TAs and EIAs seem to be improving in quality. The difficulties appear attributable more to the inherent difficulty of the task than to unscrupulous promoters. As enumerated in this section, the attempt to infer future occurrences is naturally difficult. Indeed, it proves very difficult to ascribe cause and effect relations to historical happenings. For instance, recalling the cameo on the effects of television on marital stability (p. 59), try to imagine how one might go about documenting this relationship!

An important charge against impact assessment is the lack of conceptual advance in development of methods. One of the striking findings in the Rossini et al. (1978) survey of TA practitioners was their almost universal disdain for the TA literature. If a single word were to characterize EIA methodology, it would be "pragmatism." Practitioners of EIA seem to have focused tightly on the tasks in question with minimal attention to methodological issues. We trust that this book and other recent literature present a viable counterpoint to this criticism. From our perspective, study strategies have developed to the point where useful guidance is at hand. Furthermore, specific model development and the beginnings of validation of techniques suggest a useful cumulation of methodological understanding in TA/EIA.

19.4 CRITIQUES OF TA/EIA INSTITUTIONS

Chapter 3 outlined the chain of events that led to the passage of NEPA and creation of OTA. Possibly, the rocky road that these institutions have traveled is due largely to the substantive and methodological difficulties mentioned in the previous sections. On the other hand, political infighting, jurisdictional quarrels, problematic leadership, and poor formulation of the laws may be to blame. This section explores these aspects of institutionalization of TA/EIA in the United States. We assume that the impact assessment process itself will play an increasingly important role in the policy process, but that the institutional forms may change. Hence there is merit in looking at the institutional performance.

This section does not attempt to review the performance of all institutions associated with TA/EIA. Chapter 3 dealt with the range of activities, and as indicated there, they are both extensive and diffuse. The present focus is on OTA and the EIS process established in NEPA. The OTA is in a sense a flagship for TA, and the EIS process is the mainstay of EIA.

With respect to OTA, we note three significant criticisms:

Short-range concerns have replaced long-range ones.

Jurisdictional contentions have restricted assessment activity.

Politics has harmed performance.

From its inception, OTA has been caught up in providing short-range information to Congressional committees. This was not, however, the early warning activity envisioned by the early proponents of TA. In the words of Harold Brown, former chairman of OTA's Technology Assessment Advisory Committee (U.S. Congress, OTA 1976b: 103–104):

> Inevitably there are strong pressures on the Congress as well as the Executive Branch to concentrate on immediate problems. Certainly these problems must be faced as they arise. But there needs to be a balancing effort within the Congress to foresee problems of the medium- or even the long-term future.

Jurisdictional contentions have arisen from several sources. OTA was intended as a staff arm of the Congress to help counter Executive Branch expertise. Most significantly, the Technology Assessment Board (TAB) has on occasion treated OTA as committee staff (U.S. Congress, House 1978), and some Congressional committees have viewed OTA as a competitor (Casper 1978). This attitude reflects the notion that information is power. Thus a committee would often rather remain the font of expertise on matters in its jurisdiction. Possible overlap of function between OTA and the Congressional Research Service, General Accounting Office, and the Congressional Budget Office (created in 1974) remains somewhat unsettled (U.S. Congress, House 1976).

Political intrigue has marked the early years of the OTA. The OTA staff were selected not only on the basis of merit, but on attachments to the Senators and Representatives on its Board. Office of Technology Assessment projects sometimes reflected the preoccupations of politically powerful Board members. Casper (1978) points out that political sensitivity led OTA to leave gaps in its coverage of technology-intensive policy issues. A cautious strategy of avoiding items that might prove too controversial was often followed. The review of OTA by the U.S. House Subcommittee on Science, Research, and Technology (1978) gives a qualified approval, but suggests that the TAB and the Technology Assessment Advisory Council limit their day-to-day involvement.

The OTA appears to be moving to face these criticisms. Under its second director, former governor Russell Peterson, there were indications that it would attempt to reduce the political overtones. Staff was built up and longer-range assessment priorities were discussed. Congressional resistance to this stance may have contributed to Mr. Peterson's decision to leave OTA. Hopefully, OTA can obtain the time, support, and leadership necessary to provide a true long-range assessment capability. As we went to press, OTA's third director, Dr. John Gibbons, was facing this challenging situation.

Another set of uncertainties surrounds NEPA. Some of the primary concerns include unnecessary red tape, lack of substance, poor fit with the policy process, and poor use of environmentalists' efforts.

Various arguments have been made as to the bureaucratic wastefulness of EISs. These range from concerns about the delay in project development caused by EIS preparation and review, to concern for the paper wasted. The EIS on the Alaska pipeline alone resulted in several years' delay in the project, an EIS costing some $9 million, and a 10-foot-high stack of paper. However, it also led to the redesign of a number of project features to better protect the environment and generated useful debate over the merits of the scheme. Many developments have been modified as a direct result of EIAs; more importantly, many more are presumably being modified at the planning stage in contemplation of EIA considerations.

One objection to EISs focuses on its emphasis on process rather than substance. That is, one can obtain an injunction to halt a development based on the unsatisfactory fulfillment of the process requirements (e.g., adequacy, proper hearings). One cannot stop a project based on potential environmental damage uncovered by the EIA. One could hope for greater emphasis on mitigation measures in response to identified negative impacts. However, the procedural nature of the EIS requirements may well be an enduring strength. Since an EIS does not itself lead to policy actions, the process can be somewhat standardized and maintain a less politically vulnerable posture. Indeed, it is consistent with democratic principles to allow the information uncovered by an EIA to be used by anyone attempting to influence policy through normal governmental channels. EIS information is, of course, available to the public.

To some extent, EIS requirements may be a poor vehicle to influence policy. For example, NEPA requires the stipulation of alternatives to the project, but few bureaucracies have an inclination to analyze seriously alternatives that they cannot carry out. Since any development entails dozens of decisions and decision makers, the proper focus and timing for an EIA are unclear. Issues considered at the planning stage may not be paramount later. Agency commitments tend to be ongoing—one is forever entering the fray midstream.

Although environmentalists have generally applauded the EIA institutionalization, there are counterpoints. Fairfax (1978: 743) challenged the conventional wisdom: "I suggest that NEPA does not constitute a new approach to administrative reform and is actually a poor vehicle for a reformation of agency decisionmaking." She criticized the demand on environmentalist resources required to process EIS papers. She also challenged the claim that NEPA has expanded the doctrine of standing (the right to bring suit). The courts have actually broadened standing in cases, such as the Storm King Mountain project, which predated NEPA. She also suggests that EIS preparation should include greater requirements for citizen participation.

Both presidential message and hearings (U.S. CEQ 1977) have lent strong support to the EIA process. Yet, they have also recognized merit in many of the criticisms. Specific suggestions on ways to improve the EIS process have included the following:

Make the statements shorter and provide summaries so that they will be more useful to busy policy makers and unsophisticated lay people.

Identify key options.

Include economic impacts and impacts falling outside the United States.

Perform more program EISs (i.e., more impact assessments like full-scale TAs).

The new CEQ regulations for the preparation of EISs are intended to improve the process (U.S. CEQ 1978).

19.5 CONCLUDING OBSERVATIONS

Despite its real and potential benefits, TA/EIA has some serious problems. We have mentioned many of these throughout this book, but this chapter brings together some of the more important ones. We divide these critiques into three groups: substantive, methodological, and institutional.

Substantive criticisms fall into two categories. Attitudes toward technology color one's view of TA/EIA as "technology harassment" or as wanton neglect of telling criticisms of a "technological imperative." Views of TA/EIA with respect to the political process depict it as, alternatively, antidemocratic or democratic. Others question its very relevance to policy making.

Important methodological concerns include:

sufficiently broad assessment team perspectives;

effective ways to treat value-laden aspects;

accomplishment of participatory assessment;

adequate performance of social and policy analyses;

balanced historical and futures perspectives for analysis;

adequate data and analytical techniques;

specific methodological characteristics, such as proper quantification in cost–benefit analysis or the assignment of conditional probabilities; and

development of a cumulative professional literature.

Institutional criticisms are directed primarily at OTA and NEPA. The OTA has had difficulties due to political overtones and to possible jurisdictional overlaps. A long-range research capability has yet to be established. NEPA has been criticized as being a poor vehicle for improving decision making, since it cannot force substantive consideration of negative impacts and may provide an imperfect fit with real decision processes. In addition, it can cause inefficient use of available resources. Efforts to improve OTA and EIS procedures appear somewhat promising.

Despite the importance of these criticisms, the great positive potential and the accomplishments of TA/EIA lead us to view the improvement, rather than the abandonment, of TA/EIA as the appropriate response.

RECOMMENDED READINGS

Berg, M. R. (1975) "The Politics of Technology Assessment," *Journal of the International Society for Technology Assessment* 1 (4) (December), 21–32.

A careful description of the way political structures and processes are likely to influence assessments.

Carroll, J. D. (1971) "Participatory Technology," *Science* 171 (19 February), 647–653.

Analyzes the emergence of public action at the grassroots level as a countervailing force to technological alienation.

Fairfax, S. K. (1978) "A Disaster in the Environmental Movement," *Science*, 199 (17 February), 743–748.

A negative evaluation of effects of NEPA.

White, L., Jr. (1974) "Technology Assessment from the Stance of a Medieval Historian," *American Historical Review* 79 (1) (February), 1–13.

Delightful article suggesting that TA would probably have failed to anticipate many of the most profound consequences of earlier innovations.

Winner, L. (1972) "On Criticizing Technology," *Public Policy* 20 (1) (Winter), 35–59.

A sharp warning that TA can easily lose its objectivity by being co-opted by the technological approaches it is supposed to assess.

EXERCISES

19-1. The following are listed by Kennard (1975: 43) as criticisms of TA/EIA. Organize discussion teams to debate the pros and cons of several of these points.

1. The purpose of most technology assessments is to give the go-ahead for projects that have already gone ahead.
2. Technology assessments rarely raise the important issues.
3. Any technical study that criticizes a politically potent development will not see the light of day.
4. Most cost–benefit studies don't tell who gets the benefits and who pays the costs.
5. The hidden assumption always favors the status quo.
6. Any expert who knows anything about a technological development usually works for the developer, and most of the information included in a TA comes from the developer of the technology.
7. Technologies are usually assessed by the same agencies that promote them.
8. Citizen participation, public information, and public relations are all the same to most assessment agencies.

19-2. Kennard (1975) goes on to cite several criticisms of citizen participation. Debate several of these points as in Exercise 19-1.

1. Citizen activists don't really represent anybody but themselves; they are fringe groups, by and large, and some are irresponsible to boot.
2. Citizen activists are almost always white, middle-class people who are merely defending their own selfish interests, their own special values, or are just working out their emotional hang-ups.
3. Citizen activists are either uninformed, ill informed, or misinformed. Besides, they can never be made to understand the true complexity of a technological issue.
4. Citizen activists have no right to speak for disparate, unorganized segments of the public, such as the poor.

5. Citizen activists really have no commitment to a rational planning process. They want to disrupt the planning process and to make life hell for the administrators who possess the awesome responsibility for the ultimate decision.

6. Technology assessment is an objective scientific process designed to serve the needs of policy makers for more information. Public participation is something that happens after the assessment has been completed and handed over to the responsible officials or legislative bodies.

19-3. Assume the role of a consultant to your state government. Suggest several options for institutionalizing TA. Compare the advantages and disadvantages of each, and make a recommendation.

20

Future Prospects

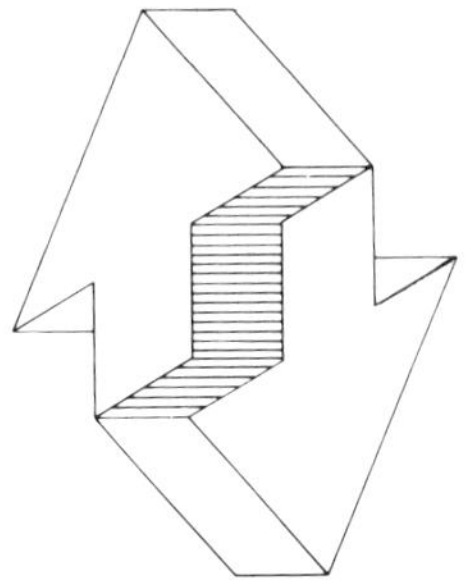

This chapter looks ahead to stimulate thoughtful consideration of the future for TA and EIA. It counterposes three brief scenarios of what might be, and adds a few final thoughts on what is and what should be.

20.1 THREE SCENARIOS

This book has attempted to present a comprehensive portrait of TA/EIA. In so doing, it has touched upon the role of technology in society and the manner in which TA/EIA can be used to trace, analyze, and evaluate the effects of that role. Specific methods and techniques for performing the assessment task have been considered and the complexity attending their application discussed.

It seems appropriate in this final chapter to reflect on TA/EIA as a social phenomenon and speculate as to its future in the policy-planning process. We believe that concern over technologies is likely to increasingly preoccupy the industrialized nations and, hopefully, the developing ones. To assist this speculation, consider the three brief scenarios that follow:

1. The scope of TA/EIA is restricted primarily to technical studies involving first-order economic, technological, and environmental consequences. Policy options to deal with these consequences are introduced and analyzed.
2. Through methodological improvements, TA/EIA is broadened to include analyses of a wide range of higher-order consequences, including the treatment of alternative intellectual and value frameworks for structuring the problem.
3. Coupled with developing communications technologies, TA/EIA is the catalyst for a wider policy process involving public participation and serious attempts at long-range planning by the private sector.

Each of these possible futures could be shaped to a degree by further development in TA/EIA itself. Some speculate that TA/EIA *as such* may disappear, while its techniques and concerns are subsumed by the broader field of policy analysis. In this case, the most visible impact would be the progressive exclusion of engineers and natural scientists from the execution of TA/EIAs, and a greater involvement of social scientists and government bureaucrats. This would imply a step backward for the involvement of engineers and scientists in the policy-planning process.

The first scenario effectively represents business as usual. Despite the lofty aspirations and generally inclusive definitions of TA/EIA, technical efficiency, economic viability, and short-range environmental effects form current policy criteria. Existing methods and techniques for studying social and institutional consequences (see Chapters

12 and 13) reflect a high degree of uncertainty. The case for this limited TA/EIA future is based on present practice, and the fact that it emphasizes the soundest TA/EIA information currently available. This is a conservative scenario that does not involve significant reshaping of social and political institutions for dealing with technology. Overall, it offers an achievable, if unimaginative, prospect for dealing with the impacts of technology.

The optimistic goals currently envisioned for TA/EIA are achieved in the second scenario. Achievement of this scenario requires improvements in the state of the art, both substantive and methodological. Necessary substantive improvements include the development of social and political theories that can be used to explain and project future societal states. In addition, a clear understanding of higher-order impacts and the process of their generation must be achieved. Methodologically, critical evaluation of the predictive validity of current techniques for understanding the future is needed. This scenario requires that the study of the future pay greater attention to the methodological sophistication that now characterizes the study of the past. It also demands a stronger institutional framework to support impact assessment. This is a scenario of institutional and intellectual evolution leading to a society that takes its future more seriously.

The third scenario can be seen either as the product of a lengthy evolution or the result of a revolution. It is easily the least plausible of the three and, therefore, the one that might be considered most surprising. In addition to the methodological and theoretical improvements needed for the second scenario, this entails an institutional change, the development of effective and powerful participatory planning institutions. Present TA/EIAs generally accept public input as information, but usually avoid having technically unsophisticated public representatives participate substantively in the study. In cases where participation has occurred, study team members generally have found it inappropriate and resented it. To achieve this scenario, the societal decision-making process must move away from the elitest approach of the past and present—a difficult task. Whereas differences of birth, race, and sex (once important determinants in the establishment of elites) are becoming less significant, differences in technical training and intellectual capacity retain their importance. Thus, for example, knowledge generation and transmission, the bases for assessment and decision making, remain among the most elite of enterprises. Major advances in communication technologies will also be required, through which public input can be incorporated into the planning process.

20.2 GENERAL OBSERVATIONS

Stepping back from the three scenarios, several factors warrant consideration. It appears likely that the private sector will increasingly embrace impact assessment, under some label or another. For one reason, EIA is mandated by law in many instances. For another, there may be public relations benefits to be gained from attention to societal interests. Most compelling, however, is the fact that TA/EIA is in a firm's self-interest. Advocacy TA can be a weapon in defense of corporate intentions and against corporate and personal liability suits. But more significantly, effective impact assessment can help anticipate future profitable developments and forestall developmental side effects that could threaten corporate profits, or even existence. "Look before you leap" makes sense for both the private and public sectors.

As outlined in Chapter 3, TA and EIA originated as public-sector institutional responses to technology-driven and technologically soluble problems. NEPA and OTA represent specific forms of this response. Although criticisms of these mechanisms were offered in the previous chapter, it appears that they will continue to exist in some form. As long as the flagship institutions remain alive and well, TA and EIA are likely to retain their identity. Insofar as strong professional associations develop, the "field" of impact assessment will be further institutionalized.

In its short history, TA/EIA has made a difference. Environmental considerations have been brought to prominence in weighing developmental decisions. TA/EIA has served to institutionalize a discussion of social impacts in the face of a long-standing American opposition to social planning. It has formalized concern for the "unintended, indirect, and delayed" effects of technological innovations. By developing analytical evaluation and planning tools, TA/EIA has supplemented intuition with valuable quantitative techniques (though their limitations must not be overlooked).

There are specific successes to which one might point. For example, the Port of New York Authority commissioned a study of the effects of extending the runways of Kennedy Airport into Jamaica Bay, heeded the findings, and dropped the project. Many projects have been altered to better their environmental implications. The Council on Environmental Quality (U.S. CEQ 1976a) reported that the EIS requirements substantially altered agency developmental procedures to include early consideration of noneconomic factors. The recently completed OTA study of solar-energy technology (U.S. Congress, OTA 1977b) has been widely circulated and given media coverage, resulting in significant broadening of dialogue concerning feasible energy scenarios. The OTA has produced quality pieces of demonstrated use to Congress.

The strongest concern expressed about TA/EIA was its potential for "technology arrestment." Certainly, EIA requirements have had short-run costs and have slowed many projects. However, as the procedures have been worked out, and as environmental considerations are being raised early in project development, this is becoming less serious. Furthermore, long-run gains from avoiding costly, late-in-the-game project abortions or unintended side-effects should outweigh the short-run inconveniences.

TA/EIA can become even more useful. Improvements in the state of the art have been made and must continue. EISs have markedly improved through several years of experience. The greatest potential improvement would be a better understanding of higher-order and unintended future consequences. Progress in impact assessment needs to be accompanied by parallel progress in the receptivity of public and private institutions to applying assessment information. Through the execution of perceptive TA/EIA studies, advances in the state of the art, and development of receptive institutional structures, society can gain the leverage necessary to magnify the positive and minimize the negative impacts of technological activities—in short, to gain some measure of control over its technological present and future.

EXERCISES

20-1. Monitor (recall Section 6.4.2) the recent developments in TA and EIA. Use such instruments as the *New York Times Index* and the *Science Citation Index* and scan journal issues (e.g., *Science*) for events with implications for impact assessment. Consider such things as Congressional attitudes toward OTA, reactions to major TA/EISs, legal changes, indications of broadening participation, and evidence for success of new assessment methodologies. Based on this information, evaluate this chapter's prognosis for its three scenarios.

20-2. Construct your own most likely scenario for the future of TA/EIA over the next 10 years. Include indications of critical events (political, social, technological, legal, etc.), and the sensitivity of your forecast to these. The scenario should be developed in considerably greater detail than those in this chapter.

References

Note: † denotes impact assessments; * marks selected methodological pieces. References to the United States National Academies of Engineering and Science, and to United States governmental units, begin "U.S." Documents associated with Congressional entities, such as OTA, are headed "U.S. Congress." Selected pieces not referenced in the text are included for possible reader interest.

Abt Associates (1975) *Water Pollution Control Act of 1972—Social Impacts, Eight Case Studies*. Prepared for the U.S. National Commission on Water Quality (NCWQ-75135). Springfield, VA: National Technical Information Service (PB 250 106).

Adkins, W. G., and Burke, D., Jr. (1971) *Interim Report: Social, Economic, and Environmental Factors in Highway Decision Making*. College Station, TX: Texas Transportation Institute, Texas A&M University.

Amara, R., and Salanik, G. (1972) "Forecasting: From Conjectural Art Toward Science," *The Futurist* (June), 112–116.

American National Standards Institute (1971) *Methods for Measurement of Sound Pressure Levels*. New York: American National Standards Institute, ANSI 51.13–1971.SI. 13–1971.

*Andrews, R. N. L. (1973) "Approaches to Impact Assessment: Comparison and Critique." Paper presented at Short Course in Water Resources Planning. Ann Arbor: University of Michigan.

*Andrews, R. N. L., Cromwell, P., Enk, G. A., Farnworth, E. G., Hibbs, J. R., and Sharp, V. L. (1977) *Substantive Guidance for Environmental Impact Assessment: An Exploratory Study*. Indianapolis: The Institute of Ecology.

Anthrop, D. F. (1973) *Noise Pollution*. Lexington, MA: Lexington Books, D. C. Heath and Company.

Appleyard, D., and Craik, K. (1976) "Berkeley Environmental Simulation Laboratory: A Tool for Transportation Planning." Paper presented at 55th annual meeting, Transportation Research Board, Washington, DC (January).

Archibald, K. (1970) "Three Views of the Experts Role in Policymaking: Systems Analysis, Incrementalism, and the Clinical Approach," *Policy Science* 1, 73–86.

Archibald, R. D., and Villoria, R. L. (1967) *Network-Based Management Systems*. New York: Wiley.

*Armstrong, J. E., and Harman, W. W. (1977) "Strategies for Conducting Technology Assessments." Report to the Division of Exploratory Research and Systems Analysis, National Science Foundation, Washington, DC: Department of Engineering—Economic Systems, Stanford University.

*Armstrong, R., and Hobson, M. (1970) "The Use of Gaming /Simulation Techniques in the Decision Making Process." United Nations Paper No. ESA/PA/MMTS/21.

Arnstein, S. R. (1973) "Citizen Participation in Airport Development." Prepared under contract with Arthur D. Little, Inc., Cambridge, MA, for the Aviation Advisory Commission Report on *The Long Range Needs of Aviation*. Vol. 1, p. I-580–590.

*Arnstein, S. R. (1975) "A Working Model for Public Participation," *Public Administration Review* 35, 70–73.

Arnstein, S. R. (1977) "Technology Assessment: Opportunities and Obstacles," *IEEE Transactions on Systems, Man, and Cybernetics* SMC-7 (August), 571–582.

Arnstein, S. R., and Christakis, A. (1975) *Perspectives on Technology Assessment*. Jerusalem: Science and Technology Publishers.

Arrow, K. J. (1963) *Social Choice and Individual Values*. New York: Wiley.

Arrow, K. J. (Chairman) (1975) "Uncertainty Analysis: Findings of the Panel on Decision Analysis Monitoring and Surveillance," in *Environmental Impact of Stratospheric Flight: Biological and Climatic Effects of Aircraft Emissions on the Stratosphere*. Washington, DC: National Academy of Sciences, pp. 291–348.

Ascher, W. (1977) "Problems of Forecasting." Paper presented to Workshop on Appraisal of Technology Assessment, Dayton, OH (December).

*Ascher, W. (1978) *Forecasting: An Appraisal for Policy Makers and Planners*. Baltimore, MD: The Johns Hopkins University Press.

*Ayres, R. U. (1966) *On Technological Forecasting*. Croton-on-Hudson, NY: Hudson Institute.

*Ayres, R. U. (1969) *Technological Forecasting and Long-Range Planning*. New York: McGraw-Hill.

Ayres, R. U., Saxton, J., and Stern, M. (1974) *Materials—Process—Product Model*. International Research and Technology Corporation. Final Report IRT-305-FR.

Bacca, R. G., Waddel, W. W., Cole, C. R., Brandsetter, A., and Cearlock, D. B. (1973) *Explore I: A River Basin Water Quality Model*. Richland, WA: Battelle-Northwest. EPA Project No. 211B00557.

Baier, K. (1969) "What Is Value? An Analysis of the Concept," in *Values and the Future* (K. Baier and N. Rescher, eds.) New York: The Free Press.

*Baldwin, M. (ed.) (1975) "Portraits of Complexity: Applications of Systems Methodologies to Societal Problems." Columbus, OH: Battelle Research Institute Monograph No. 9 (June).

Bartel, C., Coughlin, C., Moran, J., and Watkins, L. (1974) *Aircraft Noise Reduc-*

tion Forecast, Vol. 2. *NEF Computer Program Description and Users Manual.* Washington, DC: U.S. Department of Transportation, Report DDT-TST-75–4.

Battelle–Seattle (1974) Human Affairs Research Center Annual Report—Research Activities in the Behavioral and Social Sciences.

Bauer, R. A. (ed.) (1966) *Social Indicators.* Cambridge, MA: MIT Press.

Bavelas, A. (1950) "Communications Patterns in Task Oriented Groups," *Journal of the Acoustical Society of America* 22, 725–730.

Bechtel Corporation (1975) *Fuels from Municipal Refuse for Utilities: Technology Assessment.* Palo Alto, CA: Electric Power Research Institute (EPRI 261–1).

†Becker, H. S., et al. (1976) "An Assessment of Socioeconomic and Demographic Determinants of Electrical Energy and Gas Consumption in California." Glastonbury, CT: The Futures Group.

*Berg, M. R. (1975a) "Methodology," in *Perspectives on Technology Assessment* (S. Arnstein and A. Christakis, eds.). Jerusalem: Science and Technology Publ., pp. 63–72.

Berg, M. R. (1975b) "The Politics of Technology Assessment," *Journal of the International Society for Technology Assessment* 1 (4) (December), 21–32.

Berg, M. R., Chen, K., and Zissis, G. J. (1975) "Methodologies in Perspective," in *Perspectives in Technology Assessment* (Sherry R. Arnstein and Alexander N. Christakis, eds.). Jerusalem: Science and Technology Publ., pp. 21–44.

*Berg, M. R., Chen, K., and Zissis, G. J. (1976) "A Value-Oriented Methodology for Technology Assessment," *Technological Forecasting and Social Change* 8, 401–421.

*Berg, M. R., Brudney, J. L., Fuller, T. D., Michael, D. N., and Roth, B. K. (1978) *Factors Affecting Utilization of Technology Assessment Studies in Policy-Making.* Ann Arbor, MI: University of Michigan, Center for Research on Utilization of Scientific Knowledge, Institute for Social Research.

†Berkowitz, J. B., and Horne, R. A. (1976) *The Potential for the Solar Generation of Electricity and for the Solar Heating and Cooling of Buildings.* Cambridge, MA: Arthur D. Little, Inc. Prepared for the National Science Foundation under Contract NSF-C835.

*Bigelow-Crain Associates (1976) *State and Regional Transportation Impact Identification and Measurement, Phase I Report.* Transportation Research Board, U.S. National Research Council.

†Blackman, A. W. et al. (1974) "U.S. Ocean Shipping Technology: Forecast and Assessment." Final Report M-971623-16, United Aircraft Laboratories, E. Hartford, CN (February).

Bolt, Beranek and Newman, Inc. (1971) "Noise From Construction Equipment and Operations, Building Equipment, and Home Appliances." Prepared for U.S. Environmental Protection Agency, Washington, DC.

*Born, S. M., and Besadny, C. D. (1976) "The Environmental Impact Statement Content Analysis Project—Final Report." Madison: University of Wisconsin—Extension Environmental Resources Unit and Wisconsin Environmental Decade Fund.

Bragdon, C. R. (1972) *Noise Pollution: The Unquiet Crisis*. Philadelphia: University of Pennsylvania Press.

Braudel, F. (1976) *The Mediterranean*. New York: Harper.

Braybrooke, D., and Lindblom, C. E. (1963) *A Strategy of Decision*. New York: The Free Press.

Breslow, M., Brush, N., Giggey, F., and Urmson, C. (1972) *A Survey of Technology Assessment Today*. Prepared for the U.S. National Science Foundation by Peat, Marwick, Mitchell and Company, Washington, DC.

*Bright, J. R. (1972) *A Brief Introduction to Technology Forecasting: Concepts and Exercises*. Austin, TX: The Pemaquid Press.

*Bright, J. R. (1973) "The Process of Technological Forecasting—An Aid to Understanding Technological Forecasting," in *A Practical Guide to Technological Forecasting* (J. R. Bright and M. E. Schoeman, eds.). New York: Prentice-Hall, pp. 3–12.

*Bright, J. R., and Schoeman, M. E. F. (eds.) (1973) *A Practical Guide to Technological Forecasting*. Englewood Cliffs, NJ: Prentice-Hall.

*Bright, J. R. (1978) *Practical Technology Forecasting: Concepts and Exercises*. Austin, TX: The Industrial Management Center.

*Brockhaus, W. L., and Mickelsen, J. F. (1976) "A Worldwide Investigation of Writings and Publications Concerning the Delphi Methodology," *Journal of the International Society of Technology Assessment* 2, (2) (Summer), 5–34.

*Brockhaus, W. L., and Mickelsen, J. F. (1977) "An Analysis of Prior Delphi Applications and Some Observations on its Future Applicability," *Technology Forecasting and Social Change* 10, 103–110.

Brodeur, P. (1977) *The Zapping of America: Microwaves, Their Death Risk, and the Coverup*. New York: Norton.

Brooks, H., and Bowers, R. (1970) "The Assessment of Technology," *Scientific American* 222 (February), 13–21.

Bross, I. D. (1965) *Design for Decision*. New York: The Free Press.

Brunner, D. R., and Keller, D. J. (1972) *Sanitary Landfill Design and Operation*. Washington, DC: U.S. Government Printing Office.

Bruno, James E. (1974) "Monte Carlo Techniques in Educational Forecasting," in *Futures in Education: Methodologies*. (S. Hencley, and J. R. Yates, eds.). Berkeley, CA: McCutcheon, pp. 347–373.

Bunge, M. (1966) "Technology as Applied Science," *Technology and Culture* 7 (3) (Summer), 329–347.

*Burchell, R., and Listokin, D. (1975) *The Environmental Impact Handbook*. New Brunswick, NJ: Center for Urban Policy Research, Rutgers—The State University.

*Burdick, J. C., and Parker, E. L. (1971) "Estimation of Water Quality in a New Reservoir." Report No. 8, Department of Environment, Water Resources Engineering, School of Engineering, Vanderbilt University and U.S. Army Corps of Engineers.

Burger, R. M. (1975) *RANN Utilization Experience Final Report to the National Science Foundation*. Research Triangle Park, NC: Research Triangle Institute (available through NTIS).

*Burnham, J. B., Nealey, S. M., and Jones, G. R. (1974) "Quantified Social and Aesthetic Values in Environmental Decision Making." Paper presented at Symposium on Siting of Nuclear Facilities, Bordeaux, France (September 2–6).

Caldwell, L. C. (1970) *Environment: A Challenge to Modern Society.* Garden City, NJ: Anchor Book, Doubleday and Company.

*Campbell, A. (1976) "Subjective Measures of Well-Being," *American Psychologist* 31 (February), 117–124.

*Campbell, D. T., and Stanley, J. C. (1966) *Experimental and Quasi-experimental Designs for Research.* Chicago: Rand-McNally.

Caplan, N., et al. (1975) "The Use of Social Science Knowledge in Policy Decisions at the National Level." Ann Arbor: Center for Research on Utilization of Scientific Knowledge, Institute for Social Research, University of Michigan.

Carpenter, S. R. (1977) "Philosophical Issues in Technology Assessment," *Philosophy of Science* 44 (4), 574–593.

Carpenter, S. R., and Rossini, F. A. (1977) "Value Dimensions of Technology Assessment," in *The General Systems Paradigm: Science of Change and Change of Science.* Washington, DC: The Society for General Systems Research, pp. 463–469.

*Carr, A. B. (1976) "Uses of Scenario Writing in Assessing Technology," *ISTA Journal* 2 (3) (Fall), 27–30.

Carroll, J. D. (1971) "Participatory Technology," *Science* 171 (19 February), 647–653.

Casper, Barry M. (1978) "The Rhetoric and Reality of Congressional Technology Assessment," *Bulletin of the Atomic Scientists* 34 (February), 20–31.

Central New York Regional Planning and Development Board (1972) *Environmental Resources Management.* Prepared for the U.S. Dept. of HUD, NTIS PB 217–517.

Cetron, M. J., and Bartocha, B. (1973) *Technology Assessment in a Dynamic Environment.* New York: Gordon and Breach.

Cetron, M. J., and Monahan, T. I. (1968) "An Evaluation and Appraisal of Various Approaches to Technological Forecasting," in *Technological Forecasting for Industry and Government* (J. R. Bright, ed.). Englewood Cliffs, NJ: Prentice-Hall, pp. 144–179.

Chamblee, C. P., and Nehls, C. J. (1973) *SAROAD Terminal Users Manual.* Research Triangle Park, NC: U.S. Environmental Protection Agency, Publication No. AP-101.

†Changnon, S. A., Jr., Davis, R. J., Farhar, B., Haas, J. E., Ivens, J. L., Jones, M. V., Klein, D. A., Mann, D., Morgan, G. M., Jr., Sonka, S. T., Swanson, E. R., Taylor, C. R., and Van Blokland, J. (1977) *Hail Suppression: Impacts and Issues.* Urbana, IL: Illinois State Water Survey.

Chen, K., and Zacher, L. (1977) "Toward Effective International Technology Assessments." Presented Paper at Technology Assessing Conference, East–West Center (May 30–June 10).

*Chen, K., and Zissis, G. J. (1975) "Philosophical and Methodological Approaches to TA," *ISTA Journal* 1, (1) 17–28.

Chikishev, A. B. (ed.) (1965) *Plant Indicators of Soils, Rocks, and Subsurface Waters.* New York: Consultants Bureau.
Churchman, C. W. (1971) *The Design of Inquiring Systems.* New York: Basic Books.
Clippinger, J. H. (1977) "A Methodology to Assess the Psycho-Social Effects of Technological Change." Cambridge, MA: Kalba-Bowen Associates.
Clippinger, J. H. (1979) *Assessing Sociotechnological Change: A Case Study and Methods.* Cambridge, MA: Kalba-Bowen Associates.
CLM Systems, Inc. (1972) *Airports and Their Environment, A Guide to Environmental Planning.* Prepared for the U.S. Department of Transportation, Washington, DC (available as *Report # PB 219–957*, NTIS, Springfield, VA).
Coates, J. F. (1971) "Technology Assessment, The Benefits . . . the Costs . . . the Consequences," *The Futurist* 5 (December), 225–231.
Coates, J. F. (1972) "The Cost of Automobile Pollution: A First Order Analysis," *SPPSG* (Newsletter for the Science and Public Policy Studies Group) 3 (June–July), 31–34. Cambridge, MA: Massachusetts Institute of Technology.
Coates, J. F. (1973a) "Interdisciplinary Considerations in Sponsoring TA's," *Technology Assessment* 1 (2), 109–120.
Coates, J. F. (1973b) "Antiintellectualism and Other Plagues on Managing the Future," *Technological Forecasting and Social Change* 4 (3), 243–262.
Coates, J. F. (1973c) "Some Methods and Techniques for Comprehensive Impact Assessment." Conference on Preparation of EIS, Henniker, NH.
Coates, J. F. (1974) "Coates' Corner," *Technology Assessment* 2 (2) (February), 159–161.
*Coates, J. F. (1976a) "The Role of Formal Models in Technology Assessment," *Technological Forecasting and Social Change* 9, 139–190.
*Coates, J. F. (1976b) "Technology Assessment—A Tool Kit," *Chemtech* (June), 372–383.
Coates, J. F. (1977a) "Technological Change and Future Growth: Issues and Opportunities," in *U.S. Economic Growth from 1976 to 1986: Prospects, Problems, and Patterns,* (U.S. Congress, Joint Economic Committee), Vol. 9 (January 3) (95th Congress, 1st Session). Washington, DC: U.S. Government Printing Office.
Coates, J. F. (1977b) "What Is A Public Policy Issue?" Paper presented at the Annual Meeting of the American Association for the Advancement of Science, Denver, CO, February 23.
Coates, J. F. (1977c) "Life Patterns, Technology and Political Institutions," in *Changing American Lifestyles.* Valparaiso, IN: Valparaiso University.
Coates, V. T. (1972) *Technology and Public Policy: The Processes of Technology Assessment in the Federal Government,* Vols. 1, 2. Washington, DC: George Washington University Program of Policy Studies in Science and Technology.
Coates, V. T. (1976) "President's Message," *ISTA Journal* 2 (3), 4–6.
Coates, V. T. (1977a) "Technology Assessment Seeks Role in Business," *Chemical and Engineering News* (28 March), 11–13.
Coates, V. T. (1977b) "Recent TA's in Japan," *TA Update* 4 (1) (February), 4.

†Cohen, M. D. (1974) "Is TA a Social Science?" *Managing Technology Change.* Bloomington, IN: University of Indiana (February), pp. 1, 5.

Cohen, M. D., March, J. G., and Olsen, J. P. (1972) "A Garbage Can Model of Organizational Choice," *Administrative Science Quarterly* 17, 1–25.

Cole, H. S. D., Freeman, C., Jahoda, M., and Pavitt, K. L. R. (eds.) (1973) *Models of Doom.* New York: Universe Books.

Connolly, T. (1976) "The Experimenting Organization: A Decision-Focussed Framework." Presented paper, American Institute of Decision Sciences, San Francisco.

Cook, T. (1976) "Federal Agencies Impact Assessment Guidelines—Present and Future," in *Environmental Impact Analysis.* Urbana, IL: Univ. of Illinois, Dept. of Architecture Monograph Series, pp. 85–92.

*Cook, T. D., and Campbell, D. T. (1979) *The Design and Conduct of Quasi-Experiments and True Experiments in Field Settings.* New York: Rand-McNally.

*Crews, J. E., and Johnson, G. P. (1975) "A Methodology for Trade-Off Analysis in Water Resources Planning, *ISTA Journal* 1 (June), 31–35.

Daddario, E. Q. (1967) "Technology Assessment: Statement of Chairman, Subcommittee on Science, Research, and Development." 90th Congress, 1st Session.

Daddario, E. Q. (1968) Science, Research and Development Subcommittee Report, U.S. House of Representatives Committee on Science and Astronautics, 90th Congress, 1st Session.

*Dalkey, N., and Helmer, O. (1963) "An Experimental Application of the Delphi Method to the Use of Experts," *Management Science* 9 (3), 458.

Davidson, L. S., and Schaffer, W. A. (1973) "An Economic-Base Multiplier for Atlanta, 1961–1970," *Atlanta Economic Review* 23 (4), 52–54.

Davis, R. (1973) "Organizing and Conducting Technological Forecasting in a Consumer Goods Firm," in *A Practical Guide to Technological Forecasting* (J. R. Bright and M. E. F. Schoeman, eds.). Englewood Cliffs, NJ: Prentice-Hall, pp. 601–618.

*Dee, N., et al. (1972) *Environmental Evaluation System for Water Resources Planning.* Report to the U.S. Bureau of Reclamation. Columbus, Ohio: Battelle Memorial Institute.

*Dee, N., et al. (1973) *Planning Methodology for Water Quality Management: Environmental Evaluation System.* Columbus, OH: Battelle Memorial Institute.

Development Economics Group (1971) *Northern New England East–West Highway Study, Phase I.* Prepared for the New England Regional Commission and the Northern New England East–West Highway Program.

Dexter, L. A. (1977) "Court Politics—Presidential Staff Relations as a Special Case of a General Phenomenon," *Administration and Society* 9 (3), 267–283.

Dickson, D. (1974) *The Politics of Alternative Technology.* New York: Universe Books.

†Dickson, E. M. (1974) *The Video Telephone.* New York: Praeger Publishers.

†Dickson, E. M., Ryan, J. W., and Smulyan, M. H. (1976a) *The Hydrogen Econ-*

omy: A Preliminary Technology Assessment. Menlo Park, CA: SRI International. Prepared for the National Science Foundation under Grant ERP 73–02706.

†Dickson, E. M., Steele, R. V., Hughes, E. E., Walton, B. L., Zink, R. A., Miller, P. D., Ryan, J. W., Simmon, P. B., Holt, B., White, R. K., Harvey, E. C., Cooper, R., Phillips, D. F., and Stoneman, W. C. (1976b) *Impacts of Synthetic Liquid Fuel Development.* Menlo Park, CA: SRI International.

Dolbeare, K. M. (1973) "The Impact of Public Policy," in *The Policy Science Annual.* Indianapolis: Bobbs-Merrill.

Dror, Y. (1971a) *Ventures in Policy Sciences: Concepts and Applications.* New York: American Elsevier.

Dror, Y. (1971b) *Design for Policy Sciences.* New York: American Elsevier.

Duke, R. D., and Burkhalter, B. R. (1966) "The Application of Heuristic Gaming to Urban Problems." East Lansing, MI: Institute for Community Development and Services, Continuing Education Service, Michigan State University.

*Duval, A., Fontella, E., and Gabus, A. (1975) "Cross-Impact Analysis: A Handbook on Concepts and Applications," in *Portraits of Complexity* (M. Baldwin, ed.). Battelle Monograph 9, pp. 202–222.

Dye, T. R. (1972) *Understanding Public Policy.* Englewood Cliffs, NJ: Prentice-Hall.

Dye, T. R. (1975) *Understanding Public Policy*, 2nd ed. Englewood Cliffs, NJ: Prentice-Hall.

Easton, D. (1953) *The Political System: An Inquiry into the State of Political Science.* New York: Knopf.

Easton, D. (1965) *A Systems Analysis of Political Life.* New York: Wiley.

Ehrenfeld, D. W. (1976) "The Conservation of Non-Resources," *American Scientist* 64 (November/December), 648–656.

Eichholz, G. G. (1976) *Environmental Aspects of Nuclear Power.* Ann Arbor, MI: Ann Arbor Science Publishers.

Elgin, D., and Mitchell, A. (1977) "Voluntary Simplicity," *The Co-Evolution Quarterly* 14 (Summer), 4–19.

Ellul, J. (1964) *The Technological Society* (J. Wilkinson, trans.). New York: Vintage.

Enzer, S. (1970) "A Case Study Using Forecasting as a Decision-Making Aid," *Futures* 2, 202–222.

*Enzer, S. (1971) "Delphi and Cross-Impact Techniques: An Effective Combination for Systematic Futures Analysis," *Futures* 3, 48–61.

*Enzer, S. (1972) "Cross-Impact Techniques in Technology Assessment," *Futures* 4, 30–51.

†Enzer, S. (1974) *Some Impacts of No-Fault Automobile Insurance—A Technology Assessment.* Middletown, CT: Institute for the Future.

Enzer, S. (1975) "Public Analysis," in *Perspectives on Technology Assessment* (S. R. Arnstein and A. Christakis, eds.). Jerusalem: Science and Technology Publishers, pp. 107–113.

Evans, M. K. (1969) *Macroeconomic Activity: An Econometric Approach.* New York: Harper and Row.

Ezra, A. A. (1975) "Technology Utilization: Incentives and Solar Energy," *Science* 187 (28 February), 707–713.

Fairfax, S. K. (1978) "A Disaster in the Environmental Movement," *Science* 199 (17 February), 743–748.

†Farhar, B. C., Changnon, S. C., and Swanson, E. (1977) *Hail Suppression and Society*. Urbana, IL: Illinois State Water Survey.

Feldt, A. G. (1966) "The Community (Cornell) Land Use Game (CLUG)." Ithaca, NY: Department of City and Regional Planning, Cornell University.

Field, R., Struzeski, E. J., Masters, H. E., and Tafuri, A. N. (1975) "Water Pollution and Associated Effects from Street Salting," in *Water Pollution Control in Low Density Areas; Proceedings of a Rural Environmental Engineering Conference*. Hanover, NH: University Press of New England, pp. 317–340.

Findley, E. L. (1975) "Significant Ecological Considerations for Environmental Impact Analysis of Facility Development," in *Proceedings, 5th Annual Environmental Engineering and Science Conference*, Louisville, KY.

*Finsterbusch, K. (1975) "A Policy Analysis Methodology for Social Impacts," *ISTA Journal* 1 (1), 5–15.

Finsterbusch, K., and Wetzel-O'Neil, P. A. (1974) *A Methodology for the Analysis of Social Impacts*. Vienna, VA: BDM, Inc. Document BDM/W–74–049–TR.

*Finsterbusch, K., and Wolf, C. P. (eds.) (1977) *Methodology of Social Impact*. Stroudsburg, PA: Dowden, Hutchinson and Ross, Inc.

†Fletcher, K., and Baldwin, M. F. (eds.) (1973) "A Scientific and Policy Review of the Prototype Oil Shale Leasing Program Final Environmental Statement of the U.S. Department of the Interior." Indianapolis: The Institute of Ecology.

†Forester, J. W., and Rosenthal, S. (1972) "Toward a Technology Assessment: The Case of V/Stol Aircraft Systems," in *Social Change, Public Response, and the Regulation of Large Scale Technology* (T. R. LaPorte, ed.) Berkeley, CA: Institute of Governmental Studies, University of California.

Forrester, J. W. (1968) *Urban Dynamics*. Cambridge, MA: MIT Press.

Forrester, J. W. (1971) *World Dynamics*. Cambridge, MA: Wright-Allen Press, Inc.

Frank, P. M. (1978) *Introduction to System Sensitivity Theory*. New York: Academic Press.

Friesma, W., and Culhane, J. (1976) "Social Impacts, Politics, and the Environmental Impact Statement Process," *Natural Resources Journal* 16 (2), 339–356.

Fromm, G., and Taubman, P. (1968) *Policy Simulations with an Econometric Model*. Washington, DC: The Brookings Institution.

†Futures Group, The (1975) "Technology Assessment of Geothermal Energy Resources Development." Glastonbury, CT: The Futures Group.

†Futures Group, The (1976) "A Study of Life-Extending Technologies." Glastonbury, CT: The Futures Group.

Garland, G. A., and Mosher D. C., (1975) "Leachate Effects of Improper Land Disposal," *Waste Age* (March), 435–440.

Gass, S. I., and Sisson, R. L. (1975) *A Guide to Models in Government Planning and Operations.* Potomac, MD: Sauger.

Gastil, R. D. (1977) *Social Humanities.* San Francisco: Jossey-Bass.

Geiger, R. (1957) *The Climate Near the Ground,* 2nd ed. Cambridge, MA: Harvard University Press.

Gendron, B. (1977) *Technology and the Human Condition.* New York: St. Martin's Press.

*Gifford, F. A., and Hanna, S. R. (1970) "Urban Air Pollution Modeling." Paper presented to the Second International Clean Air Congress, Washington, DC: (December).

*Gilliland, M. W. (1975) "Energy Analyses and Public Policy," *Science* 189 (5) (26 September), 1051–1056.

Gofman, J. W., and Tamplin, A. R. (1970) "Epidemologic Studies of Carcinogenesis by Ionizing Radiation." *Proceedings, 6th Berkeley Symposium on Mathematical Statistics and Probability.* Berkeley: University of California.

*Gohagan, J. K. (1975) "A Practical Bayesian Methodology for Program Planning and Evaluation." Las Vegas: ORSA /TIMS Meeting (17–19 November).

Gold, R. L. (1974) "Social Impact of Strip Mining and Other Industrializations of Coal Resources," in *Social Impact Assessment* (C. P. Wolf, ed.). Milwaukee: Environmental Design and Research Association.

*Gold, R. L. (1977) "Combining Ethnographic and Survey Research," in *Methodology of Social Impact Assessment* (K. Finsterbusch and C. P. Wolf, eds.). Stroudsburg, PA: Dowden Hutchison & Ross, pp. 102–107.

Goldberger, A. L., and Duncan, O. D. (eds.) (1973) *Structural Equation Models in the Social Sciences.* New York: Seminar Press.

Goldschmidt, P. G. (1975) "Scientific Inquiry or Political Critique? Remarks on Delphi Assessment: Expert Opinion Forecasting and Group Process by H. Sackman," *Journal of Technological Forecasting and Social Change* 7 (2), 195–213.

Goodwin, R. H., and Niering, W. A. (1974) "Inland Wetlands: Their Ecological Role and Environmental Status," *Bulletin of the Ecological Society of America* 55 (2), 2–6.

Gordon, C., Galloway, W. J., Kugler, B. A., and Nelson, D. L. (1971) *Highway Noise: A Design Guide for Highway Engineers.* Prepared for the National Cooperative Highway Research Program, Washington, DC, Report No. 117NCHRP.

*Gordon, T. J. (1969) "Cross-Impact Matrices—An Illustration of Their Use for Policy Analysis," *Futures* 2, 527–531.

*Gordon, T. J., and Becker, H. S. (1972) "The Uses of Cross-Impact Approaches in Technology Assessment," in *The Methodology of Technology Assessment* (M. Cetron and B. Bartocha). New York: Gordon and Breech.

*Gordon, T. J., and Hayward, H. (1968) "Initial Experiments with the Cross-Impact Matrix Method of Forecasting," *Futures* 4, 100–116.

*Gordon, T. J., and Helmer, O. (1964) "Report on a Long-Range Forecasting Study." Rand Corporation Paper P–2982.

*Gordon, T. J., and Stover, J. (1976) "Using Perceptions and Data About the Future to Improve the Simulation of Complex Systems," *Technological Forecasting and Social Change* 9, 191–211.

†Grad, F. P., Rosenthal, A. J., Rockett, L. R., Fay, J. A., Heywood, J., Kain, J. F., Ingram, G. K., Harrison, D., and Tietenberg, T. (1975) *The Automobile*. Norman, OK: The University of Oklahoma Press.

Green, H. P. (1970) "The Adversary Process in Technology Assessment," *Technology and Society* 5, 163–167.

Green, H. P. (1975) "The Risk Benefit Calculus in Safety Determinations," *George Washington Law Review* 43 (30), 791–799.

Gross, B. M. (1966) "The State of the Nation: Social Systems Accounting," in *Social Indicators* (R. A. Bauer, ed.). Cambridge, MA: MIT Press.

Guetzkow, H., and Simon, H. R. (1955). "The Impact of Certain Communication Nets upon Organization and Performance in Task-Oriented Groups," *Management Science* 1, 233–250.

†Haas, J. E., and Mileti, D. (1976) "An Empirical Approach to Social Impact Assessment: Anticipating the Consequences of Earthquake Prediction." Ann Arbor: Second International Congress on Technology Assessment (27 October).

Hadley, G. F. (1967) *Introduction to Probability and Statistical Decision Theory*. San Francisco: Holden-Day.

Hahn, W. (1977) "Technology Assessment: Some Alternative Perceptions and Its Implications Outside the United States." Testimony before the Subcommittee on Science, Research, and Technology of the House Committee on Science and Technology (3 August) (mimeo., 56 pp.).

Hamilton, M. R. (1977) "The Use of Historical Records to Inform Prospective Technology Assessments," in *Retrospective Technology Assessment—1976* (J. A. Tarr, ed.). San Francisco: San Francisco Press.

*Hammond, K. R., and Adelman, L. (1976) "Science, Values, and Human Judgment," *Science* 194 (22 October), 389–396.

†Harkness, R. C., et al. (1976) *Technology Assessment of Telecommunications–Transportation Interactions*. Menlo Park, CA: SRI.

Harkness, T. (1976) "Visual Analysis Techniques: Outfitting Your Tool Box," in *Environmental Impact Analysis*. Urbana, IL: University of Illinois, Dept. of Agriculture Monograph Series, pp. 59–73.

Harman, W. (1976) *An Incomplete Guide to the Future*. San Francisco: San Francisco Book Company.

Harman, W. (1977) "The Coming Transformation," *The Futurist* (February), 5–11; (April), 106–112.

†Harman, W., and Reuyl, J. (1977) "Solar Energy in America's Future: A Preliminary Assessment." Menlo Park, CA: SRI International.

Harris, Louis (1975) *"The Harris Survey," The Chicago Tribune* (4 December).

†Harvey, D. G., and Menchen, R. W. (1974) *The Automobile—Energy and the Environment: A Technology Assessment of Advanced Automotive Propulsion Systems*. Columbia, MD: Hittman Associates, Inc.

Hawker, T. W. (1973) "Transportation Planning and the Environment." *Proceedings of The American Society of Civil Engineers, Journal of the Transportation Division* 99 (TE3), 499–512.

*Heer, J., and Hagerty, D. (1977) *Environmental Assessments and Statements*. New York: Van Nostrand Reinhold.

*Hencley, S., and Yates, J. R. (eds.) (1974) *Futures in Education: Methodologies*. Berkeley: McCutchan.

Henderson, H. (1973) "Ecologists versus Economists," *Harvard Business Review* (July–August), 28–36, 152–157.

Henderson, H. (1974) "Technology: A Social Cost–Benefit Analysis." Address before American Association of University Women Conference on Technology and Survival (14 February).

Henschel, R. L. (1976) *On the Future of Social Prediction*. Indianapolis, IN: Bobbs-Merrill.

*Hetman, F. (1973) *Society and the Assessment of Technology*. Paris: OECD.

Hill, S. C. (1970) "A Natural Experiment on the Influence of Leadership Behavioral Patterns on Scientific Productivity," *IEEE Transactions on Engineering Management* EM-17, 10–20.

Hiltz, S. R., and Turoff, M. (1977) "Effective Communications Structures for Technology Assessment." New Orleans: Symposium on Technology Assessment, American Chemical Society (20–25 March).

Holm, L. G., Weldon, L. W., and Blackburn, R. D. (1969) "Aquatic Weeds," *Science* 166 (7 November), 699–709.

Holzworth, G. C. (1972) *Mixing Heights, Wind Speeds and Potential for Urban Air Pollution Throughout the Contiguous United States*. Research Triangle Park, NC: U.S. Environmental Protection Agency, Publication No. AP–101.

Hornick, W. F., and Enk, G. A. (1978) *Value Issues in Technology Assessment*. Rensselaerville, NY: Institute on Man and Science.

House, P. W. (1971) "The State of the Art in Urban Game Models." Washington, DC: Environmetrics, Inc.

House, P. W., and Tyndall, G. R. (1975) "Models and Policy Making," in *A Guide to Models in Governmental Planning and Operation* (S. I. Gass and R. L. Sisson, eds.). Potomac, MD: Sauger, pp. 39–60.

*Howard, R. A. (1968) *The Foundations of Decision Analysis*. Stanford, CA: Department of Engineering—Economic Systems, Stanford University.

*Howard, R. A., Matheson, J. E., and North, D. W. (1972) "The Decision to Seed Hurricanes," *Science* 176 (16 June), 1191–1202.

Huber, W. C., and Harleman, D. R. F. (1968) "Laboratory and Analytical Studies of the Thermal Stratification of Reservoirs." Cambridge, MA: MIT, Hydrodynamics Lab Report No. 122.

Huber, W. C., Harleman, D. R. F., and Ryan, P. J. (1972) "Temperature Prediction in Stratified Reservoirs," *Journal of Hydraulics Division, ASCE, HYA* 98, 645–666.

*Huettner, D. A. (1976) "Net Energy Analysis," *Science* 192 (9 April), 101–104.

*Ignatovich, F. R. (1974) "Morphological Analysis," in *Futurism in Education: Methodologies* (S. P. Hencley and J. R. Yates, eds.). Berkeley: McCutchan, pp. 211–234.

Illich, I. (1973) *Tools for Conviviality*. New York: Harper and Row.

*Ingram, G. K., and Fauth, G. R. (1974) *Tassim: A Transportation and Air Shed Simulation Model—Volume 2: Program User's Guide*. Cambridge, MA: MIT, Final Report to U.S. Department of Transportation under Contract DOT-OS–30099.

*Institute of Ecology (1971) "Optimum Pathway Matrix Analysis Approach to the Environmental Decision Making Process: Test Case: Relative Impact of Proposed Highway Alternatives." Athens, GA: University of Georgia (mimeo.).

†International Research and Technology Corporation (1977) *An Assessment of Controlled Environment Agriculture Technology*. McLean, VA: International Research and Technology Corporation. Submitted to the National Science Foundation under Contract C–1026.

*Isard, W. (1960) *Methods of Regional Analysis: An Introduction to Regional Science*. New York: Wiley, Chapter 6.

*Jain, R. K., Urban, L. V., and Stacey, G. S. (1977) *Environmental Impact Analysis: A New Dimension in Decision Making*. New York: Van Nostrand Reinhold.

†Jamaica Bay Environmental Study Group (1971) *Jamaica Bay and Kennedy Airport*. Washington, DC: National Academy of Sciences and National Academy of Engineering.

*Jantsch, E. (1967) *Technological Forecasting in Perspective*. Paris: OECD.

†Jet Propulsion Laboratory (1975) "Should We Have a New Engine?" Vols. I, II. Pasadena: California Institute of Technology.

Johanning, J., and Talvitie, A. (1976) "State of the Art of Environmental Impact Statements in Transportation," *Transportation Research Record* 603. Washington, DC: National Academy of Science, Transportation Research Board.

Johnston, J. (1972) *Econometric Methods*, 2nd ed. New York: McGraw-Hill.

*Jones, M. V. (1971) "A Technology Assessment Methodology. Some Basic Propositions." Report MTR6009, Vol. 1, for the Office of Science and Technology. Washington, DC: The Mitre Corporation (June).

Jones, M. V. (1973) *A Comparative State-of-the-Art Review of Selected U.S. Technology Assessment Studies*. Washington, DC: The Mitre Corporation (M73–62), prepared for National Science Foundation.

†Jones, M. V., and Jones, R. M. (1975) *Scientific Earthquake Prediction: Some First Thoughts on Possible Societal Impacts*. Bethesda, MD: Impact Assessment Institute. Prepared for the National Science Foundation.

*Jones, M. V., and Jones, R. M. (1977) *Twenty-Five National Science Foundation Technology Assessment Studies: An Analytical Bibliography*. Prepared for the National Science Foundation.

Jopling, David G. (1974) "Plant Site Evaluation Using Numerical Ratings," *Power Engineering* (March), 56–59.

Kaill, W. M., and Frey, J. K. (1973) *Environments in Profile, An Aquatic Perspective.* San Francisco: Canfield Press.

*Kane, J. (1972) "A Primer for a New Cross-Impact Language—KSIM," *Technological Forecasting and Social Change* 4, 129–142.

Kantrowitz, A. (1975) "Controlling Technology Democratically," *American Scientist* 63, 505–509.

Kash, D. E. (1977) "Observations on Interdisciplinary Studies and Government Roles," in *Adapting Science to Social Needs.* Washington, DC: American Association for the Advancement of Science, pp. 147–178.

†Kash, D. E., White, I. L., Bergey, K. H., Chartock, M. A., Devine, M. D., Leonard, R. L., Salomon, S. N., and Young, H. W. (1973) *Energy Under the Oceans: A Technology Assessment of Outer Continental Shelf Oil and Gas Operations.* Norman, OK: University of Oklahoma Press.

Katz, M. (1969) "The Function of Tort Liability in Technology Assessment," *University of Cincinnati Law Review* 38 (Fall), 587–662.

Kelly, E. T. (1976) "Technology Assessment in State Government." Presented at Annual Conference, International Society of Technology Assessment, Ann Arbor, MI (26 October).

Kelly, P., Kranzberg, M., Rossini, F. A., Baker, N. R., Tarpley, F. A., Jr., and Mitzner, M. (1978) *Technological Innovation: A Critical Review of Current Knowledge.* San Francisco: The San Francisco Press.

Kennard, B. (1975) "Some Methods and Criteria for Public Participation in Technology Assessment," *ISTA Journal* 1 (3) (September), 43–46.

Kevan, D. K. (1962) *Soil Animals.* New York: Philosophical Library.

Kimball, T. L. (1973) "Why Environmental Quality Indices?" in *The Quality-of-Life Concept.* Washington, DC: U.S. EPA, Office of Research and Monitoring.

Knezo, G. J. (1973) *Social Indicators: A Review of Research and Policy Issues.* Washington, DC: Congressional Research Service.

*Knezo, G. J. (1974) "Toward Using Appropriate Social Data in Technology Assessment," *Technology Assessment* 2 (4), 273–286.

Knight, K. E. (1973) "An Application of Technological Forecasting to the Computer Industry," in *A Practical Guide to Technological Forecasting* (J. R. Bright and M. E. F. Schoeman, eds.). Englewood Cliffs, NJ: Prentice-Hall, pp. 377–403.

Koppel, B. (1977) "Technology Assessing: The Quest for Coherence." Conference held at the East–West Center, Technology and Development Institute, Honolulu, HI.

Kranskopf, T. M., and Bunde, D. C. (1972) "Evaluation of Environmental Impact Through a Computer Modelling Process," in *Environmental Impact Analysis: Philosophy and Methods.* (R. Dutton, and T. Goodak, eds.). Madison, WI: University of Wisconsin Sea Grant Prog., pp. 107–125.

*Kruzic, P. G. (1974) *Cross-Impact Simulation in Water Resources Planning.* Report 74–12 to U.S. Army Engineer Institute for Water Resources. Fort Belvoir, VA. Palo Alto, CA: SRI.

Kuchler, A. W. (1967) *Vegetation Mapping.* New York: Ronald Press Company.

*Kugler, B. A., and Piersol, A. G. (1973) *Highway Noise—A Field Evaluation of Traffic Noise Reduction Measures.* Prepared for the National Cooperative Highway Research Program, Washington, DC: Report No. 144 NCHRP.

Lagerwerff, J. V., and Specht, A. W. (1970) "Contamination of Roadside Soil and Vegetation by Cadmium, Nickel, Lead, and Zinc," *Environmental Science and Technology* 4 (7) (July).

Landis, R. C. (1971) *Mariculture (Sea Farming),* Vol. 5 of *A Technology Assessment Methodology* (M. V. Jones, ed.). Washington, DC: U.S. Office of Science and Technology. MITRE Report MTR 6009, NTIS PB 202−778.

Langham, M. R., and McPherson, W. W. (1976) (Letter to the Editor) *Science* 192 (2 April), 8.

Lasswell, H. D. (1971) *A Pre-View of Policy Sciences.* New York: American Elsevier.

Lawless, E. W. (1977) *Technology and Social Shock.* New Brunswick, NJ: Rutgers University Press.

League of Women Voters (1970) *Citizens Guide to Clean Air.* Washington, DC: U.S. Environmental Protection Agency.

*Leininger, G., Jutila, S., King, J., Muraco, W., and Hansell, J. (1975) *The Total Assessment Profile.* Toledo, OH: University of Toledo (NTIS No. N75−31919/4).

*Leontief, W. (1966) *Input Output Economics.* Oxford: Oxford University Press.

Leopold, L. B., Clarke, F. E., Henshaw, B. B., and Balsley, J. R. (1971) *A Procedure for Evaluating Environmental Impact.* Washington, DC: U.S. Geological Survey, Circ. 645.

Levis, A. H. (1973) *Users Manual, Methodology, II.* Wash-use-1 Urban Systems Engineering Demonstration Project Final Report, Systems Control, Inc. George S. Nolte and Associates, Snohomish County Planning Department.

Lewis, A. C., Sadosky, T. L., and Connolly, T. (1975) "The Effectiveness of Group Brainstorming in Engineering Problem Solving," *IEEE Transactions in Engineering Management* EM-22 (3), 119−124.

Lind, R. C. (1972) "The Analysis of Benefit−Risk Relationships: Unresolved Issues and Areas for Future Research," in *Perspectives on Benefit-Risk Decision Making.* Washington, DC: Committee on Public Engineering Policy, National Academy of Engineering.

Lindblom, C. E. (1959) "The Science of Muddling Through," *Public Administration Revew* 19 (Spring), 79−88.

Lindblom, C. E. (1968) *The Policy-Making Process.* Englewood Cliffs, NJ: Prentice-Hall.

*Linstone, H. A., and Turoff, M. (eds.) (1975) *The Delphi Method: Techniques and Applications.* Reading, MA: Addison-Wesley, Advanced Books Program.

Little, Arthur D., Inc. (1971) *Transportation and Environment: Synthesis for Action: Impact of National Environmental Policy Act of 1969 on the Department of Transportation,* Vol. 3. *Options for Environmental Management.* Report to the Secretary. Washington, DC: U.S. Department of Transportation.

†Little, Arthur D., Inc. (1975) *The Consequences of Electronic Funds Transfer.*

Cambridge, MA: Arthur D. Little, Inc. Prepared for the National Science Foundation under Contract NSF-C884.

Loucks, D. P., Bower, B. T., and Spofford, W. D., Jr. (1973) "Environment Noise Management," in *Proceedings of the American Society of Civil Engineers, Journal of the Environmental Engineering Division*, 99 (EE6), 813–829.

Love, S. K. (1961) "Relationship of Impoundment to Water Quality," *American Water Works Association Journal* 53, 559–568.

Love, S. K., and Slack, K. V. (1963) "Controls on Solution and Precipitation in Reservoirs" in Symposium on Streamflow Regulation for Water Quality Control. Cincinnati, OH: Public Health Service Publication No. 999-WP-30, pp. 97–120.

Lovins, A. B. (1977) "Cost–Risk Benefit Assessments in Energy Policy," *George Washington Law Review* 45 (5), 911–943.

Low, G. W. (1974) "National Socio-Economic Model—An Overview of Structure." Cambridge, MA: Systems Dynamic Group, Massachusetts Institute of Technology (Paper No. D-2123).

Lowi, T. J. (1964) "American Business, Public Policy, Case Studies and Political Theory." *World Politics* 16, 216–30.

Lowi, T. J. (1970) "Decision Making vs. Policy Making: Toward An Antidote for Technocracy," *Public Administration Review* 30 (May/June), 314–324.

Lowi, T. J. (1976) *Poliscide*. New York: Macmillan.

*Mack, R. P. (1974) "Criteria for Evaluation of Social Impacts of Flood Management Alternatives," in *Social Impact Assessment* (C. P. Wolf, ed.). Milwaukee: Environmental Design Research Association, pp. 175–195.

*Malone, D. W. (1975a) "An Overview of Interpretive Structural Modeling," in *Perspectives on Technology Assessment* (S. Arnstein and A. Christakis, eds.). Jerusalem: Science and Technology Publishers, pp. 229–233.

*Malone, D. W. (1975b) "An Introduction to the Application of Interpretive Structural Modeling," in *Portraits of Complexity*. Columbus, OH: Battelle Monograph No. 9.

Mantoux, P. (1961) *The Industrial Revolution in the Nineteenth Century*. New York: Harcourt.

*Martin, D. O., and Tikvart, J. A. (1968) "A General Atmospheric Diffusion Model for Estimating the Effects in Air Quality of One or More Sources." APCA Paper No. 68–148 (June). Air Pollution Control Association.

Martin, R. J., and Willeke, G. (1978) *The House That Jack Built: An Agenda for the Assessment of the Technologies of the Built Environment*. Atlanta, GA: Georgia Institute of Technology. Prepared for the National Science Foundation.

*Martino, J. P. (1973) "Trend Extrapolation," in *A Practical Guide to Technological Forecasting* (J. R. Bright and M. E. F. Schoeman, eds.). Englewood Cliffs, NJ: Prentice-Hall, pp. 106–125.

*Martino, J. P., et al. (1978) *Technology Assessment: An Appraisal of the State of the Art*. Dayton, OH: University of Dayton Research Institute.

Marx, K. (1906) *Capital*, 3rd ed. (S. Moore and E. Aveling, trans.). New York: Modern Library.

Marx, K. (1964) *Economic and Philosophical Manuscripts. Karl Marx: Early Writings* (T. B. Bottomore, trans., ed.). New York: McGraw-Hill.

Maslow, A. H. (1968) *Toward a Psychology of Being*, 2nd ed. Princeton, NJ: Van Nostrand.

Masters, G. M. (1974) *Introduction to Environmental Science and Technology*. New York: Wiley.

McCaull, J., and Crossland, J. (1974) *Water Pollution*. New York: Harcourt, Brace, Jovanovich, Inc.

McDermott, J. (1969) "Technology: The Opiate of the Intellectuals," *New York Review of Books* (13 July), 25–35.

*McGrath, J. H. (1974) "Relevance Trees," in *Futurism in Education* (S. R. Hencley and J. R. Yates, eds.). Berkeley, CA: McCutchan, pp. 71–96.

*McHarg, I. (1968) "A Comprehensive Highway Route-Selection Method," *Highway Research Record* No. 246, 1–15.

*McNamara, J. F. (1974) "Markov Chain Theory and Technological Forecasting," in *Futures in Education: Methodologies* (S. Hencley and J. R. Yates, eds.). Berkeley, CA: McCutchan, pp. 301–345.

Meadows, D. H., Meadows, D. L., Randers, J., and Behrens, W. W., III (1972) *The Limits to Growth*. New York: Universe Books.

Mesarovič, M. B., and Pestel, E. (1974) *Mankind at the Turning Point: The Second Report to the Club of Rome*. New York: Dutton.

Mesthene, E. G. (1968) "How Technology Will Shape the Future," *Science* 161 (12 July), 135–143.

Michael, D. N., Berg, M. R., and Rich, R. (1976) *Research on the Utilization of Technology Assessment Studies in Public Policy-Making and Planning*. Ann Arbor, MI: Center for Research on Utilization of Scientific Knowledge, University of Michigan. Unsolicited research proposal submitted to the National Science Foundation.

†Midwest Research Institute (1977) *Technology Assessment of Integration of the Hog-Pork Industry*. Kansas City, MO: Midwest Research Institute. Prepared for the National Science Foundation under Contract NSF-C850.

Miller, G. T. (1975) *Living in the Environment: Concepts, Problems, and Alternatives*. Belmont, CA: Wadsworth Publ. Co.

*Miller, P. D. (1976) "Technological Innovation and Assessing the Distribution of Its Social Impacts." Paper presented at annual conference, International Society for Technology Assessment, Ann Arbor, MI.

Mishan, E. J. (1969) *Technology and Growth*. New York: Praeger.

*Mishan, E. J. (1976) *Cost–Benefit Analysis*. New York: Praeger.

*Mitchell, A., Dodge, B. H., Kruzic, P. G., Miller, D. C., Schwartz, P., and Suta, B. E. (1975) *Handbook of Forecasting Techniques*. Stanford Research Institute Report to U.S. Army Corps of Engineers Institute for Water Resources, AD-A019 280; available from NTIS, Springfield, VA.

Mitroff, I. I., and Turoff, M. (1973) "The Whys Behind the Hows," *IEEE Spectrum* 10, (3) 62–71.

Moore, J. L., et al. (1973) *A Methodology for Evaluating Manufacturing Environmental Impact Statements for Delaware's Coastal Zone*. Report to the State of Delaware. Columbus, OH: Battelle Memorial Institute.

Multiagency Task Force (1972) *Guidelines for Implementing Principles and Standards for Multiobjective Planning of Water Resources.* Washington, DC: U.S Bureau of Reclamations, review draft.

Munn, R. E. (1970) *Biometeorological Methods.* New York: Academic Press.

National Environmental Policy Act (1969) Public Law 91–190 (1 January 1970), 42 U.S.C. 4321–4347.

†Nathans, Robert R., and Associates (1972) *U.S. Deepwater Port Study, Vol. IV: The Environmental and Ecological Aspects of Deepwater Ports.* U.S. Army Engineer Institute for Water Resources. IWR Report 72–8.

Noble, D. (1977) *America by Design.* New York: Knopf.

Odum, E. P. (1971) *Fundamentals of Ecology,* 3rd ed. Philadelphia: W. B. Saunders.

Oettinger, A., and Shapiro, P. D. (1974) "A Review of Edward M. Dickson, in association with Raymond Bowers, The Video Telephone, A New Era in Telecommunications, a Preliminary Technology Assessment." Available through NTIS (PB 235 145).

*Oglesby, C. H., Bishop, B., and Willeke, G. E. (1970) "A Method for Decisions Among Freeway Location Alternatives Based on User and Community Consequences," *Highway Research Record,* 305, 1–15.

*Olsen, M. E., and Merwin, D. J. (1976) *Toward A Methodology for Conducting Social Impact Assessments Using Quality of Life Social Indicators.* Richland, WA: Battelle Pacific Northwest Labs., BNWL-2084. Prepared for U.S. Energy Research and Development Administration; available through NTIS (A03/MF A01).

*Olsen, M. E., Greene, M. R., Melber, B., Curry, M. G., and Merwin, D. J. (1977) A Social Impact and Management Methodology Using Several Indicators and Planning Strategies. Richland, WA: Battelle-Pacific Northwest Labs.

*Organisation for Economic Co-operation and Development (1975) *Methodological Guidelines for Social Assessment of Technology.* Paris: OECD.

*Orlob, G. T., and Selna, L. G. (1968) "Mathematical Simulation of Stratification in Deep Impoundments, in *Proceedings of Speciality Conference on Current Research into Effects of Reservoirs on Water Quality,* ASCE, T.R. No. 17. Nashville, TN: Department of Environment and Water Resources Engineering, Vanderbilt University.

Osborn, A. F. (1957) *Applied Imagination* (rev. ed.). New York: Scribners.

*Otway, H. J., and Pahner, P. D. (1976) "Risk Assessment," *Futures* 8 (April), 122–134.

Otway, H. J., Maurer, D., and Thomas, K. (1978) "Nuclear Power: The Question of Public Acceptance," *Futures* 10, 109–118.

*Paik, I. K., Harrington, J., Jr., and McElroy, F. W. (1974) "The Integrated Multi-Media Pollution Model." Washington Environmental Research Center, prepared for the U.S. Environmental Protection Agency (EPA-600/5–74–020).

Paschen, H. (1977) "Institutionalizing the TA Function." Presented at Conference: Technology Assessing, The Quest for Coherence. Honolulu, HI: Technology and Development Institute (May 30–June 10) (mimeo., 35 pp.).

Paschen, H., Gresser, K., Friedrichs, G., and Conrad, F. (1975) "Some Problems

of Evaluations in Technology Assessment Studies" in *Methodological Guidelines for Social Assessment of Technology*. Paris: OECD, pp. 119–131.
Pavitt, K. (1972) "Analytical Techniques in Government Science Policy," *Futures* 4, 5–14.
†Pearson, G. L., Pomerov, W. L., Sherwood, G. A., and Winder, J. S., Jr. (1975) "A Scientific and Policy Review of the Final Environmental Statement for the Initial Stage, Garrison Diversion Unit (North Dakota)." Indianapolis: The Institute of Ecology.
Peat, Marwick, Livingston, and Company (1969) "*Empiric Activity—Allocation Model Study Design.*" Washington, DC: Metropolitan Council of Governments.
Peat, Marwick, and Mitchell, Inc. (1972) "Atlanta Regional Commission, Activity Allocation Mode." Atlanta, GA.
Perkins, H. C. (1974) *Air Pollution*. New York: McGraw-Hill.
Peters, B. G. (1977) "Insiders and Outsiders-Politics of Pressure Group Influence on Bureaucracy," *Administration and Society* 9 (2), 191–218.
Peterson, A. P. G., and Gross, E. E. (1972) *Handbook of Noise Measurement*. West Concord, MA: General Radio Company.
Petrie, H. D. (1976) "Do You See What I See? The Epistemology of Interdisciplinary Inquiry," *Journal of Aesthetic Education* 10, 9–15.
Platt, R. B., and Griffiths, J. F. (1964) *Environmental Measurement and Interpretation*. New York: Reinhold.
Poole, R. W. (1974) *An Introduction to Quantitative Ecology*. New York: McGraw-Hill.
*Porter, A. L., and Rossini, F. A. (1977) "Evaluation Designs for Technology Assessments and Forecasts," *Technological Forecasting and Social Change* 10, 369–380.
*Prest, A. R., and Turvey, R. (1965) "Cost/Benefit Analysis: A Survey," *The Economic Journal* 75 (300) (December), 665–735.
Preston, R. S. (1972) "The Wharton Annual Industry Forecasting Model," *Studies in Quantitative Economics No. 7*. Philadelphia, PA: Economics Research Unit, Department of Economics, Wharton School, University of Pennsylvania.
Price, D. K. (1965) *The Scientific Estate*. Cambridge, MA: Harvard University Press.
*Rao, P., and Miller, R. L. (1971) *Applied Econometrics*. Belmont, CA: Wadsworth Publ. Co.
Rein, M. (1971) "Social Policy Analysis as the Interpretation of Beliefs," *AIP Journal* (September), 297–310.
Rescher, N. (1969) "What is Value Change? A Framework for Research," in *Values and the Future* (K. Baier and N. Rescher, eds.). New York: The Free Press, 68–109.
Rescher, N. (1974) "The Environmental Crisis and the Quality of Life," in *Philosophy and the Environmental Crisis* (W. T. Blackstone, ed.). Athens, GA: University of Georgia Press, pp. 90–104.
Richardson, H. W. (1969) *Elements of Regional Economics*. New York: Praeger.

Riecken, H. W., and Boruch, R. F. (eds.). (1974) *Social Experimentation*. New York: Academic Press.

Rokeach, M. (1973) *The Nature of Human Values*. New York: Free Press.

Roper, A. T. (1976) "Technology Assessment: A Vehicle for Dialogue." Paper presented at conference, International Society for Technology Assessment, Ann Arbor, MI.

Roper, A. T., and Brophy, P. D. (1977) "An Assessment of Engineering Education for the Decade of the 1980s." Terre Haute, IN: Center for Technology Assessment and Policy Studies, Rose-Hulman Institute of Technology.

Roper, A. T., and Dekker, D. (1978) *Values: A Simulation of Values Impacts of Technological Activity*. Terre Haute, IN: Center for Technology Assessment and Policy Studies.

*Rosove, P. E. (1973) *A Trend Impact Matrix for Societal Impact Assessment*. Los Angeles: Center for Futures Research, Graduate School of Business Administration, University of Southern California.

*Rossini, F. A., Porter, A. L., and Zucker, E. (1976) "Multiple Technology Assessments," *Journal of the International Society for Technology Assessment* 2, 21–28.

*Rossini, F. A., Carpenter, S. R., Havick, J., Kelly, P., Lipscomb, M. A., and Porter, A. L. (1977) "Epistemology of Interdisciplinary Research: The Case of Technology Assessment," in *The General Systems Paradigm: Science of Change and Change of Science*. Washington, DC: Society for General Systems Research, pp. 451–498.

*Rossini, F. A., Porter, A. L., Kelly, P., and Chubin, D. E. (1978) *Frameworks and Factors Affecting Integration Within Technology Assessments*. Report to the National Science Foundation, Grant ERS 76–04474. Atlanta: Georgia Institute of Technology.

Ryan, P. J., and Harleman, D. R. F. (1971) "Prediction of the Annual Cycle of Temperature Changes in a Stratified Lake or Reservoir, Mathematical and User's Manual." Cambridge, MA: MIT, Ralph M. Parsons Lab. Report No. 137, Department of Civil Engineering.

*Sackman, H. (1974) "Delphi Assessment, Expert Opinion, Forecasting, and Group Process." Rand Corporation Report R–1283–pl (April).

*Sage, A. P. (1977) *Methodology for Large Scale Systems*. New York: McGraw-Hill.

Sartor, J. D., and Boyd, G. B. (1975) *Water Pollution Control in Low Density Areas: Proceedings of a Rural Environmental Engineering Conference*. Hanover, NH: University Press of New England, pp. 301–316.

*Sassone, P. G., and Schaffer, W. A. (1978) *Cost–Benefit Analysis: A Handbook*. New York: Academic Press.

Schaffer, W. A., and Davidson, L. S. (1972) *Economic Impact of the Falcons on Atlanta*. Atlanta, GA: The Atlanta Falcons.

Schaffer, W. A., Laurent, E. A., and Sutter, E. M., Jr. (1972) "Using the Georgia Economic Model." Atlanta, GA: Georgia Institute of Technology, College of Industrial Management.

Schick, A. (1976) "The Supply and Demand for Analysis on Capitol Hill," *Policy Analysis* 2, 215–234.
Schindler, D. W. (1976) "The Impact Statement Boondoggle," *Science* 192 (7 May), 509.
Schmookler, J. (1966) *Invention and Economic Growth*. Cambridge, MA: Harvard University Press.
Schumacher, E. F. (1973) *Small Is Beautiful: Economics As If People Mattered*. New York: Harper Torchbooks.
Scriven, M. (1967) "The Methodology of Evaluation," in *Perspectives of Curriculum Evaluation* (R. Tyler, R. Gagne, and M. Scriven, eds.). Chicago: Rand McNally.
Settle, T. (1976) "The Moral Dilemma in Political Assessments of the Social Impact of Technology," *Philosophy of the Social Sciences* 6 (4) (December), 315–334.
Sheldon, E. B., and Land, K. C. (1972) "Social Reporting for the 1970's: A Review and Programmatic Statement," *Policy Science* 3, 137–151.
Shields, M. A. (1974) *Social Impact Assessment: An Analytical Bibliography*. Ft. Belvoir, VA: U.S. Army Institute for Water Resources Available from NTIS (AD-A003 245/8).
Silverman, M. (1976) *Project Management*. New York: Wiley.
Simmonds, W. H. C. (1973) "Analysis of Industrial Behavior and Its Use in Forecasting," in *A Practical Guide to Technological Forecasting* (J. R. Bright and E. F. Schoeman, eds.). Englewood Cliffs, NJ: Prentice-Hall, pp. 215–237.
Sklarew, R. C., Turner, D. B., and Zimmerman, J. R. (1973) "Modeling Transportation Impact on Air Quality," in *Proceedings of the ASCE Transportation Division Environmental Impact Specialty Conference*. New York: American Society for Civil Engineers, pp. 127–142.
Slesser, M. (1974) "Energy Analysis in Technology Assessment," *Technology Assessment* 2 (3), 201–208.
Smith, R. L. (1974) *Ecology and Field Biology*, 2nd ed. New York: Harper and Row.
Smithsonian Institution (1974) *Planning Considerations for Statewide Inventories of Critical Environmental Areas, a Reference Guide*. Washington, DC: Center for Natural Areas, Smithsonian Institution, Report Three.
Sorensen, J. (1971) *A Framework for Identification and Control of Resource Degradation and Conflict in the Multiple Use of the Coastal Zone*. Berkeley: University of California, Department of Landscape Agriculture.
Sperry, W. C. (1968) *Aircraft Noise Evaluation*. Washington, DC: Federal Aviation Administration, U.S. Department of Transportation, FAA Report No. 68–34.
*Starr, C. (1969) "Social Benefits vs Technological Risk," *Science* 165 (19 September), 1232–1238.
*Stover, L. V. (1972) *Environmental Impact Assessment: A Procedure*. Miami, FL: Sanders and Thomas, Inc.
*Streeten, P. (1971) *Cost–Benefit and Other Problems of Method*. Paris: Mouton.
Sundaram, T. R., Rehm, R. G., Rudinger, G., and Merritt, G. E. (1971) "Research

on the Physical Aspects of Thermal Pollution." Buffalo, NY: Water Pollution Control Research, Series 16130 DPU, Cornell Aeronautical Lab., Inc.

Sylvester, R. O., and DeWalle, F. B. (1972) *Character and Significance of Highway Runoff of Waters*. Seattle, WA: Department of Civil Engineering, University of Washington.

Tamblyn, T. A., and Cederborg, E. A. (1975) "The Environmental Assessment Matrix as a Site-Selection Tool—A Case Study," *Nuclear Technology* 25, 598–606.

Tarr, J. A. (ed.) (1977) *Retrospective Technology Assessment—1976*. San Francisco: The San Francisco Press.

Thatcher, V. S. (ed.) (1971) *The New Webster Encyclopedia Dictionary of the English Language*. Chicago: Consolidated Book Publ.

Theil, H. (1978) *Introduction to Econometrics*. Englewood Cliffs, NJ: Prentice-Hall.

Thompson, G. E. (1959) "An Investigation of the Local Employment Multiplier," *Review of Economics and Statistics* 41, 61–67.

Thuesen, H. G., Fabrycky, W. J., and Thuesen, G. J. (1977) *Engineering Economy*, 5th ed. Englewood Cliffs, NJ: Prentice-Hall.

Tiebout, C. M. (1962) *The Community Economic Base Study*. Washington, DC: Committee for Economic Development.

*Tiller, J. S. (1979) "Uses of Input–Output Analysis in Technology Assessments." Unpublished master's thesis, Georgia Institute of Technology.

Toffler, A. (1969) "Value Impact Forecaster—A Profession of the Future" in *Values and the Future* (K. Baier and N. Rescher, eds.). New York: The Free Press, pp. 1–30.

Transportation Research Board (1976) *Highway Traffic Noise Prediction Methods*. Washington, DC: Transportation Research Board, Circular No. 175.

Tribe, L. H. (1972) "Policy Science: Analysis or Ideology," *Philosophy and Public Affairs* 2 (1) (Fall), 66–110.

Tribe, L. H. (1973a) *Channeling Technology Through Law*. Chicago: Bracton Press, Ltd.

Tribe, L. H. (1973b) "Technology Assessment and the Fourth Discontinuity: the Limits of Instrumental Rationality," *Southern California Law Review* 46 (June), 617–660.

Tribe, L. H., Schelling, C. S., and Voss, J. (eds.) (1976) *When Values Conflict*. Cambridge, MA: Ballinger.

Turner, D. B. (1967) *Workbook of Atmosphere Dispersion Estimates*. Cincinnati, OH: Public Health Service, National Center for Air Pollution Control, PHS Publication No. 999–AP–26.

*Turoff, M. (1972) "An Alternative Approach to Cross-Impact Analysis," *Technological Forecasting and Social Change* 3, 309–339.

United Nations Statistical Office (1976) "Report on National Practices and Plans in Reporting Statistics of Levels of Living." U.N. Document ESA/STA/AC. 4/2 (March).

U.S. Army (1975) Construction Engineering Research Laboratory "Computer Aided Environmental Impact Analysis for Construction Activities: User Manual." Champaign, IL: Technical Report No. E–50.

*U.S. Army Corps of Engineers (1970) "Application of WRE Reservoir Temperature Simulation Model NPD Computer Program 723–1C5–G067." North Pacific Division.

*U.S. Army Corps of Engineers (1972) "Matrix Analysis of Alternatives for Water Resource Development." Tulsa District draft technical paper.

U.S. Army Corps of Engineers (1973) Information Supplement No. 1 to Section 122 Guidelines CER 1105–2–105 (15 December 1972). Washington, DC.

†U.S. Army Corps of Engineers (1976) Baltimore District. "Binghampton Wastewater Management Study: Institutional Analysis Appendix."

U.S. Atomic Energy Commission (1972) Directorate of Regulatory Standards. *Guide to the Preparation of Environmental Reports for Nuclear Power Plants.* Washington, DC: Atomic Energy Commission (Issued for comment, August 1972).

U.S. Bureau of Labor Statistics (1971a) *Employment and Earnings Statistics for States and Areas 1939–1970.* Washington, DC: U.S. BLS, Bull. 1370.

U.S. Bureau of Labor Statistics (1971b) *Employment and Earnings Statistics for the United States, 1909–1970.* Washington, DC: U.S. BLS, Bull. 1312.

U.S. Bureau of Reclamation (1974) *Final Environmental Statement for the Initial Stage, Garrison Diversion Unit (North Dakota).* Washington, DC: EIS-ND-74-0058-F.

U.S. Congress, Congressional Research Service (1973a) "A Legislative History of the Water Pollution Control Act Amendments of 1972." Document prepared for Senate Committee on Public Works. Washington, DC: Committee Print, Serial No. 93–1.

U.S. Congress, Congressional Research Service (1973b) "Office of Technology Assessment." Document prepared for House Committee on Science and Astronautics. Washington, DC: U.S. Government Printing Office.

U.S. Congress, House (1966) Committee on Science and Astronautics. "Inquiries, Legislation, Policy Studies re Science and Technology, Review and Forecast." Second Progress Report of Subcommittee on Science, Research, and Development. Washington, DC: U.S. Government Printing Office.

U.S. Congress, House (1972) Committee on Rules and Administration. "Technology Assessment for Congress." Subcommittee on Computer Services. Washington, DC: U.S. Government Printing Office.

*U.S. Congress, House (1975) Committee on Merchant Marine and Fisheries. "Computer Simulation Methods to Aid National Growth Policy." Subcommittee on Fisheries and Wildlife Conservation. Washington, DC: U.S. Government Printing Office.

U.S. Congress, House (1976) Commission on Information and Facilities. "The Office of Technology Assessment: A Study of Its Organizational Effectiveness." Washington, DC: U.S. Government Printing Office.

U.S. Congress, House (1978) Subcommittee on Science, Research, and Technology. "Review of the Office of Technology Assessment and Its Organic Act." Washington, DC: U.S. Government Printing Office.

U.S. Congress, Joint Committee on Economics (1976) "Priorities and Efficiency in Federal R and D: A Compendium of Papers." Washington, DC: U.S. Government Printing Office (29 October).

U.S. Congress, Legislative Reference Service (1969) "Technical Information for Congress." Document prepared for House Committee on Science and Astronautics. Washington, DC: U.S. Government Printing Office.

U.S. Congress, Office of Technology Assessment (1974) "Hearings Before Technology Assessment Board, June 12." Washington, DC: U.S. Government Printing Office.

†U.S. Congress, Office of Technology Assessment (1975) *Automated Guideway Transit*. Washington, DC: U.S. Government Printing Office.

U.S. Congress, Office of Technology Assessment (1976a) "Annual Report to Congress." Washington, DC: U.S. Government Printing Office.

U.S. Congress, Office of Technology Assessment (1976b) "Hearings Before Technology Assessment Board: Technology Assessment Activities in the Industrial Academic, and Governmental Communities, June 8, 9, 10, 14." Washington, DC: U.S. Government Printing Office.

U.S. Congress, Office of Technology Assessment (1976c) "Summary: Hearings Before Technology Assessment Board, June 8, 9, 10, 14." Washington, DC: U.S. Government Printing Office.

†U.S. Congress, Office of Technology Assessment (1976d) "An Assessment of Information System Capabilities Required to Support U.S. Materials Policy Decisions." Washington, DC: U.S. Government Printing Office.

U.S. Congress, Office of Technology Assessment (1977a) "Technology Assessment in Business and Government: Summary and Analysis." Washington, DC: U.S. Government Printing Office.

†U.S. Congress, Office of Technology Assessment (1977b) "Application of Solar Technology to Today's Energy Needs." Washington, DC: U.S. Government Printing Office. Prepublication draft.

U.S. Congress, Office of Technology Assessment (1977c) "Program Manager's Workshop." Washington, DC: U.S. Government Printing Office (May).

†U.S. Congress, Office of Technology Assessment (1978) "A Technology Assessment of Coal Slurry Pipelines." Washington, DC: U.S. Government Printing Office.

U.S. Congress, Senate (1972) Committee on Rules and Administration. "Technology Assessment for the Congress." Subcommittee on Computer Sources. Washington, DC: U.S. Government Printing Office.

U.S. Council on Environmental Quality (1974) "Guidelines for the Preparation of Environmental Impact Statements." Washington, DC: U.S. Government Printing Office.

U.S. Council on Environmental Quality (1975) "Sixth Annual Report." Washington, DC: U.S. Government Printing Office.

U.S. Council on Environmental Quality (1976a) "Seventh Annual Report." Washington, DC: U.S. Government Printing Office.

U.S. Council on Environmental Quality (1976b) "EPA Categories for Ranking Environmental Impact Statements." 102 *Monitor* 6 (8) (September). Washington, DC: U.S. Government Printing Office.

U.S. Council on Environmental Quality (1976c) "Environmental Impact Statements: An Analysis of Six Years Experience by Seventy Federal Agencies." Washington, DC: U.S. Government Printing Office.

U.S. Council on Environmental Quality (1977) "Eighth Annual Report." Washington, DC: U.S. Government Printing Office.

U.S. Council on Environmental Quality (1978) "National Environmental Policy Act—Regulations for Implementation of Procedural Provisions," *Federal Register* 36 (November 29), Part 6, 55978–56007.

†U.S. Department of Agriculture (1975). *Minimum Tillage: A Preliminary Technology Assessment.* Washington, DC: Office of Planning and Evaluation, U.S. Department of Agriculture.

U.S. Department of Commerce (1974) Bureau of Economic Analysis. "Input–Output Structure of the U.S. Economy, 1967." Washington, DC: U.S. Government Printing Office.

U.S. Department of Commerce (1975) Bureau of the Census. *Historical Statistics of the United States: Colonial Times to 1970.* House Document No. 93–78, pp. 574–585.

U.S. Department of Health, Education, and Welfare (1973) National Heart and Lung Institute. "The Totally Implantable Heart." Washington, DC: U.S. Government Printing Office.

U.S. Department of Health, Education, and Welfare (1976) *Policy Analysis Sourcebook for Social Programs.* Washington, DC: U.S. Government Printing Office.

U.S. Department of Housing and Urban Development (1971) "HUD Noise Assessment Guidelines." Washington, DC: U.S. Government Printing Office.

U.S. Department of Housing and Urban Development (1974) "Environmental Impact Statements Format," *Federal Register* 39 (37) (22 February), 6824.

U.S. Department of Transportation (1974) *Aircraft Sound Description System, Vols. 1–4.* Washington, DC: Federal Aviation Administration, U.S. Department of Transportation, FAA Report EQ-74-2.

U.S. Energy Research and Development Administration (1975) "Total Energy, Electric Energy and Nuclear Power Projections" [WASH-1139 (75)].

U.S. Environmental Protection Agency (1971a) "National Primary and Secondary Ambient Air Quality Standards." *Federal Register* 36 (November 25), 228.

U.S. Environmental Protection Agency (1971b) *Methods for Chemical Analysis of Water and Wastes.* Washington, DC: U.S. EPA.

U.S. Environmental Protection Agency, Region X (1973a) "Environmental Impact Statement Guidelines." Seattle, WA.

U.S. Environmental Protection Agency (1973b) *The Quality of Life Concept.* Washington, DC: The EPA Office of Research and Monitoring.

U.S. Environmental Protection Agency (1973c) *Public Health and Welfare Criteria for Noise* (July 27). Washington, DC: U.S. EPA.

U.S. Environmental Protection Agency (1974a) *A Guide for Considering Air Quality in Urban Planning.* Washington, DC: U.S. Government Printing Office.

*U.S. Environmental Protection Agency (1974b) *Strategic Environmental Assessment System (SEAS).* Washington, DC: U.S. Government Printing Office.

†U.S. Environmental Protection Agency (1976) "Request for Proposal: A Technology Assessment of Energy Development in the Appalachian Region. Cincinnati" (RFP CI 76-0320).

U.S. Environmental Protection Agency (1977) *Decision-Makers Guide in Solid Waste Management,* 2nd ed. Washington, DC: U.S. Government Printing Office.

U.S. Federal Aviation Administration (1977) *Impact of Noise on People.* Washington, DC: U.S. Government Printing Office.

U.S. Federal Highway Administration (1972) *Air Quality Manual.* Washington, DC: Office of Research, Federal Highway Administration, Report FHWA-RD-72-33–40 (8 Vols.).

U.S. Federal Highway Administration (1973) *Noise Standards and Procedures.* Washington, DC: U.S. Federal Highway Administration, PPM 90-2.

†U.S. Federal Highway Administration (1974) "Final Environmental Statement for U.S. 59 in Stevens County, Minnesota" (EIS-MN-74-0180-F).

U.S. National Academy of Engineering (1969) *A Study of Technology Assessment,* Committee on Public Engineering Policy (COPEP). Document prepared for House Committee on Science and Astronautics. Washington, DC: U.S. Government Printing Office.

U.S. National Academy of Public Administration (1970) *A Technology Assessment System for the Executive Branch.* Document prepared for House Committee on Science and Astronautics. Washington, DC: U.S. Government Printing Office (July).

U.S. National Academy of Public Administration (1972) "Technology Assessment in State Government." Report of working conference (September).

U.S. National Academy of Sciences (1969) *Technology: Processes of Assessment and Choice.* Committee on Science and Public Policy (COSPUP). Document prepared for House Committee on Science and Astronautics. Washington, DC: U.S. Government Printing Office.

†U.S. National Academy of Sciences (1975) *Assessing Biomedical Technologies.* Washington, DC: Committee on the Life Sciences and Social Policy, NAS.

U.S. National Science Foundation (1974) *Program Solicitation: Technology Assessments in Selected Areas.* Washington, DC: National Science Foundation.

U.S. National Science Foundation (1977) *Program Announcement: Technology Assessments in Selected Areas.* Washington, DC (NSF77-25).

U.S. Nuclear Regulatory Commission (1975) *Reactor Safety Study: An Assessment of Accident Risks in U.S. Commercial Nuclear Power Plants.* Washington, DC: U.S. NRC, WASH-1400.

U.S. Nuclear Regulatory Commission (1976) "Final Generic Environmental

Statement on the Use of Recycled Plutonium in Mixed Oxide Fuel in Light Water Cooled Reactors." Washington, DC: NTIS (PB 256 452).
U.S. Office of Management and Budget (1974) *Social Indicators, 1973.* Washington, DC: U.S. OMB.
Victor, P. (1972) "The Macro-economics of Pollution," *Technology and Society* 7 (November), 118–121.
Vlachos, E. (1976a) "Transnational Interest in Technology Assessment." Presented at Annual Conference, International Society of Technology Assessment, Ann Arbor, MI (26 October).
Vlachos, E. (1976b) "Technology Assessment for New Water Supplies: The Case of the Arid West." Paper presented at the Second International Congress on Technology Assessment, Ann Arbor, MI (October).
†Vlachos, E., and Hendricks, D. W. (1976) "Secondary Impacts and Consequences of Highway Projects." Ft. Collins, CO: Colorado State University, Department of Civil Engineering, Environmental Engineering Technical Department (October).
*Vlachos, E., Buckley, W., Filstead, W. J., Jacobs, S. E., Maruyama, M., Peterson, J. H., and Willeke, G. E. (1975) "Social Impact Assessment: An Overview." Ft. Belvoir, VA: U.S. Army Engineer Institute for Water Resources, IWR Paper 75-P7 (December).
Wakeland, W. (1976) "QSIM 2: A Low Budget Heuristic Approach to Modeling and Forecasting," *Technological Forecasting and Social Change* 9, 213–229.
Waller, R. A. (1975) "Assessing the Impact of Technology on the Environment," *Journal of Long Range Planning* 8 (February), 43–51.
*Walton, L. E., Jr. and Lewis, E. (1971) *A Manual for Conducting Environmental Impact Studies.* Virginia Highway Research Council (available through the National Technical Information Service, PB-210-222).
Walton, W. C. (1970) *Groundwater Resource Evaluation.* New York: McGraw-Hill.
*Warner, M. L., and Preston, E. H. (1974) *A Review of Environmental Impact Assessment Methodologies.* Prepared for EPA, Office of Research and Development. Washington, DC: U.S. Government Printing Office.
*Warner, M. L., Moore, J. L., Chatterjee, S., Cooper, D. C., Ifeadi, C., Lawhon, W. T., and Reimers, R. S. (1974) *An Assessment Methodology for the Environmental Impact of Water Resource Projects.* Washington, DC: Office of Research and Development, U.S. EPA (EPA-600/5-74-016).
*Warwick, D. P., and Lininger, C. A. (1975) *The Sample Survey: Theory and Practice.* New York: McGraw-Hill.
*Water Resources Engineers, Inc. (1969) "Mathematical Models for the Prediction of Thermal Energy Changes in Impoundments." Final project report presented to the Federal Water Pollution Control Administration.
*Watson, R. H. (1978) "Interpretive Structural Modeling: A Useful Tool for Technology Assessment." *Technological Forecasting and Social Change* 11 (2), 165–185.
Watt, R. E. F. (1968) *Ecology and Resource Management, A Quantitative Approach.* New York: McGraw-Hill.

Weingart, J. M. (1977) "Transdisciplinary Science—Some Recent Experience with Solar Energy Conversion Research." AAAS Annual Meeting, Denver (February).

†Weisbecker, L. W. (1972) *Technology Assessment of Winter Orographic Snowpack Augmentation in the Colorado River Basin: The Impacts of Snow Enchancement.* Vols. I & II. Menlo Park, CA: Stanford Research Institute. Prepared for the National Science Foundation.

Weisbecker, L. W. (1974) *The Impacts of Snow Enhancement.* Norman, OK: University of Oklahoma Press.

†Weisbecker, L. W., Stoneman, W. C., Ackerman, S. E., Arnold, R. K., Halton, P. M., Ivy, S. C., Kautz, W. N., Kroll, C. A., Levy, S., Mickled, R. B., Miller, P. D., Rainey, C. T., and Vanzandt, J. E. (1977) *Earthquake Prediction, Uncertainty, and Policies for the Future.* Menlo Park, CA: SRI International. Prepared for the National Science Foundation. [Short report is *Earthquake Prediction in Society.*]

Weiss, C. H. (1972) *Evaluation Research.* Englewood Cliffs, NJ: Prentice-Hall.

Wenk, E., Jr. and Kuehn, T. J. (1977). "Interinstitutional Networks in Technological Delivery Systems," in *Science and Technology Policy* (J. Haberer, ed.). Lexington, MA: Lexington Books, pp. 153–175.

†Wertz, J. E. (1977) *Metric Transition in the United States.* Report prepared for the National Science Foundation (Grant No. GI 40445).

Wesler, J. E. (1972) *Manual for Highway Noise Prediction.* Cambridge, MA: Transportation Systems Center, U.S. Department of Transportation, Report No. DOT-TSC-FIJWA-72-1.

Western Systems Coordinating Council, Environmental Committee (1971). *Environmental Guidelines* (Robert Coe, Southern California Electric Company, Environmental Committee Chairman).

White, I. L., Chartock, M. A., Leonard, R. L., Bloyd, C. N., Gilliland, M. W., Hall, T. A., Malecki, E. J., and Rappaport, E. B. (1976) "First Year Work Plan for a Technology Assessment of Western Energy Resource Development." Document prepared for U.S. EPA (EPA-600/5-76-00).

†White, I. L., Chartock, M. A., Leonard, R. L., Ballard, S. C., Gilliland, M. W., Hall, T. A., Malecki, E. J., Rappaport, E. B., Freed, R. K., Miller, G. D., La Grone, F. S., Bartosh, C. P., Cabe, D. B., Eppright, B. R., Grossman, D. C., Lacy, J. C., Raye, T. D., Stuart, J. D., and Wilson, M. L. (1977) *Energy from the West: A Progress Report of a Technology Assessment of Western Energy Resource Development.* Washington, DC: U.S. EPA.

White, J. A., Agee, M. H., and Case, K. E. (1977) *Principles of Engineering Economic Analysis.* New York: Wiley.

White, L., Jr. (1966) *Medieval Technology and Social Change.* New York: Oxford.

White, L., Jr. (1967) "The Historical Roots of Our Ecologic Crisis," *Science* 155 (10 March), 1203–1207.

White, L., Jr. (1974) "Technology Assessment from the Stance of a Medieval Historian," *American Historical Review* 79 (1) (February), 1–13.

Willeke, G. E. (1974) "Identification of Publics in Water Resource Planning."

Atlanta, GA: Georgia Institute of Technology, Environmental Resource Center (September, ERC 1174).

Williams, A. N. (1977) "Compendium of Critiques of JPL Report SP-43-17: Automobile Technology Status and Projections Project." Pasadena, CA: Jet Propulsion Laboratory, California Institute of Technology (JPL-PUB-77-40). Also, NTIS (N77-33519/8).

Williams, R. M., Jr. (1967) "Individual and Group Values," *The Annals of the American Academy of Political and Social Sciences* 371 (May), 20–37.

Winder, J. S., Jr. and Allen, R. H. (1975) *The Environmental Impact Assessment Project: A Critical Appraisal.* Washington, DC: The Institute of Ecology (Now at Butler University, Indianapolis, IN).

Winner, L. (1972) "On Criticizing Technology," *Public Policy* 20 (1), 35–59.

Winner, L. (1977) *Autonomous Technology.* Cambridge, MA: MIT Press.

Wirth, T. J., and Associates (1972) *Report and Environmental Statement for the Lake Tahoe Planning and Effectivity Ordinances.* South Lake Tahoe, CA: Tahoe Regulatory Planning Commission.

*Wise, G. (1976) "The Accuracy of Technological Forecasts, 1890–1940," *Futures* 8 (5), 411–419.

*Wolf, C. P. (ed.) (1974) *Social Impact Assessment.* Milwaukee, WI: Environmental Design Research Association.

Wolf, C. P. (ed.) (1976—) *Social Impact Assessment* (newsletter). New York: Environmental Psychology Program, CUNY Graduate Center.

*Zajic, J. E., and Svreck, W. Y. (1975) "Environmental Impact Statement Preparation," *Journal of Environmental Systems* 5 (2), 115–120.

†Zissis, G. J., et al. (1977) *Remote Sensing: A Partial Technology Assessment.* Ann Arbor, MI: Environmental Research Institute of Michigan.

Zwicky, F. (1962) *Morphology of Propulsive Power.* Pasadena, CA: Society for Morphological Research.

Zwicky, F. (1969) *Discovery, Invention, Research: Through the Morphological Approach.* Toronto: Macmillan.

Index

Note: Only significant mentions of topics are indexed. Cameo, figure, and table notations are boldface. "U.S." is not indexed [e.g., as in (U.S.) Environmental Protection Agency]. Many entries appear under the heading TA/EIA (e.g., TA/EIA, performers).

A

Aesthetics, 319
Air, pollutants, 239–240
Analogy, use of, 314, 332
Annual equivalent, 272–274
Atomic Energy Commission, 246, 337
Automotive emissions, **106–110**

B

Bayes' Rule, 191
Bayesian statistics, 208
Benefit:cost ratio, **259,** 272, 275–276
Boom towns, 288, 318
Boulding, Kenneth, 284
Bounding, 54–55, 65–**68,** 69–70, 86, 428
Brainstorming, 6–**7, 82,** 332

C

Capitalism, 18
Cause and effect, 23, 45, 147, 316
Checklists, **83,** 105, 162, **163–165,** 166
Clean Air Act, 238–239
Coates, Joseph F., 3, 8
Communication
  patterns within team, 432–435
  of results [Chapter 16], 60, 87, 393–395, 411–417
Community, impacts, 272, 276, **277,** 278, **312–313,** 318
Computer conferencing, 434
Congress, U.S., use of TA/EIA, 406
Constant dollars, 270–271
Core team, 423, 425, 434
Cost–benefit analysis [Chapter 11], **48, 84,** 462–463
Cost–effectiveness analysis, 272
Council on Environmental Quality (CEQ), 28–29
  NEPA regulations (for EISs), 61, 67, 216–218
Cross-impact analysis, 190–203
Current dollars, 270–271

D

Daddario, Emilio Q., 30–31, 34
Data, primary and secondary, 309, **310–314**

Decision analysis, **84,** 363, 366–367, **368–370,** 375–377
Delphi, **82,** 124–125, **126–129,** 130
Determinism, soft, 16
Dimensionless scaling, 362–363, **364–365**
Discount rate, 256, 270–273, **274–275,** 276, 282, 297
  social, 271–272
Discounted cash flow, 272
Displacement and relocation, 318
Distributional impacts, 318, 358; *see also* Equity
DYNAMO simulation language, 205

E

Ecological survey, 229
Ecological systems, 222–223, 229–230
  carrying capacity, 222
  food webs, 223
  limiting factors, 222
  microclimate, 223
  triggering factors, 222
Economic analysis [Chapter 11]
Economic base models, **84,** 284, **285–287,** 288–289
Effectiveness, 254–255
Efficiency, 254–255
EIA, *see* Environmental impact analysis
EIS, *see* Environmental impact statement
Elasticity, **266–267**
Employment multipliers, 286–287
Energy analysis, 281–282
Environmental impact assessment; *see also* TA/EIA
  emergence, 3
  program, 4
Environmental impact indicators, **224–227**
Environmental impact statement (EIS); *see also* TA/EIA
  required by NEPA, 3, 29
  preparation, 29, 216–222
Environmental Protection Agency, U.S., **217,** 232, 238, 240–245
Epistemological factors, 426–427
Equity, 254–255, 318, 358; *see also* Values
Erosion, 231
Escalation rates, 270
Ethnographic field research, 309, **312–313**
Evaluation research, 443–444; *see also* TA/EIA, evaluation
Expert opinion, *see* Opinion measurement
Export base models, **84,** 284–289
Externalities, 257, 265–267

F

Feedback systems, 204–205, **206**
Forecasting; *see also* Technology forecasting; Social forecasting
  principles, 111–113

G

Gantt chart, 428, **429**
Genius forecast, 123
GESMO assessment, 93–94
Growth
  economic, 284
  population, 287, 288, 293

H

Historic preservation, 247
Historical analogy, 314, 332
Human needs, 146, 323, 325

I

Impact analysis [Chapters 9–13]
  approaches, 59–60, 86
  dissociation from forecasting, 315
Impact evaluation [Chapter 14], 60, 86–87, 248, **361**
  economic, 269–270
Impact identification [Chapter 8]
  approaches to, 58, 86, 157, 172
  economic, 258
  holistic approach to, 158–159
  institutional/political, 344, **346–348**
  legal, 337–338
  policy options, **389**–390
  reductionist approach, 158–159
Impacts
  classification, 58, 158
  environmental, **227, 228**
  higher order, 58–**59,** 77, 168, 247–248, 255, 277, 283, 301
  quantification, 183–187, **234–235,** 258–259, 264, 269, 276–**277,** 362–363
  selection of, 173–176, 316, 359–360
Industrial state, 19–20, 458
Information display, 316
Information types, 386–387
Input–output analysis, **289–290,** 291
Inquiring systems, 78, **79,** 80
Institute of Ecology, The, review of EISs, 248, 293, 451, 452

Institutional analysis [Chapter 13], 330, 339–341, **342–343,** 344, **346–347, 348**; *see also* TA/EIA, institutions
Institutions, 13, 340
Integration, of TA/EIA components, 61, 435, **436,** 437, 461–462
Interdisciplinary process, 355, 426–427, 461; *see also* Integration
Internal rate of return, 272–273, **275**
Interpretive structural modeling, **71–73,** 82
ISM, *see* Interpretive structural modeling
Iteration, 248, 432

K

Keynes, John Maynard, 283
KSIM, 196–203

L

Labor-force participation rate, 287
Land use, 230
Least-squares fit, *see* Regression analyses
Legal analysis, 330, 334–335, **336,** 337, **338–339, 340**
Liability, 335, **336,** 337
Life-cycle costing, 273
Life-styles, 319; *see also* Quality of life
Light intrusion, 247
Literature search, 110
Local–national divergences, 318

M

Macroassessment, **49, 53**
Macroeconomic, 283
Markov chains, 207
Marx, Karl, 18, 22, 283
Maslow, Abraham, *see* Human needs
Matrices, 165–168
Matrix
  cross-effect, **83,** 187–190, 362–363, **364–365**
  cross-impact, 190–195, 219
  nonoccurrence, 192–193
Measurement techniques, **224–227,** 229, 232, 236, 238, 240–244
Microassessment, **49, 53,** 69–**70**
Microwaves, effects, 247
Milestone chart, 428, 430
Miniassessment, **49, 53**
Mitigation of impacts, **227,** 241–242, 244, 247
Models
  criticism, 184, 186, 203, 205, 207
  definition, 182–183
  economic base, **84,** 284–289
  gaming, 204
  hierarchy of, 183–184
  input–output, **289–290,** 291
  land use, 203–204
  physical, 203
  planning, 203–204
  policy process, 382–384
  as process, 187, **436,** 437
  qualitative, 332
  simulation, **83,** 195–196
    air, 240
    water, 232, **234–235**
  social impact, 306–**307**
  transportation, 203–204
  uses, 177, 186, 219, 222
Monitoring, **53,** 113–**114**
Monsanto, 37
Monte Carlo simulation, 194, 207
Morphological analysis, 172, **174–175**
Multipliers
  economic base, 284–288
  employment, **286–287,** 288
  income, **285**

N

National Environmental Policy Act, provisions, 3, 27–29, 216–218, 319, 451, 466–467
National Historic Preservation Act of 1966, 247
National Register, 247
National Science Foundation, U.S., 35
NEPA, *see* National Environmental Policy Act
Net energy analysis, *see* Energy analysis
Net present value, 272–273, **274,** 297
Network scheduling, *see* PERT
Noise
  description, 242–244
  effects, 242
Noise Control Act of 1972, 242
NSF, *see* National Science Foundation

O

OECD, *see* Organization for Economic Cooperation and Development
Office of Science and Technology Policy, U.S., establishment, 35–36

Office of Technology Assessment, U.S., 33–34, 464–466
Oklahoma group, approach, 394–395, **413–416**
Opinion measurement [Chapter 6], **82,** 110–111, 122–130, 316–318, 332; *see also* Delphi; Survey; Panels
  assumptions, 123
  genius forecast, 123
Organization for Economic Cooperation and Development, 38
Organizational factors in TA/EIA, 424
OTA, *see* Office of Technology Assessment

P

Paired comparison, 278
Panels, 123–124
Participation, *see* Public participation
Parties at interest, identification, 55, 147
Payback period, 272–273, **275,** 276
PERT, 430, **431,** 432
Peterson, Russell, 465
Plutonium recycle, *see* GESMO assessment
Policy
  evaluation, 392, 393; *see also* Evaluation research
  impact, 201
  options, 69, 388–390
  process models, 382–384, 388
  recommendations, 394–395, **416**
Policy analysis [Chapter 15], 60, **68,** 87, 156–157, 176–177, **304–305**
Policy capture, **84,** 367, 371, **372–374**
Policy community, 388, 413
Policy sciences, 381
Policy sector, 60, 388
Policy studies, 381
Policy system, 388
Political analysis, 339–341, **342–343,** 344, **348,** 385–386
Pollutants, **224–227,** 230–232; *see also* Air, pollutants
Pollution control, implications, 283–284
Private goods, 257, 269
Probabilistic techniques, **83,** 207–208
Probabilities
  conditional, 191–195, 463
  marginal, 191–194
Problem definition, *see* Bounding
Profiling, *see* Social description
Project management [Chapter 17]
Project team characteristics, 425–427; *see also* Core team
Projection, *see* Social forecasting
Psychological impacts [Chapter 12], 301
Public goods, 268–269
Public opinion, obtainment, *see* Opinion measurement
Public participation, 33, 356–357, 391, 393, **407,** 408–**409,** 427, 472–473

Q

QSIM2, 202–203
Quality of life, 142–143, 323, **324**
Queueing theory, 207

R

Radiation
  effects, 246
  measure of dose, 245
Radioactive wastes, disposal, 246–247
Regional economics, *see* Economic base models
Regression analyses, 116–117, 291–294
Relevance trees, **83,** 168–169, **170–171,** 172
Request for proposals, 422
Resource Conservation and Recovery Act of 1976, 230–231
Retrospective analysis, 386, 458
Reverse adaptation, 17
RFP, *see* Request for proposals
Risk
  decision under, 366–367, **368–370**
  estimation, 279–**280,** 299
  evaluation, 279–**280,** 299
Risk–benefit analysis, 255, 278–281

S

Scenarios, **82,** 150–151, **152,** 315, 332, 472–473
Scheduling, project, 428, **429,** 430, **431,** 432
Schumacher, E.F., 22
Sensitivity analysis, **83,** 208–209, **211**
Shadow price, 268–269
SIA, *see* Social Impact Assessment
Simulation models, *see* Models, simulation
Social description [Chapter 7], 58, 86, 308
Social forecasting [Chapter 7], 58, 86, 136, 146–148, 314–316
Social impact assessment [Chapter 12], 426; *see also* Social impacts
  neglect of, 462
Social impacts, 301, **303,** 318–319; *see also* Social impact assessment
Social indicators, 58, 140–142, 148, 309

Soil, pollutants, 231–232
Solid waste, disposal, 230–231
Solid Waste Disposal Act of 1965, 230
Supply and demand, 266–267, 297–**298**
Survey, 123–**125**, 229, 299, 309
Systems analysis, ecological, 222–223
Systems dynamics, 204–205, **206**, 207
Systems model, of policy, 383

T

TA, *see* Technology assessment
TAAC, *see* Technology Assessment Advisory Council
TAB, *see* Technology Assessment Board
TA/EIA; *see also* Technology assessment; Environmental impact analysis
  advocacy, 46, 357, 474
  balance, 45
  components, 54–**55**, **56–57**, 58–60, **68**; *see also* TA/EIA, steps in process
  communicability, 47
  credibility, 46–47
  critiques [Chapter 19]
  differences, 3–4, **48–49**, **50**–51
  evaluation [Chapters 18, 19], **53**, 451; *see also* TA/EIA, objectives
  federal agencies, 34–36
  institutions, critique, 464–467; *see also* Institutional analysis
  international, 37–40
  objectives, 5, 43, 74
    improving methodology, 43, 47
    utility, 43, 46–47, 444, **445**, 446; *see also* TA/EIA, uses
    validity, 43, 45–46, 444, **445**, 446
  performers, 5, 34–40, 421, 422, **423**, 424–427
  philosophical approaches, *see* Inquiring systems
  private sector, 36–37
  problem oriented, 3, 32, 51, 103
  procedural flexibility, 76
  project assessment, 3, 51, 103–104, 360
  public participation, *see* Public participation
  political implications, 459–460
  relevance, 46; *see also* TA/EIA, utility
  reports, *see* Communication of results
  similarities, 3–4, **48–49**, **50–51**
  sources of input, 69, 356–357, 367
  spatial extent of coverage, 67, **68**, 75
  sponsor's perspective, 421–422, **423**
  steps in process, 32, 302; *see also* TA/EIA, components
  study resources, 76–77
  summaries of selected assessments, 91–96
  techniques, 81, **82–84**, **85**
    credibility, **410**, 427, 464
    environmental issues, **224–226**
    evaluation, 85, **172**–173, 186, **410**, 444–**445**, 462–464
    scanning, 81, 159–168
    tracing, 81, 159–162, 168–172
    use, 427
  technological alternatives, 75
  technology oriented, 3, 32, 51–52, 103, 360
  time horizon, 67, 76, 101, 386
  timeliness, 46
  type of technology, 69, 96, **103**, 104, 387
  types, 3, 51–52, 103–104
  users, 5, 69, 401–402, 404, **405**, 407
  uses, 4–6, 403–404, **407**
  value issues, 14–15, 21, 45, 77, 351, 355–359, 392–393, 459, 461
TDS, *see* Technological delivery system
Techniques, *see* TA/EIA techniques
Technological development, **11**–13, 101, **102**
Technological analysis, 333–334
Technological delivery system, 43, **44**, 58, 136–138, 156, 331, 340, **341**
Technological forecasting, *see* Technology forecasting
Technological imbalance, 333
Technological imperative, 17, 457
Technological optimism, 17, 33, 457–458
Technological parameters, 105, 115–119
Technological pessimism, 17, 457–458
Technological readiness, 333
Technology, 11–14
  appropriate, 22–23
  definition, 14
  neutrality, 14–15
  social interaction, 14–23
  value-free analysis, 14–15
Technology assessment, definition, 3; *see also* TA/EIA
Technology Assessment Advisory Council, 33–34
Technology Assessment Board, 33–34
Technology description [Chapter 6], 55, 86, 99–111
Technology forecasting [Chapter 6], 55, 86
  exploratory, 112, 315
  normative, 112, 315
  principles, 111–113, 115
TF, *see* Technology forecasting
Trend extrapolation, **82**
  compound parameter, 115–119

Trend extrapolation [cont.]
envelope curves, 119–122
single parameter, 115–119
S-shaped curves, 119–122
Trend impact analysis, **148–149**
Technology-values interaction, 320–321, 322

U

U.S. Army, Corps of Engineers, 302
Utility, 357–359, 362

V

Values, 143, 147, 319, 352, **353**–355; *see also* TA/EIA, value issues
change, 20, **21,** 139, 143–146, 320, 321, 322, 353
economic, 254
impact on, 320, **321,** 322–323
religious and technology, 21–22
reverence, 359
spiritual, 358–359
utilitarian, 357–359, 362

W

Water, associated impacts, **228**
Water quality
models, **234–235,** 237–238
parameters, 232–**233,** 236

X

X rays, effects, 247